A COMMONWEALTH OF KNOWLEDGE

A Commonwealth of Knowledge

Science, Sensibility, and White South Africa 1820–2000

SAUL DUBOW

OXFORD
UNIVERSITY PRESS

OXFORD
UNIVERSITY PRESS

Great Clarendon Street, Oxford OX2 6DP

Oxford University Press is a department of the University of Oxford.
It furthers the University's objective of excellence in research, scholarship,
and education by publishing worldwide in

Oxford New York

Auckland Cape Town Dar es Salaam Hong Kong Karachi
Kuala Lumpur Madrid Melbourne Mexico City Nairobi
New Delhi Shanghai Taipei Toronto

With offices in

Argentina Austria Brazil Chile Czech Republic France Greece
Guatemala Hungary Italy Japan Poland Portugal Singapore
South Korea Switzerland Thailand Turkey Ukraine Vietnam

Published in the United States
by Oxford University Press Inc., New York

First published 2006

British Library Cataloguing in Publication Data
Data available

Library of Congress Cataloging in Publication Data
Data available

Typeset by Newgen Imaging Systems (P) Ltd., Chennai, India
Printed in Great Britain
on acid-free paper by
Biddles Ltd., King's Lynn, Norfolk

ISBN 0–19–929663–4 978–0–19–929663–7

Published in Southern Africa by Double Storey Books,
a division of Juta & Co. Ltd, Mercury Crescent, Wetton, Cape Town.

ISBN 1–77013–119–1 978–1–77013–119–4

1 3 5 7 9 10 8 6 4 2

Preface and Acknowledgements

This study examines the intellectual underpinnings of white South African identity and power by foregrounding scientific and social knowledge in the process of national self-understanding. It develops more or less directly out of previous work. In *Scientific Racism in Modern South Africa* (1995) I explored the role of social Darwinism in underwriting ideologies of white supremacy during the age of segregation and apartheid. While completing that research I became interested in the broader relationship between social and scientific thought, national identity, and political power. A stimulating spell in 1995 as a fellow at the ANU Humanities Research Centre, Canberra (then under the inspiring directorship of Iain McCalman), gave me the opportunity to pursue further these still barely formulated questions.

Keith Hancock, the Australian and Commonwealth historian, who also wrote a definitive study of Smuts and whose presence at the Australian National University was still tangible, offered a natural point of entry. In his historical essay, *Australia* (1930), Hancock used the striking term 'independent Australian Britons' to characterize the emergence of colonial nationalism—a tradition that he saw as being neither assertively 'nativist' nor imitatively 'English'. At about the same time I came across Carl Berger's classic study of Canadian nationalism, *The Sense of Power* (1970). In it Berger showed how Canadian pro-imperial sentiment could be understood as a form of Canadian nationalism, rather than as its negation. This insight seemed novel to someone who had grown up thinking about identity in binary, exclusive terms, and in a society where Afrikaner nationalists laid first claim to 'being South African'.

Historians are used to analysing national identities as constructed, rather than pre-existing phenomena. Many excellent studies have approached the study of Afrikaner and African ethnic groupings in this way. But there has been a tendency to take the country's status as a unitary nation-state for granted and to assume, in the words of the Freedom Charter, that the country 'belongs to all who live in it'. The story of belonging is of course profoundly historical—as well as political—and raises important subsidiary or prior questions: how did understandings of the term 'South Africa' develop? What were considered to be its defining problems? Who laid claim to membership of the national community, and when? Answers to these questions, which have a fresh contemporary political relevance, demand a thorough engagement with the realm of ideas. They need to be posed over a broad timespan and ought to be addressed with reference to a wide range of source material.

This in turn raises problems about what to include or exclude. In a field where many of the institutions and personnel may not be familiar to readers, a balance has to be struck between the need to provide detail, and the desire to maintain the momentum of the argument. Where choices have to be made I have tended to favour the latter imperative. Fortunately, South African history is rich in biography and there are many useful accounts of institutions to be drawn on. Some of this material is narrowly focused and insufficiently critical, a good deal of it is antiquarian, but without such scholarly work it would have been impossible to undertake a project of this scale. I have also drawn freely on, and revised, my own published work, which is listed in the Bibliography.

In respect of chronology I have sought to cover the greater part of the nineteenth and twentieth centuries. Each chapter therefore deals with a roughly equivalent period of time. However, the focus of investigation, the approach, and the material chosen to elucidate the arguments varies considerably. The second half of the last century is dealt with rather more schematically and speculatively than the earlier sections, and the final chapter is therefore designed as an extended post-script. Its intention is to frame the book's central thrust on 'South Africanism' rather than to explicate Afrikaner and African nationalist views of scientific knowledge in detail.

South Africanism took many forms and resists easy definition. It may be understood as a version of the patriotic colonial or dominion nationalisms that developed in Australia, New Zealand, or Canada. (In Africa, 'Algerianism' offers interesting parallels.) It was pre-eminently the expression of a developing settler society, and as such marginalized or denied the rights of indigenous African peoples. Given the fact that whites in South Africa were so much more in the minority than in other comparable settler societies, the pathology of denial was all the more acute. The paradox is clear when one considers that South Africanism steadfastly disavowed the politics of 'racialism' (the 'colour question' was something else entirely): its proponents professed their commitment to ameliorating tensions between Afrikaners and English-speakers by stressing common bonds of patriotism.

The primary focus of South Africanism was on the country's relationship to the Empire/Commonwealth and in this sense it was international. South Africanism was expressed more as a creed or an ethos than an ideology. Indeed, it purported to be non-ideological, and therefore wrapped itself in the apparently neutral virtues of reason, progress, and civilization. This is why knowledge, and specifically scientific and technical knowledge, was so valued by its adherents. To understand the reasoning of those responsible for creating knowledge-based institutions is to cast light on their motives and to broaden our understanding of the fabric of power and collective self-belief.

Although South Africanism has attracted some scholarly attention in recent years, it has not been given the attention it deserves. As a centrist tradition designed to hold the ring between imperialism and Afrikaner nationalism, it defined itself as much in terms of what it was not than what it was. Its weakness and ultimate political failure has more or less been taken as a given in many teleological readings of modern South African history.

This book advances a somewhat different view. It maintains that South Africanism, while always an ideology of compromise, was a major—even dominant—political force from the moment of Union to the advent of the Nationalists in 1948. It argues, too, that South Africanism developed out of a prior sense of colonial identity, namely, that which developed in the Cape from the early years of British occupation at the turn of the nineteenth century. We therefore take the story of South Africanism back to the colonial politics of the nineteenth century and trace its deep foundations in the institutions and associational life of Cape colonial culture, intellect, and politics.

The history of ideas in South Africa remains poorly developed, although it is treated with somewhat less suspicion than was the case just a decade ago. Intellectual history is not by its nature democratic or popular, since it tends to focus on the thoughts and concerns of well-educated, articulate men (and sometimes women). In this regard the history of ideas is different from much social or cultural history, which proceeds from the experience of 'ordinary' people. And yet, the difference need not be so much as first appears.

Many cultural and social historians, seeking exemplars from the past to illuminate current political issues, are drawn to the lives of extraordinary, if marginalized or forgotten people, whose views and values marked them out as exceptional or extreme. This study, by contrast, focuses on the thinking of influential intellectuals who may have been part of a minority by virtue of their access to the world of ideas, but who were otherwise often conservative in their outlook, predisposed towards the politics of the middle-ground, and therefore in their own way quite ordinary. Unlike many intellectual historians, I am not much concerned with original and great ideas or with the analysis of foundational texts. The focus here is on an intelligentsia whose ideas were frequently derived from elsewhere, and the institutions they built to sustain their authority. My interest is in how ideas were selected and adapted to suit local conditions and contexts. Appraisal of their effectiveness and impact invites a re-examination of key political, social, and ideological themes in South African historiography.

As always in the field of ideas-based history, problems of representivity and reception are difficult to address: significance is easier to assert than to demonstrate (though this is surely also the case in most non-quantitative forms of historical explanation). Not all readers of this book will be sympathetic to, or convinced of, the value of this enterprise, especially those who consider that this is not the appropriate moment to write about white elites in South Africa. I differ from this viewpoint for a number of reasons. How would European historiography look, for example, if accounts of the aristocracy or the middle class were declared out of bounds? At the very least I hope that readers of this book will be inclined to view key problems in South African and imperial history—new as well as old—from a fresh perspective.

The support of the Leverhulme Trust, which granted me a Fellowship in 1998–9, allowed this project to get under way. The British Academy, which granted me a two-year Readership in 2004–5, gave me the space and time to reacquaint myself with my research and bring the manuscript to conclusion. Without the support of these two institutions it is difficult to see how this book could have been completed, and I am indebted to those anonymous reviewers and committee members who decided to back my work. The School of Humanities at Sussex University provided additional research funding which allowed me to collect material for the concluding chapter.

I am profoundly grateful to the individuals—and there are rather a lot of them—who, over the course of almost a decade, commented on parts of the evolving manuscript and advised on sources. My Sussex colleagues, Donald Winch and Brian Young, pointed out fruitful directions of research. Paul Betts commented on large sections of the manuscript, encouraged me to look beyond the familiar, and provided wise counsel. Brian Austin, Andrew Bank, William Beinart, Vivian Bickford-Smith, Jane Carruthers, Elizabeth Green Musselman, Eduard Fagan, Wieland Gevers, Albert Grundlingh, Patrick Harries, Nancy Jacobs, Alan Lester, Russell Martin, Shula Marks, Lance van Sittert, Peter Vale, and Brian Warner answered queries and commented on specific aspects of the manuscript. Richard Wilson proved as shrewd a critic as he is a generous friend. In Johannesburg, Neil McCarthy was often the first to hear about new discoveries and to indulge (in) my enthusiasms. He also took some of the photographs. The entire manuscript was read by Jeremy Krikler and Andrew Thompson, both of whom devoted enormous care and attention to the task. Their detailed commentary and imaginative criticisms have undoubtedly made this a better book.

Most of the research, and certainly the most pleasurable aspects of it, was conducted in South Africa. The professionalism and efficiency of librarians at Rhodes, Cape Town, and Wits universities, as well as the Killie Campbell Library in Durban, is much appreciated. I am particularly grateful to Margaret Northey (Wits Africana); Lesley Hart (UCT manuscripts), Sue Ogterop and Allegra Louw (UCT Africana), and Laureen Rushby (UCT Govt. Publications) for their help and engagement.

Writing this book has, for the most part, been a selfish pursuit as well as a passion. Bryony Mortimer understood this and did much to facilitate the early phases of my project. Two concentrated years of research leave, subject only to the welcome interruptions of my daughters Talitha and Bethany and the enlivening presence and love of Signe Gossmann, made the latter stages of completion most pleasurable. This book started in Cape Town, which is where I grew up and where my parents, Neville and Rhona, still live. They have always taken the keenest interest in my academic work. This book is for them.

Contents

List of Illustrations

List of Abbreviations

BAAS	British Association for the Advancement of Science
BJHS	*British Journal for the History of Science*
CLJ	*Cape Law Journal*
CMM	*Cape Monthly Magazine*
DSAB	*The Dictionary of South African Biography* (Pretoria, 1968–)
EPMM	*Eastern Province Monthly Magazine*
JMH	*Journal of Modern History*
JSAS	*Journal of Southern African Studies*
RDM	*Rand Daily Mail*
SAAAS	South African Association for the Advancement of Science
SAC	South African College
SAHJ	*South African Historical Journal*
SAJS	*South African Journal of Science*
SALS	South African Literary Society
SAM	South African Museum
SANG	South African National Gallery
ZAR	South African Republic (Transvaal)

Introduction

THEMES

Since the end of apartheid the Cape has become a prime international destination. Tourists who marvel at the majesty of Table Mountain, travel through the winelands, appreciate the distinctive architecture of Cape Dutch homesteads, are startled by the richness of the Cape floral kingdom, or applaud the conscious multiculturalism of its cosmopolitan society, all participate knowingly or unknowingly in the re-enactment of core themes in Cape identity. Many of these tropes were carefully cultivated by nineteenth- and early twentieth-century writers and publicists who imagined the Cape as a unique offshoot of Europe, located at the tip of Africa.

The material culture of the Cape is reassuringly familiar to visitors from the northern hemisphere, and all the more impressive when bathed in the Cape's mediterranean light. Classic examples of British colonial architecture are to be found in the old town centre as well as the newly reconstructed waterfront. The more discerning eye will note the lingering Dutch colonial presence, typically signalled by ornate gables, fanlights, white stucco, and brick verandas or 'stoeps'. In the course of a brief stroll through the old Company Gardens, first laid out under Dutch occupation, the tourist may encounter the National Library of South Africa, the Iziko South African Museum and National Gallery, as well as Parliament and the cluster of buildings that housed the country's first institutions of higher learning.

It is easy to overlook the underlying messages and sinews of association that link these fine buildings. Taken together, they bear testimony to a set of overlapping, interlinked networks of power and authority that significantly shaped the Cape's distinctive colonial identity. Within their portals a considerable repository of scientific and societal knowledge was assembled and conserved. They all exemplify, in subtly different but complementary ways, the ethos of progress and improvement that took root from the beginning of the nineteenth century.

When the Cape passed into British hands at the turn of the nineteenth century the primary purpose of the new occupiers was to hold it for military and strategic reasons while the Napoleonic threat remained real. The new colony was thereby drawn into an empire that was vigorously rebuilding a cordon of anti-revolutionary conservatism. British settlement and anglicization was duly encouraged, but the new settlers were not always deferential to authority and they soon began to claim the rights of free-born Englishmen. By the 1820s strains of radicalism were clearly evident. Settlers bristled at the powers of an autocratic and reactionary governor who was viscerally alert to any hint of 'jacobinism'. Their resentment at their treatment and neglect by the colonial government was a ready source of anger; liberal and emancipationist ideas provided inspiration. Dissent was expressed through the establishment of a free press and the creation of a public sphere in which ideas could be exchanged. Libraries, journals, bookshops, coffee-houses, and scientific societies came to constitute an integral part of associational and sociable life in a developing civic society.

Once the foundations of middle-class life were established in the 1830s and 1840s there were few artificial constraints—other than the scale of the Cape's economy—to limit the growth of the mercantile and agrarian middle classes. No powerful aristocracy impeded it from above or, indeed, from below, since the imprint of a slave society and the colour-coded pattern of employment meant that workers and artisans with white skins could easily aspire to middle-class status. Although rivalries and political antagonisms between Dutch- and English-speakers were keen (issues like slavery and free trade affected Dutch agriculturalists and urban-dwelling English traders and financiers very differently), there were compelling reasons for a middle class to coalesce across the ethnic divide and establish a shared sense of colonial identity. The imperative to shore up an evolving racial order in an expanding colony of settlement was the most constant underlying factor, even if it was not always immediately to the fore in political discourse.

Metropolitan disdain of colonial life channelled assertion into the quest for middle-class respectability and recognition. Many immigrants came to the Cape to improve their circumstances and to remake themselves in a new context. Awareness of the relative status of other colonial societies (Australia, most notably) provoked a keen sense of competition. Radicalism, in so far as this was expressed with regard to class or colour, was evident in the humanitarian and emancipationist lobby of the early nineteenth century. However, this was a declining force by mid-century. The reverberations of the 1848 uprisings in Europe were experienced at the Cape, not by Chartist demonstrations, but by an upsurge of righteous popular anger at a government that was attempting to land convicts at Table Bay harbour. The problem of unfree labour, whether in the form of slaves or felons, was viewed as a serious affront to the dignity of the Cape, just as it was in New South Wales. Campaigns to defend colonial rights and reputations were thus closely linked.[1]

Knowledge-based institutions and voluntary societies were one means by which the colonial middle class established its public presence. These organizations, with their constitutions, rules of procedure and etiquette, committee structures, and voluntary subscriber lists, operated as fine-grained indicators of hierarchy and achievement. At one level they proclaimed collective advance. At another, they offered a source of public recognition for individuals. The South African Library,

[1] K. McKenzie, *Scandal in the Colonies* (Melbourne, 2004), 6.

a locus of considerable colonial pride, was a particularly important and enduring institution. From 1840 leading citizens, often the colonial governor himself, were invited to deliver its annual subscribers' address. The occasion, which was widely reported, helped to cement mutual trust and regard amongst representatives of the intellectual, social, and political elites.

Ordinary citizens who were not so much in the public eye could also participate in these improving activities. A dispersed community of collectors, fossil hunters, and amateur botanists, pursuing their enthusiasms in solitary isolation or in the course of sociable activities (like picnics, meetings, or outings) were likewise involved in the quest for recognition and collective upliftment. By having a rare specimen or object of curiosity accepted for exhibit in the South African Museum, the social status of the collector or donor as well as the esteem of the Colony was enhanced.[2]

Self-regard and public esteem fed into the quest for political recognition. Thus, evidence of the Cape's maturing civic culture was increasingly tied to demands for wider constitutional rights. By the 1850s colonial nationalist assertion had reached the point where a significant measure of self-government was ceded to the citizenry of the Cape. The public institutions of the Cape expanded during this period and the emergent political culture encouraged formalization of working agreements that had evolved between Dutch- and English-speakers over two generations. By the 1870s the Cape parliament, now a respected as well as a self-important institution, served as a visible example of functioning constitutional government. Closely linked was the institution of Roman-Dutch, or 'Cape Law', whose survival and growth as a body of jurisprudence bore testimony to the underlying compatibility of English and pre-Napoleonic continental traditions. The evolution of this durable mixed legal system gave considerable scope for local intellectual initiative and supported networks of authority and power that were to have implications for the development of the future South African nation-state.

The development of an anglophone colonial intelligentsia from the mid-1850s was immeasurably stimulated by the creation of the *Cape Monthly Magazine*. Modelled on British periodical literature, the *Monthly* served as a vital medium for the interchange of ideas. Its constant watchwords, of progress and improvement, were directed to the formation and expansion of literary and scientific institutions. As well as giving vent to new ideas generated in the metropole, the *Monthly* provided an opportunity for southern African-based contributors to report on local discoveries and to generate a continuous and multi-layered conversational thread. Although the size of its readership (perhaps only a few hundred) cannot be ascertained for certain, the *Monthly*'s impact can be assessed in terms of the debates it generated, its longevity, and the prominence of those who wrote for it. A number of contributors went on to publish pioneering studies in fields ranging from botany and geology to anthropology, archaeology, linguistics, and history.

[2] The story of how cabinets of curiosities gave rise to 'curio shops' selling indigenous crafts to tourists, only to transmute into today's emporia of African art, has yet to be told.

Analysis of the *Monthly* yields substantial evidence of the development of a visible—and invisible—colonial college of ideas. In an age when professionalism and specialization had barely begun to develop, knowledge was freely shared, rapidly disseminated, and frankly expressed. Aesthetic responses to the landscape of the Cape colony, travel accounts, and characterizations of its peoples are all revealing of the formation of colonial identity.

Evidence of the deep time span of human habitation and accounts of more recent colonization helped to encourage narratives of national existence; these in turn stimulated a collective sense of self-awareness and posed questions about belonging, ownership, and rights. Greater awareness of human difference—encouraged by the application of evolutionary principles to anthropological, historical, and archaeological understandings of society—stimulated the framing of such problems in the language of race.

The growth of expert knowledge about the land and its peoples was closely bound up with processes of colonial self-discovery and understanding. The urge to know about others was born of intellectual curiosity and the urge to constitute a sense of collective self. It also had a more instrumental dimension, namely the power to identify, pronounce upon, and control South Africa's indigenous inhabitants. Many contributors to the *Cape Monthly* were pioneers in what would later become known as 'African Studies'. Amateur colonial experts took care to formulate and format their ideas so as to fit in with, and inform, universal western schemes of knowledge. Theirs was the knowledge *about* Africans rather than the knowledge *of* Africans themselves.

At another level, the colonial intellectuals who helped to define, delimit, and describe the indigenous peoples of the country, as well as those who spoke with authority about its natural history, were becoming African. Their expertise was closely bound up with claims to be rights-bearing citizens of a country that they were consciously making their own. To make this assertion entailed displacing, limiting, or bypassing the prior ownership rights of black and brown peoples, while simultaneously demanding the rights of colonial citizenship within a wider British world. This is precisely what colonial nationalism involved. Assertions of colonial nationalism could be more or less exclusionary. Within the rubric of its more liberal emanation—and subject always to acceptance of the principles of civilization and the sovereignty of the Crown—British settlers, Afrikaner colonists, and (in certain cases) blacks too could be accommodated.

Colonial nationalist awareness was sharpened considerably over the course of the last quarter of the nineteenth century. The mineral revolution in the South African hinterland, the final conquest of African peoples in the subcontinent, and the failed efforts by the metropole to confederate the loose assemblage of republics and colonies in a South African dominion, all contributed to this process. As a result, the colonial conversation of mid-century became ever more politically charged. Exchanges between locally based colonial experts and prominent outside visitors like James Froude and Anthony Trollope, both of whom toured South Africa in the 1870s, did much to stimulate this dialogue. In the ensuing discussion, ideas about 'South Africa' were clarified and the defining political problems of the country specified more precisely. These turned on the endemic, perhaps irresolvable, conflict between blacks and whites, Boers and British, settlers in the east and colonists in the west.

Recent generations of historians have tended to lose sight of the salience of tensions between imperialists and colonists, despite this being very much a live issue in the late nineteenth century. In the language of the time, colonialists were people who, by birth or adoption, came to identify not only with the metropole but also with the country in which they lived. In common with 'republicans', who constituted the third element in this linguistic triptych, colonialists resented overbearing imperial interference with their rights and freedoms and feared the destabilizing rapacity of imperial interests. Before the emergence of modern Afrikaner and African nationalisms the distinction between imperialism and colonialism was a key aspect of political parlance, and it is therefore worthy of re-examination. Indeed, one of the central contentions of this book is that the tradition of Cape colonial nationalism is vital to our understanding of nineteenth-century Cape liberalism as well as twentieth-century South Africanism.

Cape colonial nationalism reached maturity in the 1870s. The slowly emerging political arrangement between leading Afrikaner and English-speaking notables that culminated in the alliance between the Afrikaner Bond and Cecil Rhodes, marked the apogee of the Cape colonialist concordat. It was shattered by Rhodes himself, who reverted to the imperialist standard in 1895, having long presented himself as a plausible champion of colonialist interests. Patriotic colonialists were horrified at the ascendancy of militant imperialism and the power of capitalists. Yet, almost all English-speakers, and many Cape Afrikaners too, baulked at supporting retrogressive Boer republicanism. Instead, they rallied to defend the legal and constitutional traditions of the Cape against the depredations of Milner, Rhodes, and their ilk.

The South African War of 1899–1902 marked a further damaging defeat for the delicate inter-ethnic alliance of moderate Afrikaner nationalists, anglophone liberals, and enfranchised Africans. However, it did not altogether expunge the component elements and traditions of this Cape-centric colonial tradition. These were soon to regroup in the battle to oppose Milner's attempts to suspend the Cape parliament, which helped to revive the political process. The end of the war opened up new possibilities for the reconstitution of Cape colonial freedoms, albeit within the terms of a racially segregated white South African nation. Key aspects of the struggle to establish the 'new' South Africa were played out in the decade after 1904, when the meanings of loyalism, patriotism, imperialism, nationalism, and progressivism were tested and contested.

The ideology of South Africanism emerged to inhabit the space left by a retreating imperialism and a temporarily broken republicanism. Geared to the needs of a unified white nation-state, it stressed virtues of moderation and conciliation.

Emphasis was laid on the need to create broadly based institutions and to nurture a shared national culture as a necessary counterpart to successful state-building. The constitutional politics of the Cape and the traditions of Anglo-Dutch partnership were revived and reconfigured to take account of new realities and a much-expanded national entity. A key South Africanist assumption was the notion that national sentiment could be accommodated within a wider sense of imperial belonging. Racialism, understood as antagonism between Dutch and British, was deplored. The language of social Darwinism, which had previously insisted on the incorrigible degeneracy of Boers, now emphasized the common 'Teutonic' roots of English- and Afrikaans-speakers. Conversely, insistence on the unbridgeable difference between whites and blacks helped to rationalize the need for systematic racial segregation.

Scientific and technical agencies, as well as professional bodies, made a significant contribution to South Africanist ideology. The notion of science as transcendent truth rendered it possible to cast the language of progress and universality within the imperial 'chain of civilization'. One of the first national public institutions to be created after the South African War was the South African Association for the Advancement of Science. Just as its Australasian equivalent had been seen as a harbinger of the Australian Commonwealth, so the South African Association came to be seen as a step towards 'closer union'. The joint meeting of the South African Association with the British Association in 1905, provided an occasion to reassert imperial authority in a constructive and benevolent manner. In fields like agriculture and veterinary medicine, the demonstrable benefits of scientific cooperation were eagerly taken up by ambitious Afrikaner modernizers and supported by Louis Botha, the first prime minister of the Union of South Africa, who embodied the spirit of reconciliation.

Several leading mining magnates and politicians, admired and reviled in equal measure for their support of the imperialist cause, gave direct support to the ethos of South Africanism by promoting significant new cultural and scientific ventures. These ranged from museums and art galleries to the creation of the national Botanical Gardens at Kirstenbosch and the establishment of a national teaching university. Identification with the landscape was practically encouraged through the promotion of agricultural improvement, conservation, and irrigation. Railways enclosed the outer reaches of the country within the realms of civilization and became a preferred means of stimulating tourism and national awareness. Preservationist and conservationist movements led to the creation of the Kruger National Park and the consecration of Table Mountain as a symbol of national unity. Enthusiastic literary and photographic evocations of 'the veld' encouraged fresh appreciation of a landscape that had once been routinely dismissed as barren and featureless. These initiatives and responses, in which scientific awareness and aesthetic sensibility were freely interwoven, were all part of a concerted effort to encourage a sense of inclusive South Africanness.

The realization of political union did much to encourage the creation of national societies and institutions, just as it hastened creeping statism. This was by no means a smooth process, for the dynamic of centralization induced strong counter-forces. Provincial and regional sub-identities had to be accommodated or submerged within an increasingly assertive nation-state. The shift of political power northwards from the Cape to the Transvaal unleashed latent strains and fissures that often overlapped with pronounced ethnic divisions. Regionalism is a rather neglected aspect of the politics of unification. The hairline cracks that it left in the national edifice can be readily detected in the complex internal histories of institutions such as museums, botanical gardens, and especially universities, where conflict between 'broad South Africanism' and Afrikaner nationalism became acute. References to the disparity between 'northern' and 'southern' political traditions was a common, if geographically misleading, way of expressing such tensions.

If the South African War opened the way for South Africanism to take root as a political ideology, the First World War put it to the test. Participation on the side of the Empire reopened wounds and provided Afrikaner nationalism and republican revanchism with fresh impetus. Hertzog was a clear beneficiary, but Smuts was able to consolidate and refashion his political base as English-speakers rallied to his support. Alternative visions of South Africanism were offered by the two leaders: Hertzog stressed national determination (though stopping short of republicanism) while Smuts became increasingly dependent on the imperial connection. The slow haemorrhage of support to the Nationalists was partially compensated for by Smuts's growing international status and the transmutation of Empire into the idea of Commonwealth. Over the next quarter of a century South Africanism of the Smutsian variety would remain an enduring political ideology.

The economic, administrative, and bureaucratic reach of the South African state underwent consolidation and expansion during and after the First World War. One of the responses to wartime industrial requirements was the inauguration of state-sponsored scientific research. The growing university sector created new academic posts and encouraged disciplinary professionalization and specialization. Government interest in the social sciences reflected an international vogue for expert knowledge, quantification, and the pursuit of social 'efficiency'. Growing concern with the problem of white poverty stimulated research into education as well as large-scale programmes of comparative intelligence testing.

A striking feature of the inter-war period was the convergence of state-funded and university-based research. The Carnegie Corporation of New York, a private foundation with a distinctive view of America's place in the world, introduced a new dimension to the anatomy of public knowledge. In 1929 Carnegie initiated a major commission into the problem of 'poor whiteism'. As well as helping to establish sociology as a discipline and the social survey as a basic methodological tool, the Carnegie Inquiry firmed up links between official policy-makers, the universities, and voluntary welfare organizations linked to the Dutch Reformed Church. Operating within its broader Commonwealth remit, the Carnegie Corporation provided core funding for libraries and museums. From 1933 the Corporation made resources available for the development of government capacity in educational and social research. This initiative served as the nucleus of future state-directed research in the social sciences and humanities.

Inter-war state-sponsored programmes in the social sciences and humanities suffered from underfunding, lack of effective intergovernmental coordination, and an absence of clear objectives. Yet the ethos of South Africanism was fully reflected in efforts to disburse funds on a non-partisan basis (that is, to English as well as Afrikaner institutions) and in a growing emphasis on research with demonstrable South African relevance. Developing international scientific contacts and networks, especially in the field of comparative education and psychology, encouraged the promotion of South Africa as the international 'laboratory' for comparative investigations into race and culture. The eventual withdrawal of Carnegie funding to government departments led to the creation of a National Research Council and Board in 1938. This was launched amidst a fanfare of patriotism and hailed by J. H. Hofmeyr as a South African 'parliament of research'. The new body did not, however, deliver on its promise to increase research funding, and many of its operations were interrupted by the Second World War.

The limited success of state-managed intellectual activity was compensated for by research conducted in the university sector, as well as in semi-autonomous institutes like Onderstepoort (1908), the South African Institute for Medical Research (1912), and the Carnegie-funded Bernard Price Institute for Geophysical Research (1937). In 1929, on the occasion of the second visit of the British Association, Hofmeyr and Smuts trumpeted the 'South Africanization of science'. A great deal was made of the expansion in tertiary education and the growth of research infrastructure in the period since the British Association's previous visit in 1905. Discoveries of national—and international—significance in the field of prehistory and physical anthropology generated a great deal of publicity.

Jan Smuts was especially adept at placing South African scientific achievements in a broader perspective and harnessing them to the purposes of statecraft. His authority as an international statesman was enhanced (at least in the eyes of his followers) by the publication in 1929 of his personal philosophy of 'holism'. This effort to synthesize scientific knowledge and human potential in terms of organic unity and evolutionary progress provided a ready analogy for his views about political interdependence: in the Smutsian universe, small nation-states like South Africa could expand rather than dilute their identity by association with the Commonwealth, the League, or the United Nations.

The 1929 meeting of the South African and British Associations revealed a less deferential attitude on the part of local scholars to the Empire, as well as exposing underlying ethnic tensions. Afrikaner nationalists were angered by the way in which scientific universalism was equated with membership of the Commonwealth.

Strong exception was taken to E. G. Malherbe's presentation of poor-whiteism in terms of Afrikaners' psychological feelings of inferiority. The Afrikaner nationalist presence in science was still undeveloped, but there were plenty of intimations that this was a field in which the *volk* was beginning to think about asserting itself. It would take another political generation before the anglophone dominion of knowledge was challenged in earnest.

The Second World War unleashed a new era in the relationship between nationhood, state, and knowledge. The National Research Council languished as resources were diverted towards support of the war effort. Nevertheless, ideas about a better future and schemes for post-war social reconstruction abounded. South Africans of all colours reflected on the meaning of citizenship and nationality. The language of welfare and democratic entitlement entered public discourse, and was given qualified endorsement by a government that understood the need to specify its war objectives not merely as the defeat of fascism, but also in positive terms: raising hopes for the future was essential for the maintenance of public morale. This was especially important in the face of a resurgent—albeit fractious—Afrikaner nationalist movement that was mainly united by its implacable opposition to South Africa's involvement on the Allied side. Demands for social justice and citizenship rights were also heard from the African National Congress, the labour movement, and the Communist Party, all of which were reinvigorated by the war. The need to take account of their radical demands was another important reason why state intervention, collectivism, and planning became integral to government thinking.

Several important commissions of inquiry and intergovernmental committees gave expression to the vogue for integrated scientific and social planning, in areas ranging from agriculture and industry to urban welfare and medicine. It is well known that the Smuts government failed to deliver on its promises to make substantial social progress. But the wartime administration's success in mobilizing for war, and the long-term effects of this process, should not be overlooked. Smuts's use of academic experts and his creation of new institutions, like the Council for Scientific and Industrial Research, did a great deal to cement links between industry, the universities, and the government. This incipiently corporatist approach arose out of a confluence of military, strategic, and social objectives. The ethos of state intervention and planning survived demobilization, only to be taken up in the cause of radical social engineering by the apartheid government from the 1950s onwards. Technocratic assumptions found a ready home in the burgeoning state bureaucracy. These were increasingly put to use in the name of white and, specifically, Afrikaner supremacy.

The rise of Afrikaner nationalism to political dominance finally put paid to South Africanist sentiment by making loyalty to the country virtually coterminous with membership of the *volk*. Government attacks on the white political opposition focused on the strangulation of civic institutions like the press, universities, and churches, behind which liberals increasingly retreated. The remarkable resilience of many of these institutions has perhaps less to do with individual moral heroism (though there were many instances of this) than with the fact of their deep embeddedness in an anglophone public culture that was itself never directly threatened by Afrikaner nationalism. Thus, although the English-speaking establishment and its institutions were in reality often highly conservative during the apartheid era, these became indelibly associated with 'liberalism'—as much a term of opprobrium as approbation.[3] To fully understand this strange conflation, a broader appreciation of the development of Cape colonial nationalism, civil society, and nineteenth-century Cape liberalism is required. The evidence presented here is a contribution to understanding that interrelationship.

KNOWLEDGE AND POWER

This book focuses on the intelligentsia within South Africa's white elite, and some readers may find the choice of subject disconcerting. For at least a quarter of a century the most innovative directions in South African historiography have focused on history 'from below' and, in particular, on recovering the voices of the forgotten and the dispossessed. The history of resistance to white power, whether expressed directly or indirectly, has been a central concern. Major advances have been made in our understanding of African politics, cultures, and identities. There have, of course, been important studies of colonial rule, segregation, and apartheid, which have illuminated state power 'from above', but the broader domain of what whites—and most observers—would have understood as national politics and culture may be beginning to fade from view. In a post-apartheid world this is inevitable, even desirable. But there are attendant dangers here. Important dimensions of South African history risk being occluded or lost if the role of whites is viewed too narrowly in terms of settler colonialism and exploitation, and if resistance to apartheid becomes our only frame of historical reference.

It was only in the last quarter of the twentieth century, and more particularly after the Soweto uprisings of 1976–7, that the majority of whites and their leaders began to sense that power was slipping away from them. From the perspective of the dominant groups, periods of crisis like the mass resistance of the early 1950s or the 1960 Sharpeville massacre seemed to have been successfully weathered or contained. For most of the twentieth century politics therefore meant white politics, and South Africans were assumed to be white. It was one thing to view blacks as a latent or periodic threat, quite another to see them as citizens competing directly for power.

This assumption, however complacent and misguided, is of central importance to an understanding of white hegemony. That blacks and whites interacted in

[3] The paradoxes and ambiguities in this association became manifold in left-wing and black nationalist critiques of liberalism from the 1970s.

daily life is not in question, nor is the fact that structural interdependence was a feature of South African existence from the beginnings of white settlement. But this does not detract from the reality that, in the realm of perceptions and social existence, whites and blacks inhabited separate universes. That these boundaries were largely self-policed imparted a sense of permanence and normality to white ascendancy.

South African novelists, from Olive Schreiner onwards, have cast the most penetrating light on this state of illusion or delusion. Journalists have made their contribution too. Ironically, historians have been less successful, precisely because (ever since the pioneering work of W. M. Macmillan and C. K. de Kiewiet in the 1930s and 1940s) they have sought in the spirit of rational enquiry to demonstrate the underlying reality of racial interdependence and injustice. In doing so—and to convincing effect—modern South African historiography has been disinclined to pay serious attention to the barely conscious assumptions that sustained white power for so long.

Historians who have concentrated on the history of Afrikaner nationalism are a partial exception to this generalization, though those who have narrated the story from within are more like the exceptions who prove the rule, in that they reflect rather than reveal the inwardness of white identity. Moreover, historians of Afrikaner identity—whether sympathetic to their subject or not—have too easily viewed Afrikaner nationalism as the purest expression of white South Africanness—if only because that is what Afrikaner nationalism chose to believe so insistently. This is perhaps one of the reasons why, during much of the apartheid period, many commentators believed that political change could only come from within the Afrikaner nationalist movement. One of the implications of this study is that we need to bear in mind how much more broadly white South African identity was based.

This book has relatively little to say about Afrikaner or African nationalisms, both of which have already attracted detailed scrutiny from historians. As the major claimants for political power in the second half of the twentieth century, such attention is both appropriate and explicable. A consequence of such approaches has been to occlude the political traditions that held the centre ground.

The history of South Africanism, which held power or stood in close proximity to power for most of the first half of the twentieth century, has been especially neglected. One reason is its close association with white, anglophone identity—which until the last decade or so has not been seen as a topic worthy of serious discussion.[4] Ironically, recent generations of English-speaking historians seem to have been complicit in its erasure. Indeed, English-speaking South Africans have

[4] The exception that proves the rule in the case of revisionist historiography is B. Bozzoli, *The Political Nature of a Ruling Class* (London, 1981), which devotes considerable attention to South Africanism. Recent treatments include M. Cardo, 'Culture, Citizenship and White South Africanism, *c.* 1924–1948', (Ph.D thesis, Cambridge University (2002); J. Lambert, 'South African or British? Or Dominion South Africans? The Evolution of an Identity in the 1910s and 1920s', *SAHJ* 43

long managed the trick of defining everyone else in the country as racially or ethnically 'other'—while blithely assuming their own identity to be somehow 'normal' and therefore not suitable for deep investigation. In a post-apartheid and post-colonial world this position is becoming less and less tenable.

Another reason for the neglect of South Africanism may have to do with the fact that it was tainted by power in the first half of the last century and increasingly enfeebled in the second half. Its moderate outlook, and its moral equivocation on questions of race, have not proved appealing to historians. Unlike liberalism, with which it was closely associated (and which has generated a great deal of controversy), South Africanism is not conducive to heroic or passionate treatment. From the perspectives of the political right and the political left alike, it is a hopelessly compromised 'lost cause'.

This does not, however, negate its importance or its interest. Linda Colley, in her introduction to *Britons*, has made the point that our fascination with dissenting voices from the past should not 'drown out the other, *apparently* more conventional voices'. Those who appear to be conformists, she explains, ought to be rescued from the condescension of posterity—as well as the ignorance of historians. Patriotism was 'more often than not a highly rational response and a creative one as well'.[5] This book does something similar by viewing South Africanism as an ideology of centrism, one that was heavily dependent on claims to even-handed political neutrality and normalcy, and which cultivated a form of patriotism that was compatible with internationalism. The role of knowledge and of knowledge-centred institutions in supporting these sentiments lies at the core of our investigation.

Broadly speaking, the relationship of knowledge to power, and its role in supporting claims to national identity, went through four identifiable phases from the early nineteenth to the mid-twentieth century. The first of these covers the period from the 1820s to the early 1850s. In this era, knowledge and knowledge-based institutions formed part of a drive to establish a middle-class civic order at the Cape that resisted autocratic colonial rule and asserted the rights owing to respectable citizens in a British colony of settlement. The second phase, which took in the remainder of the century, experienced a growing convergence between knowledge-centred institutions and the constitutional prerogatives of a colony enjoying the fruits of self-government. Tensions between colonial nationalism and imperial assertion shaped this relationship until the outbreak of the South African War.

A third phase was entered into during the period of post-war reconstruction in the first decade of the twentieth century. The politics of knowledge now became closely associated with the spirit of reconciliation and the ideology of

(2000); S. Dubow, 'Scientism, Social Research and the Limits of "South Africanism": The Case of Ernst Gideon Malherbe', *SAHJ* 44, (2001); J. Hyslop, 'An Anglo-South African Intellectual, the Second World War, and the Coming of Apartheid: Guy Butler in the 1940s', in S. Dubow and A. Jeeves (eds.), *South Africa's 1940s: Worlds of Possibilities* (Cape Town, 2005).

[5] L. Colley, *Britons: Forging the Nation 1707–1837* (New Haven 1992), 4–5.

South Africanism. Scientific and cultural bodies helped to give shape to the emergent white nation-state by building up its intellectual infrastructure. The period from the outbreak of the First World War to the close of the Second inaugurated a fourth phase in the politics of knowledge. Science was used to underpin a sense of South African patriotic achievement within the broader context of Commonwealth belonging. The period also saw growing state interest in, and direction of, practical knowledge in the social and natural sciences.

The final chapter of this book sketches the lineaments of a rather different phase, which is glossed as the 'renationalization' of knowledge. Here the politics of indigenous knowledge in contemporary South Africa are touched upon. This field of study, though lying outside the scope of the present work for reasons already indicated, has attracted growing interest of late. Africanist scholars writing within the idiom of social history have made much of the exchange of knowledge as a central aspect of the colonial encounter. They are apt to stress reciprocity and permeability in the interaction of different knowledge systems. The approach has been shown to work particularly well in areas like environmental history, medicine, and field sciences such as botany and zoology.[6]

The forms of colonial knowledge considered in this study were, however, generally of a different order. Borrowings and interchange are more likely to have been common in the seventeenth and eighteenth centuries, when European travellers and scientists scoured Southern Africa to arrange its fauna and flora within emerging global typologies and taxonomies.[7] Writings about indigenous peoples at this time were generally less racialized or contemptuous than they were to become in the mid- to late ninteenth century.[8] From about the 1830s and 1840s, as scientific institutions were implanted in the country, knowledge became increasingly bounded and exclusive, even though it took the better part of a century to become thoroughly professionalized and disciplinary-based.

Colonial knowledge producers presumed the universality of 'western' scientific knowledge and sought to root its ideas, institutions, and systems in an African context. Unless local knowledge and belief systems could be translated and absorbed—with or without attribution, and always selectively—within a 'western'

[6] Amongst many recent southern African examples, see W. Beinart and J. McGregor (eds.), *Social History and African Environments* (Oxford and Ohio, 2003); N. J. Jacobs, *Environment, Power and Injustice* (Cambridge, 2003) and her article on 'The Intimate Politics of Ornithology in Colonial Africa', in *Comparative Studies in Society and History* (forthcoming); P. Harries, 'Field Sciences in Scientific Fields: Entomology, Botany and the Early Ethnographic Monograph in the work of H.-A. Junod', in S. Dubow (ed.), *Science and Society in Southern Africa* (Manchester, 2000); E. Green Musselman, 'Worlds Displaced: Projecting the Celestial Environment from the Cape Colony', *Kronos*, 29 (2003); R. Grove, *Green Imperialism: Colonial Expansion, Tropical Island Edens and the Origins of Environmentalism, 1600–1860* (Cambridge, 1995).

[7] W. Beinart, *The Rise of Conservation in South Africa* (Oxford, 2003), Ch.1. Cf Grove's analysis, in *Green Imperialism*, of the Company Garden at the Cape under Dutch occupation, which he sees as playing a conspicuous role in the global exchange of plant and medical knowledge.

[8] I. Glenn, 'The Bushman in Early South African Literature', in P. Skotnes (ed.), *Miscast: Negotiating the Presence of the Bushmen* (Cape Town, 1996), 46.

framework, these tended to be disparaged as quaint (at best) and irrelevant or irrational (at worst). The power to declare Boer or African folk knowledge as valid or invalid was assumed to be the sole prerogative of western science.

This should not come as a surprise, but it was not solely a function of racism or prejudice. Colonial patriotism and pride depended on the capacity to demonstrate that the local and the particular formed part of a larger, universal scientific scheme. There were, moreover, close links between colonial naturalism and colonial nationalism. To know the land and to conceptualize its peoples was to assert cognitive power and to proclaim a custodial or proprietorial sense of ownership. It was, in a sense, an assertion of acquired indigeneity. If this is right, it may therefore make sense to distinguish between colonial and autochthonous knowledge—and to steer clear of the problematic and essentializing category of indigenous knowledge.[9]

A further distinction requires us to separate out colonial from imperial knowledge. Studies produced within a Saidian or Foucauldian framework of reference frequently elide 'colonial' and 'imperial' in their critique of 'orientalism'. One reason for this is that Saidian-inspired critics are often not much interested in the internal dynamics and complexities of the societies that they wish to speak for. The iniquitous 'western gaze' is the primary object of their criticism. Demonstrating that knowledge about 'others' is an intrinsic aspect of, or even a necessary precondition for, imperialism is the central focus of analysis.[10]

There is undoubtedly some truth in this claim, but it vastly oversimplifies. In the case of this study, it risks overlooking the ways in which the urge to know about others was closely bound up with the process of identity formation. Colonial local knowledge was not only an instrumental resource directed to wielding power over others; it was also bound up with conceptions of self-empowerment and in demonstrating one's worth to one's peers and betters. Colonists and settlers were more than ciphers in the arithmetic of imperialism.

Colonial intellectuals were part of the network of imperial knowledge; crucially, their intellectual outlook was supra-local as well as local, which is to say that their reference points were broadly based and comparative, even if the material they considered was specific.[11] This was not always a comfortable position to adopt, and the position of colonial intellectuals in the imperial network gave rise to feelings of ambivalence, even resentment. They were flattered by the glow of metropolitan attention when their endeavours were recognized. However, as their confidence and sense of self-sufficiency developed, they became increasingly

[9] For recent treatments of indigenous knowledge in the context of developmental and anthropological debates see J.Pottier *et al.* (eds.), *Negotiating Local Knowledge* (London, 2003); M. Leach *et al.* (eds.), *Science and Citizens: Globalization and the Challenge of Engagement* (London and New York, 2005).

[10] See e.g. the powerful polemic against Saidian orientalism developed by J. M. McKenzie in *Orientalism: History, Theory and the Arts* (Manchester, 1995).

[11] I have adapted the use of local and supra-local from H. Raffles, 'Intimate Knowledge', *International Social Science Journal*, 173 (2002).

sensitive to metropolitan condescension—especially if this meant that they were treated as mere field-workers and their institutions regarded as little more than outposts.

In many ways South Africa mimicked in the world of ideas its place in the imperial economy: as a producer of raw materials for export and manufacture overseas. Yet, in other respects the relationship of colonial-based knowledge producers to the imperial centre was less dependent and more analogous to that of provincial-based scientists in (say) Bristol or Manchester, working outside of the golden triangle of London, Cambridge, and Oxford.

Arising out of this is the question of the circulation of knowledge. Historians of science have long moved away from a simple diffusionist model that treats the colonial periphery as little more than a receptor of ideas or a mere base for generating empirical facts. Carl Berger's 1982 lectures on 'Science, God and Nature in Victorian Canada' stated this with elegant simplicity:

> The implanting and growth of science in Victorian Canada was one strand in a complex fabric of transplanted British civilization overseas; like other strands in that culture it was modified and the resulting pattern was not an exact duplication. Nor was it entirely a matter of borrowing. Canadian naturalists belonged to an international community, and their contributions to science were recognized abroad long before it was ever admitted that the country had a history, still less a literature.[12]

There are more theoretical ways of saying this. Roy Macleod's metaphor of the 'moving metropolis', which he borrowed and adapted from the Australian historian Keith Hancock, is often cited in the literature. Macleod used this image to reject George Basalla's mechanistically sequential theory of the spread of western science, on account of its failure to take adequate account of the internal complexities of colonial societies or the reciprocal influence of colonies on the metropole.[13] Macleod's critique is compelling. But its strength is somewhat vitiated when one considers that it rests on the substitution of one developmental model for another, more sophisticated version.

A quarter of a century later historians of empire are less concerned with causal or developmental triggers and rather more interested in fluidity, process, and webs of interlocking influence. There are many excellent examples of this de-centred approach. Christopher Bayly's study of the 'information order' in colonial India, for example, draws attention to the intricacy of communications networks and the interplay between European and indigenous discourses and knowledge systems.[14] Richard Drayton has used the story of Kew Gardens to present a global history of botany that treats empires as 'engines which bring human

[12] C. Berger, *Science, God, and Nature in Victorian Canada* (Toronto, 1983), pp. xiii–xiv.

[13] R. Macleod, 'On Visiting the "Moving Metropolis": Reflections on the Architecture of Imperial Science', in N. Reingold and M. Rothenberg (eds.), *Scientific Colonialism: A Cross-Structural Comparison* (Washington, DC, 1987).

[14] C. A. Bayly, *Empire and Information: Intelligence Gathering and Social Communication in India, 1780–1870* (Cambridge, 1996).

communities, once separated by distance and culture, into systems of exchange and interdependence'.[15] In the South African context, Alan Lester's account of colonial expansion in the eastern Cape examines the creation of British cultural identity and settler power in terms of the region's connections with other places and spaces in a shared imperial network.[16] Kirsten McKenzie's comparative study of colonial reputation and moral worth in nineteenth-century Cape Town and Sydney proceeds from similar assumptions.[17]

The circulation of knowledge, through various nodal points, proceeded in many different ways. By the last decades of the nineteenth century the infrastructure provided by transatlantic steam shipping and the undersea telegraph meant that the Dominions and the dependent Empire were physically linked with the British metropole in ways that would have been almost incomprehensible at mid-century. One manifestation of this was the rise of what Simon Potter terms an 'imperial press system' from the 1870s, characterized by flows of shared information and journalistic interchange.[18] Well before this period, local newspapers in cities throughout the Empire revealed a remarkable awareness of developments in comparable realms of the colonial world. The constant movement of soldiers, settlers, governors, and proconsuls enabled knowledge and comparative experience to be shared and transferred from one region to another. But this interconnected perspective does not mean that the Empire was always supra-national: its spread and development was also consonant with the rise of distinctive national traditions.

Viewing the Empire as an interconnected zone constituted by multiple points of contact offers a significant advance on older, often economic-based theories of core and periphery. Andrew Thompson's recent work, *The Empire Strikes Back*, draws on a formidable range of scholarship to show how the colonial world impacted on metropolitan politics, economics, institutions, and self-image.[19] In respect of the emergence of a widely broadcast British world, such work develops the insights of writers like Linda Colley, John Mackenzie, Catherine Hall, and Norman Davies, all of whom have variously argued that British national identity was constituted, not in isolation, but as a result of conflict and interaction with other peoples and societies.[20] Sometimes outsiders see this most clearly. It was a New Zealand-born intellectual historian, J. G. A. Pocock, who wrote a powerful manifesto in 1975 that challenged A. J. P. Taylor's wilful insularity in

[15] R. Drayton, *Nature's Government: Science, Imperial Britain, and the 'Improvement' of the World* (New Haven, 2000). See also Richard Grove's *Green Imperialism*.

[16] A. Lester, *Imperial Networks: Creating Identities in Nineteenth-Century South Africa and Britain* (London, 2001).

[17] McKenzie, *Scandal in the Colonies*.

[18] S. J. Potter, *News and the British World: The Emergence of an Imperial Press System* (Oxford, 2003).

[19] A. Thompson, *The Empire Strikes Back? The Impact of Imperialism on Britain from the Mid-Nineteenth Century* (Harlow, 2005).

[20] Colley, *Britons*; J. M. Mackenzie, *Propaganda and Empire* (Manchester, 1984); C. Hall, *Civilising Subjects: Metropole and Colony in the English Imagination, 1830–1867* (Cambridge, 2002); N. Davies, *The Isles: A History* (London, 1999).

order to make the case that British history ought to be written in terms of the 'intercultural' story of 'conflict and crossbreeding between societies differently based'.[21]

This study is strongly influenced by such approaches in its treatment of the dissemination of knowledge and the plasticity of English and colonial identities. The thematic and temporal scope is broad, while remaining focused on the emergence of a definable national tradition in one particularly restive colonial component of that British world. My discussion hinges on the tension between knowledge, conceived of as a universal, modernizing force, and its realization in the particular context of a society that was increasingly divided along complex ethnic and racial fault-lines. It seeks to examine the manifold ways in which knowledge has served to support claims to nationhood and political ascendancy in modern South Africa. In so doing, it aims to pose some new questions about old problems.

21 J. G. A. Pocock, 'British History: A Plea for a New Subject', *JMH* 47 (1975), 604–5.

1

Literary and Scientific Institutions in the Nineteenth-Century Cape Colony

CONTEXT

When the Cape colony was first captured from the Dutch in 1795 it was, according to the historian de Kiewiet, 'economically more undeveloped, politically more inexperienced, and culturally more backward than any of the greater colonies of settlement. After one and a half centuries the Colony contained one town worthy of the name, and five or six little villages. The rest of the population was scattered over an immense area.'[1] Little had changed by the time of the British reoccupation of the Cape in 1806, notwithstanding some energetic efforts to improve its prosperity during the preceding three-year Batavian interlude. It is against this background that the impact of British settlement in the nineteenth century should be seen.

The British forces occupied a town that barely filled the natural basin formed by the protective arms of the mountain and the sea. Cape Town was laid out in a grid, with one axis running from the historic Company Gardens to the bay and another between the borders formed by the Buitenkant and Buitengracht arteries. The more important streets were lined with oaks and, in Dutch fashion, watercourses known as 'canals' were also in evidence. Along the shore there were double-storeyed workshops and warehouses. There were several taverns and canteens at which people could gather and talk, but sociable outlets for the expression and dissemination of ideas—such as coffee-houses, a theatre, bookshop, or newspaper—were lacking. No secular institutions devoted to the pursuit of formalized knowledge were yet in existence. There was not even a high school.[2]

[1] C. W. De Kiewiet, *A History of South Africa: Social and Economic* (London, 1941), 30. See also N. Worden and C. Crais, *Breaking the Chains: Slavery and its Legacy in the Nineteenth-century Cape Colony* (Johannesburg, 1994), 1–2, who argue that, in the context of the Atlantic world economy, the slave society of the Cape was a zone of very minor importance.

[2] V. Bickford-Smith, 'Leisure and Social Identity in Cape Town, British Cape Colony, 1838–1910', *Kronos*, 25 (1998/9), 114; R. Lewcock, *Early Nineteenth Century Architecture in South Africa* (Cape Town, 1963), 26, 394; H. Giliomee, *The Afrikaners: Biography of a People* (London, 2003), 28.

The Cape had briefly been touched by enlightenment ideas in the 1780s in the form of the Dutch 'Patriot' movement, yet its impact was transitory. Economically, the settlement had been in decline from the 1790s, so that evidence of prosperity—fine clothes and furniture, for example—was deceptive. Two hundred and fifty years of Dutch settlement were most clearly evident in the substantial material culture left behind from Company rule. Amongst the most impressive public buildings were the Reformed Groote Kerk, the old Slave Lodge, the Good Hope Masonic temple, Government House, and the Castle. Some of these bore the refined late eighteenth-century rococo contributions of Thibault and Anreith. There were also a number of fashionable residences and farmhouses with vineyards situated on the outskirts of the town. But for the most part the vernacular architecture was characterized by a mass of low, flat-roofed, square-fronted buildings, usually plastered in white and framed by green shutters and woodwork.[3]

Facing the great southern ocean on one side and with Table Mountain directly behind, Cape Town was a picturesque, if isolated, enclave, a port situated at the tip of a barely understood continent. Early nineteenth-century maps of the Cape contain only the most rudimentary detail about the interior, and mountain ranges are drawn in such a way as to emphasize the obstacles to travel. This is true even of John Barrow's ambitious attempt, at the beginning of the nineteenth century, to chart the colony's boundaries and interior. Barrow's map, which was intended to facilitate colonization and exploitation of the Cape's resources, represented a remarkable advance in cartographic knowledge and was unequalled for many years.[4] But, despite its ambitious intent, it merely served to underline the reality that the hinterland began almost immediately beyond the environs of Cape Town itself.

Prior to 1830 there was only one 'made' road in the Cape peninsula, leading southwards along the Cape flats from Table Bay via Wynberg to Muizenberg, and thereafter in rather more rudimentary form to the naval port of Simonstown. Cape Town was linked to Stellenbosch, centre of the Dutch-speaking wine-growing district, by a poor-quality road until this route was improved in the mid-1840s. The soft sand of the Cape flats and the ring of mountains north and east of the peninsula made travel beyond the immediate environs of Cape Town extremely difficult; even the immediate grain-producing regions of Caledon and Swellendam were remote. W. L. Sammons, editor of *Sam Sly's African Journal*, wryly observed in 1841: 'You must find your own way into the interior of Africa without guide posts or signposts.'[5]

Amongst historians the predominant view holds that the initial objectives of the British administration were mostly consonant with those of the Dutch East

[3] A. F. Hattersley, *An Illustrated Social History of South Africa* (Cape Town, 1969), 62; Lewcock, *Early Nineteenth Century Architecture*, 11, 14–17; N. Worden *et al.*, *Cape Town: The Making of a City* (Cape Town, 1998), ch. 2.

[4] N. Penn, 'Mapping the Cape: John Barrow and the First British Occupation of the Colony, 1795–1803', *Pretexts*, 4:2 (1993), 31.

[5] Hattersely, *Illustrated Social* History, 107–8.

India Company which it replaced, namely, that it regarded the Cape as a strategic way-station to the east and that its prime purpose was to service the needs of passing ships. Because the Cape was conquered 'for its extrinsic rather than intrinsic value',[6] holding the Castle and securing the sea route was the new British administration's first priority.[7] The successive British occupations of 1795 and 1806 may therefore be seen as incidental consequences of the Napoleonic wars, rather than the fruition of any worked-out plan to create a new colony of settlement in Africa. Only after Waterloo did it become apparent that the British occupation was to be of a permanent nature.

This interpretation has been modified in recent years by one that views the British occupation of the Cape in more purposive terms, and which is inclined to see the early decades of the nineteenth century as an era of genuine transformation. Building on Vincent Harlow's notion of the rise of a 'Second British Empire' in the period between American Independence and the Great Reform Bill, Christopher Bayly suggests that the British occupation was part of a broader imperial resurgence that was 'fired with a sharper sense of British national and imperial mission'. Its leading proponents, committed to maintaining the English *ancien régime*, 'represented a conservative philosophical cabal, as influential as that of Milner, Curzon and Baring a century later, during another period of "new imperialism" '. According to this analysis, the Cape (like India and Ireland) was subjected to a vigorous proconsular colonial despotism whose ethos was determinedly aristocratic, hierarchical, anti-Jacobin, and agrarianist.[8] While Bayly's reassessment of the period may incline one to exaggerate the immediate practical consequences of the British takeover, it would be wrong to ignore the profound shift in ideological register that accompanied British occupation or to underestimate the foundations that were laid for subsequent structural reforms. Change may have been fitful, but it was clearly discernible.

John Barrow, who served Governor Macartney's administration during the first British occupation and exerted an influence on official policy long after his return to Britain in 1803, worked hard to prepare the Cape for more thoroughgoing British colonization. His detailed mapping and statistical survey of the country has been said to have helped to place the Cape 'within the cognitive co-ordinates of the Second British Empire'.[9] During the course of his subsequent forty-year career at the Admiralty, Barrow became a key power-broker in the expanding world of imperial exploration and scientific discovery.[10] His writings about the Cape stressed the need to inculcate British traditions, institutions, and values, all

[6] Penn, 'Mapping the Cape', 21.

[7] Davenport, *South Africa: A Modern History*, 4th edn. (Basingstoke, 1991), 37; T. Keegan, *Colonial South Africa and the Origins of the Racial Order* (Cape Town, 1996), 42; R. Robinson and J. Gallagher, *Africa and the Victorians* (New York and London, 1961), 15.

[8] C. A. Bayly, *Imperial Meridian: The British Empire and the World 1780–1830* (Harlow, 1989), 134, 137, 8–9.

[9] Penn, 'Mapping the Cape', 38.

[10] R. A. Stafford, 'Scientific Exploration and Empire', in A. Porter (ed.), *The Oxford History of the British Empire, Vol. III*, (Oxford, 1999).

of which were presented as the *sine qua non* of progress and improvement—both keywords in the late Georgian lexicon. Barrow also supplied commentators and future historians with a well-stocked repertoire of anti-Dutch images and stereotypes. For example, he reviled the indolence and inertia of the Boers and deplored their cruel oppression of the indigenous Khoisan peoples, whose predicament he likened to slaves. Such ideas were soon taken up and refined for somewhat different purposes by opinion-forming philanthropists and abolitionists like John Philip.[11]

Official measures to anglicize the Cape date back to the regime of John Cradock, who expressed a desire, soon after his appointment in 1811, to 'assimilate the Institutions of this country to those of England...'.[12] Pressures for anglicization, coupled with anti-Dutch sentiment, were given further force by senior administrators like Henry Ellis, who helped to prepare the way for the 1820 settlers and whose recommendations found expression in an 1822 proclamation that sought to replace Dutch by English as the official administrative and legal language of the Cape. The governor, Somerset, arranged the importation of six Scottish Presbyterian ministers versed in the Dutch language to take up positions in the Reformed Church. He also recruited British schoolmasters to teach in government schools and established a grammar school for the instruction of children of leading Indian and Cape civil servants—not least to counter the influence of the private school established by Thomas Pringle in 1823, which Somerset famously regarded as a 'seminary of sedition' and a source of republican sentiment.[13] However, such attempts at anglicization were not implemented in any systematic fashion. Although English had become the main language of administration and political argument by the 1830s, Dutch predominated as the vernacular in the countryside, and the effect of removing it from the syllabus of government schools in 1828 led to a significant drop in attendance.[14] In the mid-1830s fully half of the Cape's twenty-two Reformed ministers were Scots, though in time most integrated themselves into local communities and came to be assimilated as Afrikaners.

The new colonial regime's reluctance to become embroiled in territorial expansion if this meant taking on additional fiscal obligations, meant that British

[11] For discussions of Barrow's writings see e.g. M. L. Pratt, *Imperial Eyes: Travel Writing and Transculturation* (London, 1992); Penn, 'Mapping the Cape'; J. M. Coetzee, *White Writing: On the Culture of Letters in South Africa* (New Haven, 1988).

[12] Cradock introduced English-language schools, attempted legal and land reforms, and declared that fluency in English should be a prerequisite for public employment. See J. Sturgis, 'Anglicisation at the Cape of Good Hope in the Early Nineteenth Century', *Journal of Imperial and Commonwealth History*, 11:1 (1982), 8–9. I use the term 'anglicize' in the sense used by Sturgis, namely, 'those processes or policies which had as their aim the assimilation of Afrikaners into a predominantly English culture' (p. 5). Sturgis points out that the word 'anglicize' was in contemporary usage along with 'anglify'.

[13] Ibid. 8–10, 14, 16. A. K. Millar, *Plantagenet in South Africa: Lord Charles Somerset* (London, 1965), 129–30; E. A. Walker, *The South African College and the University of Cape Town 1829–1929* (Cape Town, 1929), 10–11.

[14] Sturgis, 'Anglicisation at the Cape of Good Hope', 17; Giliomee, *The Afrikaners*, 196–7.

institutions were most often integrated with existing Dutch systems rather than replacing them altogether. Traditions of patronage and clientelism, which so characterized Company rule, largely went unchecked during the first two decades of British occupation. Indeed, the centralized authority of Company rule was matched, even exceeded, by the autocracy of the British. Rule by gubernatorial proclamation was the norm and perquisites continued to be tolerated.[15]

Until mid-century the Cape was typically governed by a senior army officer and veteran of the Peninsular wars. The upper echelon of Cape Town society was dominated by a tightly knit military–civil establishment, centred on Government House, and supported by a small number of leading English merchants and Dutch notables (such as the Truters, van Rynevelds, and Cloetes).[16] Support for the wine industry strengthened connections with the Afrikaner gentry in the countryside. A combination of intermarriage, patronage, and two-way cultural assimilation soon gave rise to an identifiable Anglo-Dutch colonial oligarchy. Within this upper stratum of society, ethnic differences were often transcended by mutual economic, family, and political interests. The persistence of prominent family names is as good an indication as any of the way in which networks of colonial society were sustained and reproduced.

The tenure of Lord Charles Somerset, who governed the Cape from 1814 to 1826, marks an important moment in the Cape's transition from a small trading outpost, ruled by a coterie of military and administrative officials, to an expanding British colony with an increasingly vocal settler population. This outcome was not necessarily his intention, for Somerset's instincts were those of an arch-Tory and a reactionary. While committed to anglicization in respect of fashion and style, he was distinctly averse to social change. A scion of the powerful duke of Beaufort and descendant of the Plantagenet kings of England, Somerset presided over an administration that is generally considered to have been capricious and authoritarian.[17] Like the Regency set in Brighton, with whom he was closely associated, Somerset was fond of pageantry and ostentatious display. He was passionately devoted to such gentlemanly pursuits as hunting and shooting, and served as an enthusiastic patron of the South African Turf Club, established by officers of the first British garrison. Aristocratic, instinctively anti-Whig, and viscerally opposed

[15] J. B. Peires, 'The British and the Cape, 1814–1834', in R. Elphick and H. Giliomee (eds.), *The Shaping of South African Society 1625–1840*, 2nd edn. (Cape Town and London, 1989), 490–3; Keegan, *Colonial South Africa*, 43, 47. Walker, *The South African College*, 9, waspishly characterized the procession of colonial governors as a 'Horse Guards dynasty'.

[16] Worden *et al.*, *Cape Town*, 91, estimate that more than three-quarters of Cape Town civil servants were from Dutch families in 1820.

[17] A sympathetic biography of Somerset is provided by Millar in his *Plantagenet in South Africa*, but the general effect is to confirm rather than reverse the tenor of Somerset's critics. Recent literature is inclined to focus less on Somerset's personal failings and to locate him more as a victim of changing circumstance and as a symbol of a passing age. As Jeff Peires puts it in 'The British and the Cape', 472: 'It was his misfortune to govern the Cape at the time when the new social forces generated in a rapidly industrialising Britain engulfed the colony, sweeping aside not only Somerset but the entrenched power of the local oligarchy and the established rhythms of the local economy.'

to pressures for parliamentary reform at home, Somerset feared that events like Peterloo might spark radical insurrection in the colonies.[18]

These fears were not altogether misplaced. Strains of radicalism and demands for legal and political rights came to the Cape with the 1820 settlers, whose aspirations as free-born Englishmen reflected those of the educated urban, professional, artisanal, and mercantile middle classes in Britain; their grievances at having been dumped without adequate support on an isolated and hostile frontier 500 miles from Cape Town were real enough. Somerset met such pressures with undisguised antipathy, just as he resisted the humanitarian impulses of evangelical missionaries associated with the London Missionary Society and the anti-slavery movement. In his desire to govern the Cape like a private estate, Somerset resisted pressures for economic, social, and political reform, whether these emanated from England or Cape Town. But, although able to control the pace of change to an extent by accommodating the new forces that threatened him from within and without, Somerset was eventually overwhelmed by events and compelled to return to England in 1826, where he resigned his office.

A concerted challenge to Somerset came in the form of Lord Liverpool's reforming Liberal Tory administration, noted for its recognition that industrialization was inevitable, rather than aberrant, and for attempting to devise a coherent theory of economic policy.[19] Liverpool's government despatched a Commission of Inquiry to the Cape in 1822 as part of a broad review of Britain's new colonial acquisitions, and in order to open up the Indian Ocean to free trade. The commissioners' report, submitted to Bathurst in 1826, recommended far-reaching administrative, legal, and institutional changes. Amongst the most important of these were the creation of an efficient civil service and the abolition of old offices of government, the separation of judicial and executive functions of officials in the countryside, the establishment of a Supreme Court and appointment of judges from the British bar, and an end to institutional patronage and corruption. Philosophically, the assumptions of the Commission were underpinned by a firm conviction that the restrictions of the mercantilist system should give way to the liberal principles associated with free enterprise and good government. Politically, they were directed against the established Anglo-Dutch ruling oligarchy.[20] Thus, the reformist measures recommended by the Commission were directed as much against the 'old corruption' of Somerset's conservative administration as they were against the extant Dutch establishment. In Wayne Dooling's judgement, the Commission 'broke the corrupt cabal that was the colonial government'.[21]

Modernization and anglicization were strong features of the Commission's recommendations, but many were diluted in practice by a mixture of resistance

[18] Hattersley, *An Illustrated Social History of South Africa*, 115; Millar, *Plantagenet in South Africa*.

[19] Dror Wahrman, *Imagining the Middle Class: The Political Representation of Class in Britain, c.1780–1840* (Cambridge, 1995), 233–4.

[20] Peires, 'The British and the Cape', 495–9; Keegan, *Colonial South Africa*, 96, 100–2.

[21] Wayne Dooling, *Slavery, Emancipation, and Colonial Rule in South Africa* (forthcoming), ch. 2.

and inertia. Compromise, rather than wholesale assimilation of English jurisprudence and traditions, was the result. Notably, Roman-Dutch law was maintained as the legal bedrock of the Cape, with long-run implications for the constitution of colonial society. Nevertheless, modernization of the ramshackle Dutch colonial regime proceeded with gathering pace, spurred by Somerset's successor, Richard Bourke, a Whig reformer who carried through many of the commissioners' recommendations in addition to reforms of his own devising.[22]

Some three decades after its incorporation into the British Empire, the Cape was therefore conspicuously engaged in a process of development which, however slow and uneven, saw its transformation from an agrarian-based Dutch colonial slave society to an extensive and expanding English colony in which commerce and internal trade were assuming increasing importance. Progress was the watchword of this new regime. As Robinson and Gallagher pointed out over forty years ago, it was the 'spirit of progress no less than the surpluses and shortages of the industrial community that drove Britons outward'.[23] The concept of improvement had also been a ubiquitous feature of the early modern Atlantic world of British settlement. As well as helping to gain approval in the metropole by countering assumptions of endemic colonial idleness, the language of improvement and civilization served to legitimate the colonial enterprise in the eyes of colonists themselves, and thereby assisted them in establishing a 'corporate sense of identity'.[24]

The arrival of the 1820 settlers had the effect of extending the dominion of the colony eastwards, quickening the pace of economic activity, and heightening tensions with Xhosa-speaking Africans on the hotly contested frontier. By about 1840 the total population of the Cape had doubled to approximately 150,000, amounting to two persons per square mile of territory.[25] The population of Cape Town in 1840 approached 20,000, about half of whom were ex-slaves or 'coloured'. Not counting the imperial garrison, the white population of the town therefore stood at about 9,000. The majority of these were Dutch. A German visitor at this time commented that there were few cities in the world in which so many different nationalities were concentrated.[26]

Economic reforms were a central concern of the British administration. Many of the commercial restrictions that characterized Dutch Company rule were eased and, from the 1820s especially, concerted efforts were made to introduce a freer labour market. Stimulated in part by the introduction of preferential rates on

22 Keegan, *Colonial South Africa*, 100–1.

23 Robinson and Gallagher, *Africa and the Victorians*, 2.

24 J. H. Elliot and Jack P. Greene, in N. Canny and A. Pagden (eds), *Colonial Identity in the Atlantic World, 1500–1800* (Princeton, 1987), 9–11, 228–9.

25 D. Hobart Houghton, *The South African Economy*, 4th edn. (Cape Town, 1976), 5; R. Elphick and H. Giliomee, 'The Origins and Entrenchment of European Dominance at the Cape, 1652–*c.*1840', in Elphick and Giliomee (eds.), *The Shaping of South African Society*, 524; R. Ross, *Beyond the Pale: Essays on the History of Colonial South Africa* (Johannesburg, 1993), part 1.

26 E. A. Walker, *The South African College and the University of Cape Town 1829–1929* (Cape Town, 1929), 4–5; Worden *et al.*, *Cape Town*, 89.

wine exports to Britain, the agrarian-based economy made steady, if modest, progress. The abolition of slavery had a major political and economic impact on the structure of colonial society.[27] Payment of compensation to slaveholders by the British government led to a significant inflow of capital which stimulated the development of commercial and public institutions. However, efforts to supplant a century and a half of mercantilism with the spirit of free enterprise and commercialism took time to have an effect. Sources of capital were still restricted and Cape merchants remained heavily dependent on their London partners or backers. The first private Cape bank was only established in 1837.[28] Between 1807 and 1835 the balance of Cape trade was overwhelmingly negative, the deficit being largely offset by payments to support the military and civil administration. Commercial prosperity, still focused on Cape Town, was considerably affected by the size of the garrison and the continued presence of the British civil administration, both of which attracted substantial inflows of capital and funds from London.[29] Cape society therefore remained heavily biased towards the needs of its military and administrative elite.

Traces of anglicization soon became apparent in the physical environment of Cape Town (more so in country towns like Grahamstown, Stellenbosch, and Graaff-Reinet), but here too the results were happily inconclusive. Spare Georgian architectural tastes modified, albeit without transforming, the 'true baroque landscape' left behind by the Dutch. St George's, the imposing Anglican church located at the bottom of Government Avenue, was funded by the issue of interest-bearing shares and completed in 1834. A Wesleyan chapel was built in 1822, and the Presbyterians built St Andrew's (generally known as the 'Scotch' church) in 1829.[30] The Commercial Exchange, a vaunting single-storey neoclassical building with a Corinthian portico, was built in 1822 on the western flank of the Grand Parade, and funded by subscription shares. The use of private funds for investment in public institutions was a conspicuous feature of British rule, testimony both to the enhanced role of the market and the emergence of a commercially oriented middle class.

The construction of the Royal Observatory, which was completed in 1828, sounded a note of direct imperial interest. Crucial support for the decision to locate the Observatory at the Cape came from John Barrow, now secretary of the Board of the Admiralty, and Joseph Banks of the Royal Society and the Colonial Office. The Cape Observatory was the colony's first major scientific institution, if measured only by the scale of investment in buildings, salaries, and equipment.[31] It also possessed considerable cultural authority. Situated on a low hill below the

[27] For a fresh statement on the transforming effects of British colonialism and slave emancipation on the agrarian economy of the nineteenth-century western Cape, see Wayne Dooling's forthcoming book *Slavery, Emancipation, and Colonial Rule in South Africa*.

[28] Ross, *Beyond the Pale*, 35.

[29] Ibid. 23; J. L. Meltzer, 'The Growth of Cape Town Commerce and the Role of John Fairbairn's *Advertiser* (1835–1859)', MA thesis, UCT (1989), chs. 2, 3.

[30] Hattersley, *Illustrated Social* History, 122, 125.

[31] The South African Library and the South African Museum should also be noted (see below).

mountain's sublime Devil's Peak, a few miles outside the town near the Salt and Liesbeek rivers, the Observatory is today lost from view amidst highways and overgrown bush. Yet it occupied a dominant position when it was built.

Designed in the mode of the Greek Revival by the neoclassicist architect John Rennie, the Observatory's stark stucco surfaces, lead roof, and careful mathematical proportioning served clear symbolic as well as functional purposes. It was all the more imposing for its restrained, rational lines. The Time Ball which it dropped to allow mariners to synchronize their ships' chronometers was clearly visible from Table Bay, and served as a powerful reminder of ordered British rule. As we shall see, the Cape's royal astronomers—Fearon Fallows, Thomas Maclear, and David Gill among them—played leading roles in the intellectual and associational life of nineteenth-century Cape Town, variously lending their support to its developing scientific and educational institutions.[32]

From the ordinary colonists' perspective, some of the most important physical registers of progress during the second British occupation were macadamized and paved street surfaces. Malodorous canals were filled in and piped water provided, along with street taps at regular intervals.[33] Civic and domestic space was also affected by English precedents. A more open civic environment was created through the laying out of new formal public squares.[34] Spatial boundaries between the public and private were redefined in this process. For example, many of the elevated 'stoeps' or porches that rose via a flight of stairs from the street—and whose low brick seats were used by householders for smoking and casual observing of the passing parade—were either dismantled or enclosed, as if to emphasize the separate propriety of private home and public street.[35]

FAIRBAIRN, PRINGLE, AND THE *SOUTH AFRICAN JOURNAL*

Early attempts at anglicization, whether of a conservative and surface nature (Somerset) or else directed to deeper structural reform (Barrow, Ellis, and the Commission of Inquiry), were driven from above and without. From the early 1820s, however, a new phase of anglicization became evident. Significantly, this was led from within and below. A wave of immigrants and colonists, with no official ties to the colonial government, claimed the citizenship rights of free-born Englishmen either for themselves or for others, sometimes both. In the case of those abolitionists, Christian evangelicals, and humanitarians who fought to secure rights for slaves and indigenous inhabitants, success was considerably

[32] Brian Warner, *Royal Observatory, Cape of Good Hope, 1820–1831: The Founding of a Colonial Observatory* (Dordrecht, 1995); Lewcock, *Early Nineteenth-Century Architecture*, 251–4.

[33] Hattersley, *Illustrated Social History*, 74–6; Worden *et al.*, *Cape Town*, 113.

[34] Lewcock, *Early Nineteenth-Century Architecture*, 392, 394–5.

[35] Ross, *Status and Respectability*, 81.

reliant on the capacity to mobilize the support of sympathetic networks in Britain. By contrast, those who claimed the rights and freedoms due to colonists, worked to build and defend locally based political and social institutions, often by reference to comparable developments in other parts of the Empire. Some of these activists were disposed to liberal notions of freedom and subscribed, at least in theory, to the notion of a 'colour-blind' society. Others, especially eastern Cape frontiersmen based in Grahamstown, combined an explicitly anti-humanitarian platform with a strong assertion of British settler identity. As Alan Lester shows, this mind-set of fear was steeped in the anxieties of domination.[36]

Opposition to the humanitarian lobby, as well as advocacy of self government, was also voiced by *De Zuid-Afrikaan* (est. 1830), which spoke on behalf of Dutch-speaking slave-owners in the western Cape and flourished long after the demise of its anglophone equivalent, the *South African Commercial Advertiser*. Yet a shared hostility to interfering liberals was not sufficient to draw together pro-British settlers and Dutch-speaking slave-holders. As the humanitarian impulse faltered from the 1840s, common political ground was instead forged between Dutch and English colonists around the need for citizenship rights and representative institutions.[37] It is through the interplay of the struggle for citizenship rights and colonial freedoms that the beginnings of colonial identity can be discerned at the Cape.

In countless historical overviews and school textbooks the brave efforts of the Scots immigrants John Fairbairn and Thomas Pringle to establish 'freedom of the press', in defiance of Somerset's strictures, mark the beginnings of the colonial desire to win and protect political liberties. The determination of these pugnacious colonists to publish the *Commercial Advertiser* (with the printer George Greig) was rewarded in 1828 when the secretary of state, William Huskisson, removed the publication of newspapers from direct executive control. Historians writing in the South African liberal tradition have celebrated the epic struggle of Pringle and Fairbairn, linking it to the campaign by the missionary emancipationist John Philip to improve the legal and social rights of the indigenous Khoi population, the fight against slavery, and the battle to establish representative political institutions at the Cape.[38] It is around these related causes that the core foundational myths of South African liberalism have been constructed.

Radical and Marxist critics of the liberal school, by contrast, are more inclined to focus on the links between Fairbairn's advocacy of colonial liberties and his role as a spokesperson for the Cape's nascent mercantile bourgeoisie.[39] Timothy Keegan, one of the more subtle proponents of this view, dismissively refers to

[36] A. Lester, *Imperial Networks: Creating Identities in Nineteenth-Century South Africa and Britain* (London, 2001), ch. 3; A. Bank, 'Liberals and their Enemies: Racial Ideology at the Cape of Good Hope, 1820–50', Ph.D thesis, Cambridge University (1995), introduction.

[37] Giliomee, *The Afrikaners*, 117; Bank, 'Liberals and their Enemies'.

[38] See e.g. H. C. Botha, *John Fairbairn in South Africa* (Cape Town, 1984); Hattersley, *Illustrated Social History*, 68. Cf. A. D. Hall's 'Pringle, Somerset and Press Freedom', *English Studies in Africa*, 3:2 (1960).

[39] Meltzer, 'The Growth of Cape Town Commerce'.

Fairbairn as the 'chief propagandist for the humanitarian–mercantile lobby', adding that his ideological message reflected an 'uneasy coexistence of philanthropic and more overtly utilitarian thinking'.[40] This analysis is supported by the fact that, as the century progressed, humanitarian impulses were increasingly resolved in favour of the economic and political self-interest of the colonists. Whereas Fairbairn strove constantly to reconcile and advance the common interests of Dutch- and English-speaking colonists, his record as a supporter of the rights of indigenous South Africans wavered, and during the 1840s he became distinctly more conservative on issues of colour.[41]

Fairbairn's wavering commitment to humanitarianism has perplexed and disappointed twentieth-century historians, who have looked to him as a standard-bearer of modern liberal values. Due allowances for changes in liberal sensibility and doctrine over the course of more than a century are not always made. Even so, the shift in his outlook becomes more explicable—and less contradictory—when one treats Fairbairn in terms that may have made sense to him: namely, as a proto-colonial nationalist who, in the name of civilization and progress, consistently promoted the economic, social, and political interests of the Cape's aspiring middle classes, while seeking, wherever possible, to establish common ground between English- and Dutch-speakers. Fairbairn performed this function as the untiring and irrepressible progenitor of improving public institutions, ranging from societies concerned with the advancement of science and literature to organizations dedicated to the promotion of agriculture, temperance, and life-assurance. Voluntary societies like these helped to comprise the rich 'associational world' that Peter Clarke describes as having come of age in Britain and the English-speaking world by 1800, and which went on to 'reach full maturity as a powerful social institution' in the nineteenth century. Such organizations were 'increasingly perceived as central, not only to urban sociability, but also to public advancement and communal identity'.[42]

Fairbairn's brand of liberalism was primarily concerned with securing political, economic, and institutional freedoms for the middle classes in the face of imperial despotism; the mix of humanitarian, philanthropic, and egalitarian ideals that

[40] Keegan, *Colonial South Africa*, 98, 99.

[41] On Fairbairn's changing views and the tensions within and between evangelical, humanitarian, and 'political economy' liberalism, see A. Bank, 'Liberals and their Enemies'; also A. Bank, 'Losing Faith in the Civilising Mission: The Premature Decline of Humanitarian Liberalism at the Cape, 1840–60', in M. Daunton and R. Halpern (eds.), *Empire and Others: British Encounters with Indigenous Peoples, 1600–1850* (London, 1999).

[42] P. Clarke, *British Clubs and Societies 1580–1800* (Oxford, 2000), 139–40, 273. In the USA, as Clarke shows, the American Revolution led to an upsurge of voluntary associations. Their spread to the Empire was more patchy. Clarke includes the Cape in his 'associational world', but because his study ends at the turn of the nineteenth century he is able to note only the first examples of such institutions (e.g. the Sick and Burial Society, the Turf Club, and masonic lodges), several of which which were decidedly the domain of the elite (pp. 95, 423–4). For the later period see R. J. Morris, 'Clubs, Societies and Associations', in F. M. L. Thompson (ed.), *Cambridge Social History of Britain 1750–1950* (Cambridge, 1990).

inspired representatives of the evangelical liberal tradition (most notably his father-in-law, John Philip) were supported by him with somewhat lesser conviction. His unshakeable belief in progress and in the desirability for civilization to prevail over savagery was mitigated by an ingrained sense of justice and fairness. But, at a time when the Cape colony was rapidly encroaching onto lands occupied by Africans in order to expand its agricultural base and pool of available labour, it is undeniably the case that Fairbairn became more a champion of the colonizers than the colonized.[43]

Fairbairn's close collaborator Thomas Pringle shared many of his objectives, though his political trajectory was somewhat different. Pringle was initially concerned with the rights of settlers on the frontier, and he was primarily motivated by anger against the colonial government. However, after meeting Philip and Stockenstrom in 1825, Pringle became increasingly attuned to the injustices perpetrated against the country's indigenous peoples, and he left the Cape for London, where he devoted his energies to the cause of abolitionism until his death in 1834.[44] Pringle's most distinctive contribution to the creation of colonial consciousness was as a romantic poet in the mould of Wordsworth and, especially, Walter Scott, who served as his patron and mentor.[45] The Scottish moral philosophers Dugald Stewart and Adam Ferguson were early influences.

As a lowland Scot displaced onto the eastern Cape frontier, Pringle was 'precariously situated on the threshold of culture and nation' in South Africa as well as Scotland. For this reason he was, as Damian Shaw argues, a 'Borderer' in a double sense. Pringle's attempts to invest the exotic eastern Cape landscape with aesthetic and moral meaning, and his efforts to envision Africa within a recognizably European optic, point to the indeterminacy and ambivalence that so characterized nascent anglophone colonial identity.[46] His sense of liminal identity—a central theme of his poetry—was made all the more real by the multiple tensions that existed between his position as leader of a vulnerable Scottish settler party in the midst of a hostile frontier, his conviction that the settlers were being poorly

[43] Cf. Meltzer, 'The Growth of Cape Town Commerce', 184–7. See also S. Trapido, 'The Emergence of Liberalism and the Making of "Hottentot Nationalism", 1815–1834', in *Societies of Southern Africa: Collected Seminar Papers*, Institute of Commonwealth Studies, vol. 17 (1992). Alan Lester, while agreeing that Fairbairn's politics became much more conservative during the 1840s, says that he was loathed by eastern Cape settlers in the 1820s and early 1830s and that he adopted a sympathetic line towards the Xhosa during the 1834–5 frontier war (personal communication).

[44] D. J. Shaw, 'The Writings of Thomas Pringle', Ph.D thesis, Cambridge University, (1996), 120, 122–4.

[45] Pringle met Walter Scott in 1816. It was through Scott's influence that he was awarded land as a settler in Glen Lynden, and Scott was also instrumental in his appointment to the South African Library.

[46] Shaw, The 'Writings of Thomas Pringle', 64. Also J. M. Coetzee, *White Writing*, 45–48, 163–4; G. Hutchings, ' "Toughly Disputing the Right of a Poem to Possess it": Landscape in South African Colonial Poetry', *English in Africa*, 24: 2 (1997); A. E. Voss, 'Thomas Pringle and the Image of the "Bushmen" ', *English in Africa*, 9: 1(1982).

treated by the colonial government, and his growing understanding that settlers were not the only victims of colonial oppression and unfreedom.[47]

As Scots (or 'North Britons'), Pringle and Fairbairn brought a fresh perspective to colonial society. Drawing on their formative experiences of Edinburgh (which, like Cape Town, was at once a capital and a provincial city), they sought to import the political and cultural values inherited from the Scottish enlightenment. Edinburgh, with its vigorous civic and radical political culture centred on literary and debating societies, clubs, periodicals, and other improving institutions, constituted an obvious point of reference; evangelicalism, utilitarianism, and romanticism provided the interconnected set of ideas that framed their thinking. While firm in their opposition to slavery, Pringle and especially Fairbairn were mostly concerned with the rights and status of enlightened colonists, Dutch as well as British.[48] Scotland's deep involvement with empire, the close affinities between Scottish and Cape Dutch legal traditions, and the connections between the Scottish and Dutch Reformed churches all provided natural entry-points into colonial society.

It was by fashioning a literary and scientific culture that adapted ideas of universal progress to local contexts and sensitivities that Pringle and Fairbairn sought to position the Cape within the wider Empire, as a dynamic new colonial outpost populated by proud and respectable citizens, whose commitment to social progress, improvement, and civilization entitled them to assert the common rights of free-born Englishmen (or Britons). Condescension from the metropole was an especially sensitive matter to this fledgling colonial community. Thus, they rounded on the 'Contempt' displayed towards 'Cape people' (in this case, remarks by Barrow, now in a position of influence at the Admiralty) in their capacities as editors of the *South African Commercial Advertiser*:

Whatever we are, whether born in the Northern or Southern hemisphere, in England, in Holland, or in Africa, if we have made Africa our home, and feel a common interest in the prosperity of the Colony, we are all Africans.

If we do as we ought—we will love, respect, and promote the welfare, first of our families, friends, and connections—secondly, that of the Colony we inhabit—next of the Empire of which it forms a part—and lastly of the whole human race.[49]

[47] Shaw demonstrates that Pringle's political awareness developed only gradually and that it was only towards the end of his stay in South Africa that he began to focus on injustices perpetrated against indigenous South Africans rather than against settlers and himself. It was as secretary of the Anti-Slavery Society from 1827 until his death in 1834 that his commitment to abolitionism became a dominant aspect of his life. See 'The Writings of Thomas Pringle', 120 ff.

[48] Ibid. 75–7. As McKenzie points out, Fairbairn carefully modulated his criticisms of the institution of slavery and avoided the use of emotive terms in order not to antagonize the Dutch colonial constituency on whose support he depended. See K. McKenzie, 'The *South African Commercial Advertiser* and the Making of Middle Class Identity in Early Nineteenth Century Cape Town', MA thesis, University of Cape Town (1993), 145.

[49] *South African Commercial Advertiser*, 17 Mar. 1824. This hierarchy was a standard Scottish trope and features prominently in Adam Smith's *Theory of Moral Sentiments*. It acquired political resonances when radicals wanted to alter the ordering to place love of humanity above love of country or Burke's 'little platoons'. My thanks to Donald Winch for this insight.

The editors' broad definition of colonial identity to include all (presumably, white) inhabitants of the Cape as 'Africans' is striking.[50] A little over a century after the drunken Hendrik Biebouw protested that he was an 'Afrikaander'—the first recorded use of this term[51]—two serious-minded Scottish immigrants were declaring their own adopted African affinities. Also worth noting is the hierarchy of loyalties and affinities that they sketch out: beginning with kith and kin and expanding outwards to embrace the whole of humanity.

This expansive evocation of a colonial moral community may be compared with the narrow remit of the *South African Commercial Advertiser's* founding editor, printer George Greig, who envisaged the newspaper principally as 'a medium for business transactions and advertising'.[52] Under Fairbairn and Pringle's editorship the *Commercial Advertiser* considerably expanded its horizons. Kirsten McKenzie neatly characterizes its 'underlying agenda' as 'the creation of a rational public sphere within the colonial context, out of which a distinctive middle class identity might be formed and which might allow representative government to be established at the Cape'.[53] Members of the Cape Dutch community who shared these enlightened goals were included in this programme. By emphasizing the creation of a colonial middle-class identity, and by drawing attention to the notions of masculinity and respectability that underpinned it, McKenzie moves well beyond simplistic liberal/idealist or materialist/instrumentalist readings of the *Commercial Advertiser*. Yet even this conception does not take full account of its editors' ambitions. Still more attention ought to be placed on the Pringle–Fairbairn collaboration as part of a dual-pronged attempt to fashion the moral and cultural politics of the Cape.

This dimension becomes clearer when we recall that Pringle and Fairbairn began their journalistic careers in South Africa not with the *Commercial Advertiser*, but with a literary periodical entitled the *South African Journal*. Their campaign to publish without pre-censorship began in 1823, in association with Abraham Faure, a fitfully anglophile ... Dutch minister who was responsible for producing the *Journal*'s vernacular equivalent, *Het Nederduitsch Zuid-Afrikaansche Tijdschrift*. In a memorial presented to Governor Somerset in February of that year, Pringle and Faure specifically requested permission to publish a 'South African Magazine' in order to serve the needs of Dutch as well as English inhabitants of the colony and to 'promote the improvement of South Africa'. The petitioners' emphasis on the mutual interests of Dutch and English colonists is key. So, too, is their stress

50 McKenzie, 'The *South African Commercial Advertiser*', suggests that the use of 'African' in preference to 'native' amounts to a form of appropriation whereby 'Africa was being reinvented by the colonists' and turned 'into a domain of rational enlightenment' (p. 188). There is something to this analysis, but the choice of 'African' over 'native' was surely also conditioned by the editors' recognition that many of their reading public, like themselves, were recent immigrants.

51 Giliomee, *The Afrikaners*, 22–3.

52 Shaw, 'The Writings of Thomas Pringle', 72. Shaw adds that Greig was 'resolutely determined to avoid personal controversy or any subjects relating to colonial government'.

53 McKenzie, 'The *South African Commercial Advertiser*', 4.

on the benefits to be derived from the 'general diffusion of knowledge, Piety and civilization throughout this extensive Colony', and their argument that a periodical literature enjoying governmental support was a feature of every other significant British colony as well as the more advanced parts of Europe.[54]

Pringle evidently had in mind a literary journal akin to Blackwood's *Edinburgh Monthly Magazine*, of which he had briefly been a founding joint editor and to which his friend and partner Fairbairn had contributed various literary pieces.[55] Constable's *Edinburgh Magazine and Literary Miscellany* (which Pringle edited after leaving Blackwood's) and the intellectually powerful *Edinburgh Review* provided other models. Publications such as these were intended to be active agencies of political enlightenment and social criticism, in addition to their concern with literary and scientific matters. The central theoretical and practical issue that the *Edinburgh Review* sought to address has been neatly summarized by the question: 'on what basis could one hope to design political institutions suitable for modern commercial society?'[56] Fairbairn and Pringle were equally concerned with this question. In the repressive context of the early nineteenth-century Cape, raising such questions inevitably brought them into conflict with the authorities and cast them as advocates of colonial freedoms.

Other colonial precedents—the American one in particular—were also to the fore. As Fairbairn wrote in accepting Pringle's invitation to join him in Cape Town: 'Your hint about magazines and newspapers pleases me exceedingly. What should hinder us from becoming the Franklins of the Kaap?'[57] Fairbairn's invocation of the American printer—and his use of the vernacular Dutch to describe the Cape colony—bears comparison with Benedict Anderson's reminder, in the context

[54] Letter from Faure and Pringle to Somerset, 3 Feb. 1823, in G. M. Theal, *Records of the Cape Colony from August 1822 to May 1823*, vol. 15 (Cape Town, 1903), 263.

[55] J. Meiring, *Thomas Pringle: His Life and Times* (Cape Town, 1968), 14–15; Botha, *John Fairbairn*, 9–10. *Blackwood's* was established by James Blackwood in 1817 under the joint editorship of Pringle and J. T. Cleghorn, in order to appeal to Scottish Tories and to act as a counter to the *Edinburgh Review*. Pringle and Cleghorn were fired after only six numbers, before Blackwood himself took over. According to *The Wellesley Index to Victorian Periodicals 1824–1900,* vol. 1, 'Pringle and Cleghorn would seem to have been generally incompetent' (p. 7). This seems unfair. G. Pottinger, *Heirs of the Enlightenment: Edinburgh Reviewers and Writers, 1800–1830* (Edinburgh, 1992), 168–9, explains that they were 'not the right men to take charge of the magazine', and that they were replaced in part because they failed to make much of an impact on the literary scene, but largely because, instead of attacking the *Edinburgh Review*, they published a laudatory essay on one of its founders, Francis Horner. Pringle's politics were certainly Whiggish, though, as Damian Shaw's dissertation on the 'Writings of Thomas Pringle' shows, he adopted an apolitical stance in his early journalistic years.

[56] B. Fontana, *Rethinking the Politics of Commercial Society* (Cambridge, 1985), 10. On the culture of periodical literature in Edinburgh at this time, see J.Shattock, *Politics and Reviewers: The Edinburgh Quarterly and the Quarterly in the Early Victorian Age* (Leicester, 1989); Pottinger, *Heirs of the Enlightenment.* Wahrman, in *Imagining the Middle Class*, 237, argues that the *Edinburgh Review* played a crucial role in the 1820s and 1830s 'in asserting the link between the "middle class" and the beneficial consequences of social progress'.

[57] *CMM* 7: 42 (June, 1860), 375; Kirsten McKenzie, 'Gender and Honour in Middle-Class Cape Town: The Making of Colonial Identities 1828–52', D.Phil. thesis Oxford University (1997), 71–2. The example of Benjamin Franklin was one that Fairbairn was later to invoke in his role as editor of the *South African Commercial Advertiser.*

of his classic discussion of the relationship of print-capitalism to the emergence of imagined national communities, of Benjamin Franklin's indelible association 'with creole nationalism in the northern Americas'.[58]

In March 1824 the first issue of the *South African Journal* appeared, in a run of 1,000 copies. The venture did not survive past the first two issues (though Faure's Dutch equivalent continued to publish without interference until 1843, apparently because it steered clear of confrontation with officialdom). Particular offence was caused by an article written by Pringle which was critical of Somerset's treatment of eastern Cape settlers. Pringle also published 'Afar in the Desert', in which he addressed the problems of spiritual displacement and exile faced by a native Scot in a new and exotic land. This much-anthologized poem, which drew high praise from Coleridge, is commonly regarded as the first authentic example of South African versifying. A. M. Lewin Robinson, who observes Pringle's willingness to use Cape Dutch words and 'assimilation of local idiom' into his poetry, extends this interpretation to the *Journal* itself, locating its importance 'in the fact that it set out to be South African and not merely an English imitation . . . '.[59]

The other major contributor to the *Journal* was Fairbairn, who revealed his youthful attachment to romanticism by way of a lengthy appreciation of Wordsworth's poetry. Fairbairn's account of 'Literary and Scientific Societies', and their role in the diffusion of knowledge, proved more controversial. This drew directly on his formative intellectual experience as a member of the Literary and Philosophical Society of Newcastle from 1818 to 1823; it amounted to a thinly disguised manifesto for the creation of a similar institution at the Cape, with all the connotations of radical nonconformism associated with provincial or peripheral 'Lits and Phils'.[60]

The suggestion was received without favour by Somerset, whose instinctive response was to regard Fairbairn and Pringle as agents of Jacobinism and radicalism—just as they would have been regarded by others of Somerset's social class and outlook in Britain.[61] It is hardly surprising, therefore, that the attempt to create a South African Literary Society was suppressed soon after eleven men attended its first meeting in 1824. Here, Fairbairn spoke in barely coded terms about how such societies could help to advance scientific enquiry and encourage

[58] B. Anderson, *Imagined Communities* (London, 1983), 62.

[59] A. M. Lewin Robinson, 'The English Periodical Literature of the Cape Colony from its Beginnings (1824) to 1835', Ph.D thesis, University of Cape Town (1961), 56, 57–8, 75. See also Meiring, *Thomas Pringle*, 62: His poems 'indicate how deeply he absorbed the atmosphere of South Africa with all its colourful and expressive Dutch, Hottentot and Kaffir associations, and, by introducing them into his writing, he became the first colonial writer to make use of indigenous words for effect'.

[60] Botha, *John Fairbairn*, 6. Fairbairn moved to Newcastle in 1817 or 1818. He worked as a teacher in Bruce's Academy, having studied medicine at Edinburgh University. He did not complete his degree. On the culture of provincial Lits. and Phils., see e.g. Arnold Thakray, 'Natural Knowledge in Cultural Context: the Manchester Model', *American Historical Review*, 79: 3 (1974); I. Inkster and J. Morrell, *Metropolis and Province: Science in British Culture, 1780–1850* (London, 1983).

[61] R. D. Altick, *The English Common Reader* (Chicago, 1957), 77.

a taste for reading and enquiry amongst wider sections of the population.[62] At least fourteen of the sixty-three names associated with the South African Literary Society had earlier signed a petition in favour of press freedom. This was sufficient to condemn it under the 1800 anti-Jacobin proclamation, designed to control radical clubs and societies.[63]

Fairbairn's conception of scientific knowledge was closely modelled on the upwardly mobile, utilitarian and dissenting traditions of the Newcastle Lit. and Phil. It was concerned more with improving the political and cultural standing of its members, and changing the balance of social relations, than it was devoted to the pursuit of knowledge for its own sake. A substantial literature has developed which shows how, in the period 1815–48, British scientific institutions and other agencies of specialist knowledge were often used by liberals and supporters of manufacturing and commerce to attack the inherited privileges of the traditional ruling classes. As Morris notes, urban elites in Britain commonly looked to science as 'the basis for cultural assertion' . . . 'Science was built into middle-class and elite urban identity as part of a bid for legitimacy and power'.[64]

Although the *South African Journal* managed only a short public existence, it was a significant venture. Not only did the *Journal* help to inaugurate the battle for press freedom, it also served as the prototype of later literary magazines in the colony that sought to create a distinctive colonial identity at the Cape (see Chapter 3). In seeking to define an intellectual public existing independently of established networks of power and patronage, the *Journal* served notice of the tentative emergence of a self-conscious English-speaking colonial community whose interests were increasingly at variance with those of the governing elite. Here was an excellent example of an initiative to establish the 'literary public sphere' which Jürgen Habermas sees as one of the institutional pillars of bourgeois civil society.[65]

This interpretation, stressing the quest for individual and collective moral and cultural authority, allows us to see beyond the liberal view that the struggle for

[62] Lewin Robinson, 'The English Periodical Literature', 82; L. Crawford, 'The South African Literary and Scientific Institution, 1832–1857', *Transactions of the Royal Society of South Africa*, 22 (1934), 314. Crawford records that Fairbairn said that it would be prudent to exclude from the discussions of the Literary Society topics of great interest and importance relating to the conduct of government, controversial theology, and slavery. Somerset was nonetheless outraged at reports of the meeting (which had resolved to invite him to act as patron and to grant them a licence), summoning John Fairbairn and John Philip before the Fiscal and charged them with holding an illegal meeting. The first annual general meeting of the South African Literary Society took place only in 1829.

[63] L. D. Bregman, ' "Snug Little Coteries": A History of Scientific Societies in Early Nineteenth Century Cape Town, 1824–1835', Ph.D dissertation, University of London (2003), 56, 58. Bregman's detailed history of the SALS amply confirms and expands the point that Fairbairn saw scientific societies as 'a means to secure the social and political legitimacy' of their middle-class membership, and that he saw such organizations as the most effective way to counter gubernatorial autocracy (see p. 23, and ch. 2), also, Clarke, *British Clubs and Societies*, 440.

[64] Morris, 'Clubs, Societies and Associations', 410; Inkster and Morrell (eds.), *Metropolis and Province*, 57.

[65] C. Calhoun (ed.), 'Introduction', in *Habermas and the Public Sphere* (Cambridge, Mass., 1992).

press freedom was an end in itself (an abstract love of liberty), or the instrumental materialist view that it was merely a means to an end (ideological legitimation of the interests of the mercantile bourgeoisie). Rather, it draws attention to the importance of visible demonstrations of the maturity and worth of the Cape's aspirant citizenry. Proof of worthiness and respectability was essential in order to combat disparaging metropolitan attitudes towards 'colonials', to attract investment, and to validate claims to political freedoms and autonomy.[66]

Such a perspective invites us to expand our understanding of the 'needs' of the rising middle classes—whose ranks included merchants, professionals, civilian doctors, churchmen, civil servants, and farmers—and to appreciate that they were by no means solely focused on economic success. Status was equally salient. As in Scotland, where the notion of the 'middling ranks' had been used by David Hume and Adam Smith to link commerce with good government and liberty (a theme later taken up by James Mill and Thomas Macaulay in the first decades of the nineteenth century), commercial growth was closely associated with civic virtue and firmly linked to political and social progress. Middle-class identity thus depended not only on economic position but also on shared systems of political and moral values—of which education, taste, refinement, progress in the arts and sciences, and the development of political consciousness, were all important components.[67] Schools, churches, organizations, and networks of marriage were all central to the developing bourgeois culture of the Cape's new middle-class society. So, too, were institutions devoted to the cultivation of science and the intellect.[68]

THE SOUTH AFRICAN LIBRARY, COLLEGE, MUSEUM, AND INSTITUTION, 1822–1833

In light of Somerset's hostility to press freedom and the South African Literary Society, it is notable that the reactionary governor was prepared to support the creation of the South African Library, established by proclamation in 1818 and opened to the public in 1822. Somerset may well have envisaged it as a genteel

[66] Shaw, 'Writings of Thomas Pringle', 77–9. For an account of prejudice and disdain towards 'colonials' on the part of the British governing classes see T. E. Kirk, 'Self-government and Self-defence in South Africa: The inter-relations between British and Cape Politics, 1846–1854', D.Phil. thesis, Oxford University (1972).

[67] B. Fontana, *Rethinking the Politics of Commercial Society*, 156–7; S. Collini *et al.*, *That Noble Science of Politics: A Study in Nineteenth Century Intellectual History* (Cambridge, 1983), 27–9; see also Wahrman, *Imagining the Middle Class*.

[68] Bregman, '"Snug Little Coteries"', 10–11, records the creation or proposal of no fewer than eleven societies and institutions broadly devoted to the pursuit of science in the decade from 1824—where previously there had been none. These include the Cape of Good Hope Horticultural Society (1826–8); the South African Medical Society (est. 1828); and the Cape of Good Hope Agricultural Society (est. 1831).

club which would constitute a lesser political threat than other public or voluntary associations.[69] In its first years the Library was funded by an ingenious fiscal arrangement that exemplified the higher virtues of trade: a tax of one rix dollar on every cask of wine, brandy, and vinegar brought into Cape Town. The core of the Library's holdings was the impressive private collection of over 4,500 books assembled by the eighteenth-century German settler, Dutch East India Company official, gentleman antiquarian, and private trader Joachim Nicholas von Dessin. This he bequeathed on his death in 1761 to the Dutch Reformed Church (then the only learned body in existence at the Cape) in anticipation of a dedicated public library. Other parts of the von Dessin collection—coins, medals, and miscellaneous 'curiosities'—subsequently formed key elements of the South African Museum and the National Gallery.[70]

The fact that the Library's holdings derived from the Dutch period served as a salutary reminder of the broadly based origins of the Cape's distinctive colonial culture. Yet the Library was most closely associated with the rising anglophone middle class. Between 1828 and 1860 it was housed in the Commercial Exchange. This substantial neoclassical building aptly symbolized the merchant community's sense of self-importance and aspiration. The Exchange served as a brokerage in the broadest sense, for in addition to functioning as a centre of business activity and link to London-based finance houses and politicians, it also provided a venue for social gatherings and public meetings. Fairbairn was a leading member of the Exchange, and the *Commercial Advertiser* was afforded its strong support. The Exchange was thus a further expression of the closely intertwined struggle by the colonial middle class to establish commercial, cultural, and political institutions independent of official control.[71]

The origins of the South African Museum likewise go back to the time of Somerset, who announced its purpose in 1825 as 'the reception and classification of the various objects of the Animal, Vegetable and Mineral Kingdoms which are found in South Africa, whereby an opportunity will be opened to the colonists of becoming acquainted with the general and local resources of the Colony'.[72] The honorary superintendent was Dr Andrew Smith, an upwardly mobile medical graduate of Edinburgh University, who arrived at the Cape in 1821 as a young army surgeon in pursuit of adventure and to further his prospects. Smith's medical training was wide-ranging. He developed extensive scientific interests while living and working on the frontier, in particular in the fields of natural history and anthropology. He also gained a notable reputation as a scientific explorer of the

[69] It is worth noting that Thomas Pringle served as one of the first librarians until he resigned the post in 1824 in protest at Somerset's suppression of the press. See W. H. P. A. Tyrrell-Glynn, 'The History of the South African Public Library, Cape Town, 1830–1961', Ph.D thesis, Stellenbosch University (1983), 1.

[70] R. F. H. Summers, *A History of the South African Museum* (Cape Town, 1975), 2–3.

[71] McKenzie, 'Gender and Honour', 78; R. F. M. Immelman, *Men of Good Hope 1804–1954* (Cape Town, 1955); Meltzer, 'The Growth of Cape Town Commerce', 30–2; Worden *et al.*, *Cape Town*, 101.

[72] Summers, *History of the South African Museum*, 5.

South African interior (doubling as a secret intelligence gatherer for the colonial government), during which time he made contact with African leaders like Dingane, Moshoeshoe, and Mzilikazi.[73] His most ambitious scientific project was undertaken in 1834–6, when he led a major scientific expedition to the northern Cape and as far as the Limpopo. He was supported in this venture by the newly created South African Literary and Scientific Institution, as well as by a specially constituted committee of prominent and influential Cape Town citizens which raised money for the trip by issuing subscription shares. It was portentously styled the Cape of Good Hope Association for Exploring Central Africa.[74]

As well as being an accomplished naturalist Smith was an energetic curator and an assiduous collector. Within a year of the Museum's establishment he compiled a *Descriptive Catalogue* devoted to mammalia. Smith developed the museum's holdings by means of public appeals for interesting specimens, building on the donations of amateur collectors and antiquarians like von Dessin and another noted German immigrant, the apothecary and horticulturalist 'Baron' Carl von Ludwig. Smith also worked with local providers of live and stuffed animals like C. M. Villet, who catered to the overseas market's demand for zoological specimens.

The French firm of taxidermists Maison Verreaux established a branch in Cape Town in 1818 and maintained a close relationship with the Museum. One member of the family, Jules Verreaux, combined his professional work with acting as a curator of the Museum from 1829 to 1838. During its first years the Museum accumulated ethnographic and natural history collections, principally for scientific study. There were in effect four collections making up its holdings: Smith's, von Ludwig's, and Verreaux's (which were private and on loan), and the Institution's contents which were referred to as the Government Collection. This arrangement was not unusual. In Britain, conditions of 'shared proprietary ownership' between individual collectors and embryonic natural history museums or societies were common at the time.

In its early days the general public was not afforded easy access to the Museum, which was regarded as a private club for knowledgeable gentlemen. One important visitor who is said to have inspected its contents was the 27-year-old Charles Darwin, who spent time with Smith when the *Beagle* called in at the Cape in

[73] P. R. Kirby, *Sir Andrew Smith—His Life, Letters and Works* (Cape Town and Amsterdam, 1965). Smith served from 1821 to 1825 as a military doctor on the eastern frontier—arriving soon after the departure of the anatomist Robert Knox, who had been similarly engaged. Smith may well have been influenced by Knox in his choice of career as an army doctor, for they had overlapped briefly as students at Edinburgh. Another contemporary of his was the well-known Cape society and army doctor James Barry.

[74] The prospects for trade and colonization clearly whetted the appetites of those who were not necessarily scientifically inclined. Smith's 1836 *Diary of the Expedition* is now an Africana classic, as are his pathbreaking taxonomic studies of South African natural history, published in four volumes in 1849 as *Illustrations of the Zoology of South Africa*. See Summers, *History of the South African Museum*, 6 ff.; Kirby, *Sir Andrew Smith*; id., 'Andrew Smith, M.D., Founder of the First South African Museum', *Annals of the South African Museum*, 36: 1 (1942), 4.

1836. The two evidently held each other in high regard, for they maintained a friendship and correspondence which lasted until Smith's death.[75] Darwin did not, however, find the Cape physical environment attractive, and nor was he overly impressed by it as a place of scientific interest.[76]

The contents of the embryonic Museum were removed into storage in 1827, when the building housing it and the South African Library appeared to be on the point of collapse. Just as its creation was dependent on private patronage and commercial enterprise, so its demise reveals the fragility of civic and intellectual culture at this time. Yet the story of its creation is strongly suggestive of the importance of natural history collections and embryonic museums as manifestations of local civic pride and cultural erudition in provincial British towns and imperial entrepôts.[77] The Museum's next incarnation occurred in association with the creation in 1829 of the South African Institution, which benefited from the active patronage of its 'privileged member', the governor Sir G. Lowry Cole. The Institution was intended to stimulate investigation of the geography, natural history, and general resources of the country. Andrew Smith served as joint secretary, and his friend and associate Fearon Fallows, the astronomer royal at the Cape, served as one of the vice-presidents. Jules Verreaux served as acting curator of the Museum in Smith's absence, and probably succeeded him.[78]

According to Bregman, the South African Institution was geared more directly than any other organization to the pursuit of original scientific research, and its membership was closely aligned to the governing elite, unlike Fairbairn's literary societies, which interpreted science in utilitarian terms and whose conception of knowledge was implicitly opposed to the existing order.[79] One of the Institution's first initiatives was to launch the *South African Quarterly Journal*. This publication, which ran intermittently from 1830 to 1837, served as an outlet for Smith's scientific papers on topics ranging from South African birds and poisonous snakes to pioneering ethnographic essays on the character and history of the 'Hottentot race'. The Institution also pressed for the establishment of a scientifically oriented botanical garden, and it assumed responsibility for the residual remains of the original museum in 1830.[80] When Smith left the Cape in 1837 to mount a major exhibition in London, arising out of his 1834–6 expedition to the 'Interior of

[75] Summers, *History of the South African Museum*, 8–13, 15, 16–17; K. H. Barnard, 'History of the South African Museum', unpublished TS, South African Museum, pp. 21–6; Kirby, *Sir Andrew Smith*, 223, 266. Darwin proposed Smith as a Fellow of the Royal Society in 1857.

[76] W. S. Barnard, 'Darwin at the Cape', *South African Journal of Science*, 100 (2004), 243–48.

[77] S. Alberti, 'Placing Nature: Natural History Collections and their Owners in Nineteenth-century Provincial England', *BJHS* 35 (2002), 292.

[78] Smith first met Fallows on board ship to the Cape in 1821, and they had both petitioned Somerset in 1825 in favour of the creation of a Literary Society.

[79] Bregman, '"Snug Little Coteries"', 26 and *passim*. This is one of Bregman's more original claims, and it is substantiated in exhaustive detail. Yet he may draw too strong a distinction between these organizations, based on analysis of their membership lists.

[80] Summers, *History of the South African Museum*, 13; J. C. Adamson, 'The South African Institution', *CMM* 7: 39 (1860), 153–7; Warner, *Royal Observatory*, 177–8. Kirby, *Sir Andrew Smith*, 88.

Africa', he took with him his private collection from the Museum. Von Ludwig likewise removed his collection of birds and insects. Here again was a reminder that embryonic civic institutions like the Museum were dependent on private goodwill, and that they were heavily reliant on a tiny network of individuals. It was probably on the authority of the Institution that the remnants of the Museum were given on loan to the newly established South African College and used as part of its teaching collection.[81]

The South African College (SAC) had been founded in 1829, initially without government assistance and with the frank intention of facilitating the emergence of an Anglo-Dutch male colonial elite. The *Commercial Advertiser* welcomed it as a 'popular institution altogether formed by the People', which was 'fashioned and so far completed by the patriotic exertions of the Colonists alone'. Although avowedly non-partisan in political and religious matters, women and non-whites were excluded.[82] The SAC attracted a number of distinguished Dutch- and English-speaking scholars as professors at this time, including Abraham Faure, James Adamson, James Rose Innes, and A. N. E. Changuion, all of whom contributed significantly to the intellectual and cultural and civic activities of Cape Town. Conceived as a local institution, it functioned as an embryonic stem-cell, thickening the tissues of colonial nationalism. Many SAC graduates went on to fill important positions within colonial society, especially as openings were created with the attainment of self-government in 1853. Notwithstanding its uncertain beginnings in the old orphanage, hampered by years of financial uncertainty, fluctuating student numbers, and political infighting, the College was an enduring institution with a significant educational legacy. It eventually transmuted into the University of Cape Town.[83]

The year 1829 also saw the revival of the hitherto suppressed South African Literary Society. Fairbairn was again the key mover. At its peak the new Literary Society attracted 104 members, the majority of whom were drawn from the Dutch-speaking colonial community. Sir John Truter and Andries Stockenstrom

[81] Summers, *History of the South African Museum*, 17–18.

[82] W. Ritchie, *The History of the South African College 1829–1918*, vol. 1 (Cape Town, 1918), 51. In acknowledgement of the College's opening to women students a new verse was added to its anthem in 1887: 'Some students are girls, | South African pearls, | And ladies of high degree; | They blend sweetness with light, | mingle meekness with might, | And add grace to the grave Q.E.D.' Another verse captured the College's non-sectarian aspirations: 'Our blood varies much, | We are English, French and Dutch, | And German in pedigree, | But whatever our race, | Racial feuds we efface | In love for the old S.A.C.' See ibid. 303, 304. Keegan, *Colonial South Africa*, 98, says of the South African College: 'It was a monument to the humanitarian-mercantile alliance that dominated Cape Town politics at the time—although it is indicative of the fundamentally conservative nature of that alliance that people of colour were not regarded as yet ready for admission.' This point is taken but, considered from the point of view of the aspirant white colonial citizenry fighting to secure their own civil rights, the term 'the people' was not entirely misplaced.

[83] Walker, *The South African College*; W. K. Durrill, 'The Rising Colonial Elite in Nineteenth Century Cape Colony: Students, Violence and Social Relations at South African College', South African Historical Society Conference, July 1997; Hattersley, *Illustrated Social History of South Africa*, 133–4.

were elected as the first joint presidents, Revd W. Wright and Fairbairn as joint chairmen. Bregman draws attention to the high proportion of colonial-born members on the committee. His detailed analysis of the Literary Society's membership provides further evidence of the emergence, over a relatively short period, of an increasingly confident middle class that was reformist and often humanitarian in political outlook and reasonably inclusive in terms of ethnic and social composition.[84]

There is indeed good evidence to see the post-Somerset era, a period of major social and political realignment, as a time of cultural vibrancy and civic assertion.[85] The year 1829, in which the ordinance guaranteeing freedom of the press was passed, saw an overall expansion of the associational world of voluntary societies. The foundation of the South African College, Institution, and the Literary Society in 1829 has already been noted.[86] 1829 also saw the formation of a civilian theatre company, which put on several Shakespeare plays in its inaugural season.[87] Access to the old African Theatre, originally an elite preserve, had become more open following the addition of a pit, but reaction set in during the 1820s: measures to exclude theatregoers of 'doubtful appearance' were put in place, and in 1829, following a brawl, slaves and 'free blacks' were barred from entry.[88] If anything can be concluded from this incident, it is that in the process of redrawing the domain of legitimate public activity and sociability—particularly as social hierarchies were becoming more relaxed—measures of inclusion and exclusion went hand-in-hand.

In light of the small size of the Cape's intellectual community, as well as the degree of shared membership between the Literary Society and Andrew Smith's South African Institution, these two organizations amalgamated in 1832 to become the South African Literary and Scientific Institution, which claimed eighty-nine ordinary members and six subscribers the following year.[89] The members of the combined Institution constituted a roll-call of Cape Town's great and good:

[84] Bregman, '"Snug Little Coteries"', ch. 3. The Literary Society had ambitious intellectual and scientific aims, though most projects seem not to have been fulfilled. Bregman draws attention to Fairbairn's emphasis on utilitarian science and the particular emphasis he placed on agricultural improvement.

[85] This was true of Dutch- as well as English-speaking society.

[86] In 1829, too, Baron Von Ludwig began to lay out his gardens. See below.

[87] D. Johnson, *Shakespeare and South Africa* (Oxford, 1996), 36.

[88] Bickford-Smith, 'Leisure and Social Identity', 114. The first theatrical performances were put on by the members of the British garrison. The civilian African Theatre, pet project of Governor George Yonge, was built during the first British occupation. Sited like a 'Temple of Dionysus' in the middle of Hottentot (Riebeek) Square, it was an all-box affair funded by hefty subscriptions on the part of the upper English and Dutch elite. A pit was added later, though social status was preserved by graded ticket prices.

[89] Crawford, 'The South African Literary and Scientific Institution', 316. See Bregman, '"Snug Little Coteries"', chs. 4, 5, for full discussion of the South African Institution and the Literary and Scientific Institution. He argues that the latter organization carried through the traditions of Smith's South African Institution, with its more establishment base and elite view of science, and that Fairbairn's Literary Society with its more diverse intellectual agenda and humanitarian/colonial nationalist orientation, was largely effaced as a result of the merger. This analysis, though plausible, is rather too narrowly based to be fully convincing or indicative of broader trends.

men such as Adamson, Borcherds, Breda, Cloete, Ebden, Fairbridge, Jardine, Murray, Neethling, Stockenstrom, Smuts, Silberbauer, Tredgold, Truter, Versfeld, and Watermeyer were leading representatives of the legal, official, merchant, and professional groupings that comprised Cape Town's emerging Anglo-Dutch urban middle class. Dr Philip was a member of the Institution's Council and Thomas Pringle became an honorary member (having left the colony in 1826).[90] Pringle's name represented a direct link with the struggle waged a decade earlier to create the essential infrastructure of civic society. Fairbairn, however, was notably absent.[91]

In 1834 the Institution elected Sir John Herschel, the recently knighted astronomer and leading member of the Royal Society, as its president. His presence had a galvanizing effect on the organization's activities and drew its work closer to metropolitan scientific concerns. Herschel had arrived at the Cape that same year with his young family. He soon established himself at Feldhausen, a Dutch country estate in the fashionable rural surroundings of Claremont. Here Herschel built an observatory and reflecting telescope in order to undertake the monumental task of cataloguing the clusters, nebulae, and double stars of the southern skies.[92] Herschel was not officially connected with the Royal Observatory, but he maintained a close personal and professional association with its incumbent, Thomas Maclear. One project in which they cooperated (and which grew out of the activities of the SA Institution) was the inauguration of systematic meteorological observations at the Cape. Herschel also indulged in a number of other scientific pursuits at this time, ranging from experiments in physics and chemistry and pioneering work in photography to significant work on the unique floral kingdom of the Cape.[93]

Unconstrained by any official position, financially independent, and with versatile scientific interests and abilities, Herschel was the epitome of the independent gentleman scholar. Lady Margaret, an accomplished botanical illustrator, added lustre to her husband's stellar reputation.[94] During their four-year sojourn in the colony the Herschels exerted an enlivening influence on its social, cultural, and artistic life and Feldhausen became a byword for gracious living. The Herschels afforded easy hospitality to many of the leading citizens and dignitaries of the day with little regard to protocol or social exclusivity. They maintained a wide social circle, while keeping their distance from the conservative elite set

[90] Crawford, 'The South African Literary and Scientific Institution', 317.

[91] Bregman, '"Snug Little Coteries"', 169.

[92] Herschel's achievement may be seen as an extension, to the southern hemisphere, of his father William's great project of surveying the northern celestial sphere.

[93] G. Buttman, *The Shadow of the Telescope: A Biography of John Herschel* (London, 1974), 112; B. Warner (ed.), *John Herschel 1792–1992: Bicentennial Symposium*, Royal Society of South Africa (1992).

[94] On Lady Margaret Herschel, see N. Erlank, 'Circulating in Cape Town: Material Culture and Social Relations in the Life of Lady Margaret Herschel', BA (Hons) thesis, University of Cape Town, (1993).

surrounding Government House. Amongst the many local and overseas visitors to Feldhausen were the liberal missionary John Philip and the Griqua leaders Adam Kok and Andries Waterboer, who was shown round the Observatory in 1834 by Maclear. Charles Darwin, who regarded Herschel as a hero—and saw the Cape as one of the many colonies around the world in which 'little embryo Englands are hatching'—was another appreciative guest at Feldhausen.[95]

Darwin's observation neatly captures the role of science in enlarging the networks of imperial influence. In a recent study, Steven Ruskin locates Herschel firmly in the Humboldtian tradition of scientific explorers like James Cook and Joseph Banks. Already a highly respected scientific figure before arriving at the Cape, Herschel became an iconic national hero on his return to England, and was much fêted for advancing the glory of the British Empire. At a dinner in his honour held at the Athenaeum in London in 1838, several notable speakers likened Herschel's scientific achievements to those of explorers and conquerors, while noting the importance of astronomical science to the navy. The marquis of Lansdowne, who spoke on behalf of Cambridge University, linked Herschel's role in 'annexing' a new scientific hemisphere to the 'empire of knowledge' with the consolidation of British rule at the Cape.[96]

Herschel, who insisted on funding his voyage to the Cape himself and who steadfastly declined official patronage, did not explicitly associate himself with an expansionist empire, and was more directly concerned with the advancement of science. But he undoubtedly supported the orderly progress of civilization and viewed astronomy in terms that underlined the need to devise improved forms of imperial governance. Elizabeth Green Musselman's striking analysis of the scientific metaphors used by Herschel shows how his view of astronomy as a celestial 'hunt' was modified by his experience at the Cape. Herschel's exposure to the reality of colonial violence, she argues, prompted a shift of register such that he increasingly likened astronomy to the more pacific process of 'harvesting' associated with that higher stage of production, agriculture.[97]

The Herschels were rapidly sensitized to colonial social realities. In 1834, the year of their arrival, they were witness to the financial wrangling and political uproar surrounding the emancipation of slaves. They recoiled at the outbreak of the Sixth Frontier War, which was accompanied by a significant hardening of racial attitudes and resulted in the colonial annexation of Xhosa territory

[95] P. Moore and P. Collins, *The Astronomy of Southern Africa* (Cape Town, 1977), 60–3; B. Warner, 'Sir John Herschel at the Cape', in Warner (ed.), *John Herschel*, 50–1. Darwin had long wanted to visit Africa. He held Herschel in the highest esteem and, while at Cambridge, had been greatly influenced by his *Preliminary Discourse on the Study of Natural Philosophy* (1831). A. Desmond and J. Moore, *Darwin* (London, 1991), 91, 184; Kirby, *Sir Andrew Smith*, 222–3; Barnard, 'Darwin at the Cape'.

[96] S. Ruskin, *John Herschel's Cape Voyage: Private Science, Public Imagination and the Ambitions of Empire* (Aldershot and Burlington, 2004), 63–5.

[97] Elizabeth Green Musselman, 'Swords into Ploughshares: John Herschel's Progressive View of Astronomical and Imperial governance', *British Journal for the History of Science*, 31 (1998).

between the Kei and Keiskamma rivers.[98] Such was Herschel's disapproval of the war that he chose to leave town in the company of the Maclears in order to avoid attendance at a dinner marking Governor Benjamin D'Urban's return from the frontier. Yet, Herschel did not support the Xhosa unreservedly. Rather like George Grey a generation later, his view was that Africans would have to learn to live peaceably under colonial rule and that their social system had to be reformed.[99] Referring to the frontier conflict, Margaret Herschel commented in a letter that they were 'delighted' by the Cape—with the exception of the 'tyrannical Spirit which Colonies are apt to engender among their Inhabitants towards the Natives'. Her husband, she said, 'is much interested in the causes of this war & throws the weight of his name & character into the side which humanity & justice dictate, & which is guarded by a mournful minority at the Cape . . . '.[100]

An enthusiastic champion of social and scientific reform, Herschel was a leading proponent of working-class education in England. In his (unsuccessful) campaign to become president of the Royal Society in 1830 Herschel had advocated the need to substitute the 'aristocracy of rank and power' with 'the prouder aristocracy of science'.[101] Unlike the desire of utilitarians such as Henry Brougham, whose desire to diffuse knowledge amongst the masses was infused with paternalism and didacticism, Herschel accepted that many aspirant working-class readers looked to literature for diversion and entertainment, not just for instruction and improvement.[102] These principles he helped to put into practice. Working in collaboration with John Fairbairn and John Bell, Herschel proved instrumental in persuading Governor Napier to institute an advanced system of state-funded education from 1839. This was notably independent of the churches, oriented towards the pursuit of general and scientific knowledge (rather than the classics-based curriculum of the elite English public schools), and without the overt stress on anglicization that marked the educational initiatives pursued by Somerset.[103] In the Cape, as elsewhere in the world, the creation of

[98] On the effects of the 1834–5 frontier war on racial attitudes, see Bank, 'Liberals and their Enemies', esp. chs. 4 and 8.

[99] Elizabeth Green Musselman, 'John Herschel and Imperial Governance', 428–9.

[100] D. S. Evans *et al.*, *Herschel at the Cape: Diaries and Correspondence of Sir John Herschel, 1834–1838* (Austin, Tex., 1969), 167, Lady Herschel to Caroline Lucretia Herschel, 19 May 1835.

[101] Herschel, cited in T. L. Alborn, 'The Business of Induction: Industry and Genius in the Language of British Scientific Reform', *History of Science*, 34 (1996), 94.

[102] Altick, *English Common Reader*, 96, 105, 137, 140, 371. Tyrrell-Glynn considers that Herschel was in part responsible for persuading Jardine, librarian at the South African Library, to stock fiction. See his 'The History of the South African Public Library', 15; cf. Johnson, *Shakespeare and South Africa*, 25–7, 30, which is rather less generous to Herschel.

[103] E. G. Pells, 'Sir John Herschel's Contribution to Educational Developments at the Cape of Good Hope', in *Quarterly Bulletin of the South African Library*, 12: 2 (1957); W. T. Ferguson and R. F. M. Immelman, *Sir John Herschel and Education at the Cape 1834–1840* (Cape Town, 1961); P. E. Spargo, 'Foundations Strong and Lasting—Herschel's Work in Education at the Cape', in Warner (ed), *John Herschel*. What came to be known as the 'Herschel System' involved a two-tier form of education: free government-supported elementary schools with teaching either in English or Dutch, and government-aided fee-paying 'classical' or higher schools. The system was overseen by a superintendent-general of Education; James Rose Innes, appointed in 1839, was its first incumbent.

a colonial school system proved to be a key component in the development of youthful colonial nationalist consciousness.

PUBLIC AND CIVIC INSTITUTIONS, *c.*1835–1860

By the mid-1830s an interconnecting set of public and voluntary institutions can be said to have taken root at the Cape. These were guided by the spirit of improvement, civic virtue, scientific enquiry, and reasoned debate that marked the emergence of an increasingly confident and predominantly English-speaking urban middle class in a new colony of settlement. To a significant degree they reflected the compound of evangelical and utilitarian impulses that was so conspicuous a feature of political and social life in Britain at the time.[104] The constellation formed by the Library, Museum, College, and Institution (all bearing the title 'South African') as well as the Royal Observatory, was comprised of bodies whose orbits were not yet fully defined. The individuals who shaped their trajectories had barely attained critical mass (there were probably fewer than 1,000 permanent administrators, professionals, and merchants in Cape Town at this time and a similar number attached to the garrison). Yet, the framework of a knowledge-based civic society was identifiably in place, stimulated by growing mercantile prosperity and steady, though unplanned, British immigration.

A number of short-lived periodicals, such as the *South African Journal* (1824), the *South African Chronicle* (1824–6), the *South African Quarterly Journal* (1830–6/7), and the *Cape Literary Gazette* (1830–5), gave voice to these developments. Modelled on British examples, and with a high proportion of local content, they shared a common commitment to the encouragement of reading and the diffusion of 'useful' knowledge.[105] The South African Library functioned as key intellectual and cultural brokerage centre, as well as serving as a clublike meeting place for the exchange of news and views. The connections between the Library, the Museum and the South African College were close, and the membership of all these institutions overlapped to a significant extent.

During the tenure of Alexander Johnstone Jardine, who succeeded Fairbairn as librarian in 1824 and remained in post until 1845, the Library's holdings expanded significantly from around 10,000 volumes to something over 30,000. After 1827, when it lost its revenue from wine taxation and was converted to a subscription library, the Library's finances underwent rapid improvement. In 1830 it was designated as the library of the Cape Colony, thereby anticipating its subsequent claim to be a 'national' institution. Typical of the many favourable testimonials was that of George Greig, who declared it the 'pride and boast of the

104 See e.g. B. Hilton, *The Age of Atonement: The Influence of Evangelicalism on Social and Economic Thought, 1795–1865* (Oxford, 1988); E. Stokes, *The English Utilitarians and India* (Oxford, 1959).

105 For more details see Lewin Robinson, 'The English Periodical Literature'.

Colony' in his *South African Almanac* of 1832. The Library also enjoyed an excellent reputation amongst overseas visitors, a significant component of whom were the much-prized Anglo-Indian visitors (or 'Hindoos')—senior officials and military officers with high disposable incomes who elected to spend their leave at the Cape rather than return to Britain. One of the Library's boasts was that it was recommended to the Bengal civil and military services in 1830 for the excellence of its collection.[106] Herschel called the Library 'the bright eye of the Cape'.[107]

Jardine was a prominent local worthy who enjoyed a reputation as an amiable man of sophisticated tastes. He was much respected for his professionalism and the breadth of his scholarly interests. Visitors to the Cape, especially those from India, were often directed to Jardine at his desk in the Library, where he acted as an informal interlocutor between colony and home country. A Scottish borderer, like his predecessor Pringle, Jardine was also interested in literary matters, combining his librarianship duties with journalism and editing the *South African Chronicle* and the *Cape of Good Hope Literary Gazette*. The former occupied the space left vacant after the demise of the *South African Journal*—much to Pringle's annoyance.[108] Although more circumspect in politics than Pringle, Jardine was a fervent abolitionist and a Whig who supported the 1832 Reform Bill. He was likewise committed to the development of colonial institutions and did much to extend the provision of libraries and book societies in country towns. One of Jardine's initiatives was the creation of a 'popular' library designed to cater for the (presumably white) working population who could not afford the South African Library's subscription charges. But he was less apparently well regarded amongst Afrikaners as a consequence of his attacks on the newspaper *De Zuid-Afrikaan* and his condescending attitudes towards the 'country Dutch'.[109]

This deserves further comment. English attitudes towards the less progressive Dutch 'section' were often scornful and dismissive. Many influential members of the anglicized Dutch-speaking elite were disposed to accommodate themselves to the new colonial order and the economic opportunities it offered. As Hermann Giliomee explains, this stratum (they were sometimes called 'Anglomen') was 'increasingly attracted to the idea of liberal progress, with its emphasis on the non-racial franchise, representative democracy, separation between church and

[106] This paragraph relies on material in Tyrrell-Glynn, 'History of the South African Public Library', 47–53. As well as being a vital source of information and statistics, the *Cape Almanac* served as a register of all forms of colonial progress and improvement.

[107] *Proceedings of the 25th Anniversary Meeting of the Subscribers to the Public Library, Cape Town, Saturday 6th May, 1854* (Cape Town, 1854), 20.

[108] Tyrrell-Glynn, 'History of the South African Public Library', 16–17, 47–51; A. M. Lewin Robinson, 'The History of the South African Public Library', in C. Pama (ed.), *The South African Library: Its History, Collections and Librarians 1818–1968* (Cape Town, 1968), 4; A. M. Lewin Robinson, 'The English Periodical Literature', 146, 153–4.

[109] Tyrrell-Glynn, 'History of the South African Public Library', 52; Lewin Robinson, 'The English Periodical Literature', 153–4. But see Jardine's 1827 *A Fragment of Church History at the Cape of Good Hope* (repr. Cape Town, 1979), pp. iv, 13, 40, which stresses the affinities between the Presbyterian and Dutch Reformed traditions and defends the character of 'Dutch boers'.

state, secular education, and the value of scientific enquiry'.[110] But liberal nostrums and English cultural arrogance provoked a reaction amongst other Dutch-speaking intellectuals and churchmen, who sought, in the 1820s and 1830s, to revive links with the Netherlands and to defend their ethnic and cultural distinctiveness. Opposition to Ordinance 50 of 1828, which abolished status distinctions based on colour, and above all to slave emancipation provided a vital political rallying point for such dissentients.

The reassertion of Dutch intellectual and cultural values at this time led to a questioning or rejection of anglophone cultural and political hegemony, though not necessarily of the concept of liberal progress itself. This cultural revival was expressed in the theatre and in journalism as well as in efforts to retain the use of the Dutch language in the Reformed Church.[111] Examples include Abraham Faure's *Nederduitsch Zuid-Afrikaansche Tijdschrift*, C. E. Boniface's satirical plays, the anti-humanitarian newspaper *De Zuid-Afrikaan* with its splenetic attacks on liberal 'humbugs', Joseph Suasso de Lima's literary ventures, and the *Maatschappij ter Uitbreiding van Beschaving en Letterkunde* (est. 1833), which worked on a history of the Cape. Yet the extent to which this literary outpouring signified a concerted attack on British values, much more a rejection of British authority, is open to question; in the more liberal environment of the post-Somerset era, cultural and intellectual activity was able to flourish in many diverse ways. We should also remember that men like Faure, de Lima, and Boniface were personally invested in more than one culture.[112]

A more uncompromising rejection of the modernizing colonial order was evident in the 'countryside conservatism' propounded by the doughty Paarl minister G. W. A. van der Lingen. As Jean du Plessis explains, van der Lingen fought a lonely rearguard battle from the 1830s to the 1860s against the forces of liberalism and progress. He detected such influences in the theology of the Dutch Reformed Church, in linear theories of historical development (he favoured a cyclical view of historical change), and in the construction of a railway link between Cape Town and Paarl—all of which he opposed. Van der Lingen's influence over his countrymen was considerable, but he was increasingly viewed by fellow Afrikaners as an anachronistic voice.[113]

[110] Giliomee, *The Afrikaners*, 128, 198.

[111] S. Trapido, 'Van Riebeeck Day and the New Jerusalem: Identity, Community and Violence in the 18th and 19th Century Cape', seminar paper, Institute of Commonwealth Studies (1992–3), 15–18.

[112] Ibid. See also Giliomee, *The Afrikaners*, 198–9; Worden *et al.*, *Cape Town*, 133, 157–8. Faure was partly educated in England; Boniface was born in France; De Lima was an Amsterdam-born Portuguese Jew.

[113] Jean du Plessis, 'Colonial Progress and Countryside Conservatism: An Essay on the Legacy of van der Lingen of Paarl, 1831–1875', MA thesis, University of Stellenbosch (1988). Interestingly, du Plessis argues that van der Lingen was strongly opposed to a local Dutch-based nationalism. He looked instead to reinvigorate the *Nederduitsche* character of the Cape, and it is therefore mistaken to see him as an early Afrikaner nationalist.

It would therefore be mistaken to overestimate, as some historians have tended to do, the extent or long-term significance of the anti-English reaction on the part of elite colonial Afrikaners. Notwithstanding intense anger at slave emancipation and the broader reaction against humanitarian ideals and practices which helped to spur the Voortrekker movement, the attempt at a Dutch cultural and intellectual renaissance was losing momentum. By the 1840s, as du Toit and Giliomee have pointed out, Cape Afrikaners were largely 'accepting [of] British rule, positively identifying themselves with the norms of British political history and taking their stand on their rights as British subjects'.[114]

The Library

As noted already, the South African Library was a bastion of anglophone cultural and intellectual influence. Its status was underlined by the tradition, beginning in 1840, of an annual public address to Library subscribers which helped to cement links between members of the intellectual, social, and political elites. The annual address was typically delivered by the governor or a leading political personage. Speeches were given prominent attention in the local press and the event itself formed part of Cape Town's social calendar. By convention, the Library address was an occasion to reflect on matters relating to the Colony's intellectual, cultural, and educational progress; a recurring theme was the utilitarian role of knowledge in fostering social advancement, moral instruction, and mental improvement.

The inaugural Library address was delivered by the attorney-general, William Porter.[115] An Irish Presbyterian of progressive social views, Porter was considerably influenced by the New Light movement. He was a driving force in securing a socially inclusive colour-blind franchise at the Cape with a low property qualification, and is often quoted approvingly by historians for his aphorism: 'I would rather meet a Hottentot at the hustings voting for his representative than meet the Hottentot in the wilds with his gun upon his shoulder.'[116] Porter was also a long-standing member of the Library committee and an enthusiastic supporter of its broader educational aims. Recalling his first visit, Porter said he had felt as though he was standing 'within the precincts of a Library collected in some

[114] A. du Toit and H. Giliomee, *Afrikaner Political Thought: Analysis and Documents 1780–1850* (Cape Town, 1983), 249; Giliomee, *The Afrikaners*, chs. 4, 5.

[115] William Porter (1805–80) came to the Cape in 1839 as attorney-general and held this influential post for 26 years. He played a major role in the creation of the first Cape parliament and was active in civic life—in particular through his involvement with the Library, the University of the Cape of Good Hope, the South African College, the Literary and Scientific Institution and the South African Infant Schools. He deplored contemptuous attitudes towards Dutch-speakers and, on his death, left money to facilitate women's education at the University of Cape Town. See J. L. McCracken, *New Light at the Cape of Good Hope: William Porter, the Father of Cape Liberalism* (Belfast, 1993).

[116] S. Marks and S. Trapido (eds.), *The Politics of Race, Class and Nationalism in Twentieth Century South Africa* (Harlow, 1987), 5.

great metropolis of Europe, rather than in one which had grown up in a few years in the little capital of a thinly populated Colony at the farthest part of Africa!' He called it 'the Library of every man in the community', arguing that it was essential to the 'instruction of the people—of the whole people—of the people of every age, and sex, and class, and colour.'[117]

Porter's pride in a local institution worthy of a much larger European city was widely shared. Yet his vision of the Library as a beacon of enlightenment and education for all was by no means universally accepted. For example, Porter's use of the term 'the people' was much more socially inclusive than it was in the hands of more conservative liberals like John Fairbairn, for whom it was coextensive with members of the male middle-class.[118] Arguments about the purpose and function of the Library were a matter of long-running public debate from the mid-nineteenth century, and they are revealing of different views about the desirable constitution of colonial society. Recurring areas of controversy included whether the Library should be a lending institution or merely a resource for reference and research; whether it should stock novels and 'works of imagination' or restrict itself to 'serious' literature; whether it should aim to serve the people of Cape Town or else concentrate on becoming a 'national' institution like the British Library; and whether it should be kept open at night to increase access, notwithstanding the extra expense and risk of fire that this involved.

Bearing on all these issues of access was the Library's complex financial situation, given that it benefited from a government grant in addition to being funded by private subscription. The mechanism of subscription was a familiar and well-understood way of raising money for voluntary organizations and societies. In the British context, Morris uses the term 'subscriber democracy' to characterize this widely used institutional mechanism. While outwardly a measure of public-spiritedness, the subscription model was also an unmistakable register of social status. With their elaborate hierarchies (comprising patrons, vice-presidents, trustees, and so on) and rules of association, voluntary societies dependent on subscription membership at once proclaimed the respectability and status of their citizen members while at the same time confirming existing social gradations and boundaries between them.[119]

In the view of many critics, the South African Library was an elite club geared towards satisfying the narrow interests of its subscribers, whose numbers were

[117] *Proceedings at the 11th Anniversary Meeting of the Subscribers to the Public Library, Cape Town, Tuesday, 28th April, 1840* (Cape Town, 1840), 4–5, 6. See also Tyrrell-Glynn, 'History of the South African Public Library', 41–2.

[118] Porter also had advanced views on the education of women. On his death he bequeathed £20,000 towards the establishment of a reformatory for boys, adding that if this was not established a proportion should go towards women's higher education at the Cape University. He also left sums to the South Africa Library and every other public library in the country. McCracken, *New Light*, 132.

[119] Morris, 'Clubs, Societies and Institutions', 412–13; also 'Voluntary Societies and British Urban Elites, 1780–1850: An Analysis', *Historical Journal*, 26: 1 (1983), 101; McKenzie, '*The South African Commercial Advertiser*', 75.

lamentably small.[120] It was frequently accused of being an 'aristocratic' institution whose conditions of membership and opening hours were designed to exclude the 'working man'. Several prominent members of the library committee, and indeed two of its founding librarians, were committed to the idea of popular adult education, viewing libraries as important instruments in the process of enlightenment and social advancement. Thomas Pringle had helped to establish a small circulating library in the late-1820s at Glen Lynden on the eastern Cape frontier.[121] Alexander Jardine's 'Popular Library', created in 1834, was designed to cater for the needs of mechanics, labourers, and apprentices who could not join the South African Library proper.[122] John Herschel was a leading supporter of this venture and lent his considerable status to the cause of popular education on the occasion of its anniversary meeting in 1835. The astronomer argued that, in order to give the noble invention of printing

> its full effect as a civilising agent, its inestimable products must not be confined, like the luxuries of life and gifts of fortune, to a few, but must be made to circulate among the many; not regarded as an adjunct of superior station and wealth, and limited in their use to persons of the higher classes—the cream of society—but diffused throughout the great subordinate mass—and brought into contact, if possible, with every individual of a state.[123]

The Popular Library survived, on a shaky financial basis, for some thirty years until it was finally forced to close. Attempts to export to the Cape the tradition of adult education and self-improvement associated with the rise of a skilled working class in early nineteenth-century Britain were also largely unsuccessful.[124] The consensus of the few writers who have studied the history of mechanics' institutes in nineteenth-century South Africa is that, in contrast with Britain and other colonies of settlement (like Australia, Canada, and New Zealand), such organizations failed to take root because they proved unable to adapt themselves to

120 Hattersley, *Illustrated History*, 141.

121 F. G. van der Riet, 'An 1820 Settler Circulating Library at Glen Lynden, Eastern Province', *South African Libraries*, 19: 4, (1952). Pringle had served as leader of a small party of Scottish settlers at Glen Lyden. Little is known about the library, as van der Riet explains. It was probably formed from a nucleus of books collected by Pringle after his return to Scotland in 1826.

122 The Popular Library was in effect an extension of the South African Library and operated from the same building. It may be that Jardine promoted the idea of the popular library in part in order to retain the exclusivity of the South African Library.

123 Cited in D. H. Varley, 'The Nineteenth Century and the "Common Reader" ', *South African Libraries*, 19: 4 (1952), 105.

124 From the late-1820s there are several examples of attempts to create such organizations: a Mechanics Institution was created in Cape Town in 1828 but survived only for a year. In 1845 two rival organizations were created: the Institute for the Diffusion of Useful Knowledge and the Cape Town Literary and Debating Society. Both soon sank into oblivion. The Port Elizabeth Mechanics' Institute and Library, formed in 1849 by an immigrant English ironworker, appears to have enjoyed more success, as did the Book Society in Pietermaritzburg established in 1846. A revived Mechanics Institute, formed in Cape Town in 1853, survived into the 1870s before dissolving. See P. C. Candy, 'Mechanics and Literary Institutes: A Neglected Foundation of Adult Education in Colonial South Africa' (unpublished paper).

South African conditions. Most were formed and dominated by prominent middle-class citizens whose attitudes towards the skilled working class were tinged with more than a little paternalism and sanctimonious didacticism.

The self-conscious and relatively cohesive constituencies of white, English-speaking, nonconformist workingmen from which the Victorian movement for adult education and self-improvement typically arose were largely absent in pre-industrial South Africa. The endemic lack of skilled white labour meant that the social and geographical mobility of newly arrived English-speaking artisans did not depend on the assistance of mutual improvement associations.[125] Had the 'great subordinate mass' referred to by Herschel been thought of as suitable targets for the diffusion of useful knowledge, the story might have been different. But this mass was overwhelmingly not white, and when it was, it was often not English-speaking.

If Whitelock is correct in his claim that the mechanics' institutes which 'spread throughout the empire like a cultural epidemic' during the nineteenth century 'contributed more to the sense of imperial community than all the rhetoric in the press and the stately tours by proconsuls', their relative absence in South Africa may help to explain the fragmented nature of English-speaking identity and its lack of a popular base in this part of the nineteenth-century Empire. Certainly, South Africa could not compare with Canada, which boasted over 100 mechanics' institutes in Ontario alone by 1880, or with Australia where there were over 1,000 by 1900.[126]

Pressure on the South African Library to cater for the needs of the 'common reader' came to a head in 1848, and it is inviting to see this both as a reverberation of the revolutionary spirit sweeping Europe at the time, as well as a prelude to the franchise debates of the early 1850s. Speaking at the annual meeting of the Library that year, the South African College mathematics professor James Adamson declared that his sympathies 'were all with the movements which provided, not an overabundance for the few, but much for the many'. He was ostensibly referring to the provision of books and the need to expand the Library's membership, but his phrasing indicates wider concerns. The *Cape Town Mail* called for evening opening and the creation of a cheap third-class rate of subscription as

[125] Here I am largely following the interpretation given by Candy, ibid. See also Varley, 'The Nineteenth Century'. A contemporary commentator, 'J.S.H.', in 'Mechanics' Institutes and their Social Relations', *Cape Monthly*, 8: 8(1860), offered as one of three reasons for the specific failure of mechanics' institutes in the Cape the paucity of potential members: amongst the 'heterogenous mass' of 'Christians of various denominations, Mahommedans of contending sects, and heathens sunk into the lowest state of ignorance and degradation', 'the Anglo-Saxon element is the only one that troubles itself about the spread of knowledge in this particular way; so that if we were as active in the support of Mechanics' Institutes as the population of an English town (and this is no very forcible comparison), instead of representing the mechanical influence which is supposed to exist in a town of nearly thirty thousand inhabitants, we should only represent that of about one sixth of this number'.

[126] D. A. Whitelock, *The Great Tradition: A History of Adult Education in Australia* (Brisbane 1974), 86–7, 127.

another way in which the Library, 'as a public institution, and in terms of the moral and intellectual aims of its founders, could be more beneficial especially to the working classes and the less wealthy'. (There were, at that time, a mere 200 paid subscribers.)[127] The leading figures in the 1848 campaign for third-class £1 subscriptions were William Porter, Richard Solomon, and J. A. Merrington. This democratic objective they successfully achieved, and the Library's finances improved as a result. However, Solomon was summarily removed from the committee in 1850 by elements who continued to resent the wider membership base he had engineered. Typical of these was Sir John Wylde, the hard-drinking and scandalous first chief justice of the Cape, who accused the 'lower orders' of filling the reading room with foul cigar smoke.[128]

Discussions about access and exclusivity were inflected by other considerations too. Thus, champions of education such as Fairbairn, Adamson, and Langham Dale used the occasion of the annual address to press the Library's higher function as a centre for scholarly research and reference, arguing against stocking works of amusement such as novels and questioning the Library's role as a lending or 'circulating' institution. Their claims were strongly reminiscent of British utilitarians and evangelicals who, reasoning from secular and religious premises respectively, found common cause in the view that habits of reading should be instilled amongst the working classes for the purposes of instruction and enlightenment, but disdained the notion that literature should also cater for entertainment or leisure.[129] On the other hand, Governor Harry Smith, speaking at the Library's annual general meeting in 1848, was inclined to support 'the lighter literature of the age' (i.e. novels), in keeping with the Library's expressed commitment to 'Home Education' and broader knowledge.[130] Perhaps the gallant governor and military hero was indeed being true to the interests of the privileged and leisured class with which he was associated, for much of the debate about the worth of novels and the question of the Library's core function was directed to the reading habits of 'ladies' and the role of the Library in providing entertainment.

The argument against 'circulation' was pressed hard in the 1870s and was linked to the idea that the Library's primary purpose was to serve the interests of 'the nation' by acting, like its British equivalent, as a repository of great books. Governor Barkly, speaking in 1871, tried to get around the vexed issue by

[127] *Proceedings at the 19th Anniversary Meeting of the Subscribers of the Public Library, Saturday 29th April* 1848 (Cape Town, 1848); Tyrrell-Glynn, 'History of the South African Public Library', 57–8.

[128] Wylde was sternly rebuked for this intemperate attack in the *Cape of Good Hope Observer*, which insisted that the higher subscribing classes were in fact to blame. See Tyrrell-Glynn, 'History of the South African Public Library', 60–5.

[129] Altick, *The English Common Reader*, 99–100, 129–30. See J. Rose, *The Intellectual Life of the British Working Classes* (New Haven and London, 2001), for a forthright defence of the classic literary canon in the education of ordinary readers.

[130] Tyrrell-Glynn, 'History of the South African Public Library', 59–60.

Fig 1. View of the South African Library and Museum with statue of Sir George Grey in foreground, from the Botanical Gardens, 1866. Lithograph by Thomas Bowler

recommending that the committee should use a word like 'lending department' or 'issue branch'; the term 'circulating library', he explained, was apt to conjure up the image of 'Lydia Languish, in her curl papers, receiving a lot of trashy romances from her waiting woman'.[131] But, as the *Cape Times* pointed out, the education of women was a 'topical' movement; books were vital for the 'advancement' of women: the Library should therefore retain its circulating function even if this meant that its stock would become dog-eared.[132]

This brief glimpse into the micro-politics of reading, seen in the context of divisions by class and gender, is suggestive of how wider political debates and discussions about the composition of colonial society were played out within the institutional framework of the Library. They cast light on questions of social mobility and status, social and moral worth, as well as how conceptions of private pleasure conflicted with those of the public good. Similar tensions can be observed in two other institutions that formed part of the constellation of civic institutions referred to earlier, namely the Botanical Gardens and the South African Museum.

[131] Tyrrell-Glynn, 'History of the South African Public Library', 124–8; also 138–40, 159–60. See also Johnson, *Shakespeare and South Africa*, ch. 1.

[132] Tyrrell-Glynn, 'History of the South African Public Library', 160.

The Gardens

Laid out in a long rectangle in the city centre, the Dutch East India Company's Garden was synonymous with the Cape's original *raison d'être:* it was established to provision passing ships with fresh fruit and vegetables. As this function became redundant, so the Company Garden fell into neglect. When the British first arrived at the Cape they immediately expressed interest in rehabilitating the Company Garden. Governor Macartney and his secretary John Barrow were both keen on botany and were regular visitors to Kew. Like other imperial administrators at the beginning of the nineteenth century, they saw the Company Garden in terms of agricultural improvement and regarded it as a visible representation of good government.[133] Real progress took rather longer to achieve. In 1848 a commission composed of five prominent Cape citizens (Adamson, Arderne, Fairbairn, Pappe, and Rutherfoord) persuaded the governor to proclaim part of the Garden a botanical preserve. The Garden was thereby reconstituted as a site of scientific significance while serving, too, as a place for public procession and display. Like the Library, the Garden was funded through a mixture of government grant and private subscriptions. In common with other botanic gardens established throughout the Empire in the nineteenth century, the avowed purpose of those at the Cape was to further scientific inquiry, naturalize and distribute commercially useful plants and trees, and provide training for horticulturalists and botanists.

A secondary—but by no means less significant function—of the Botanical Gardens was to provide a fashionable venue and place of entertainment for the subscribing elite, who were promised privileged access at certain times, admission to the highly popular concerts put on by military bands, and the perk of claiming free plants up to the value of their subscriptions. Richard Drayton considers that the Cape's Garden was 'more of an urban playground than a scientific centre', but this recreational function should not be underestimated.[134] Thomas Bowler's watercolour paintings of citizens listening to military bands in the Garden brilliantly evoke its role as a setting for public promenading, a space in which women as well as men were able to disport themselves. There were 300 founding subscribers or ticket-holders, but their number fell off sharply from the 1860s as the middle classes migrated away from the city to the expanding suburbs of Cape Town.[135]

Given its reliance on public funding, there was a strong current of feeling against the privileges enjoyed by ticket-holders. *The Cape Monitor*, for example, combined an attack on the credentials and competence of the Garden's management committee with a declaration that it ought, as a recipient of public funding, to exist for 'the recreation and instruction of the whole people, and not one

[133] R. Drayton, *Nature's Government: Science, Imperial Britain, and the 'Improvement' of the World* (New Haven, 2000), 121, 183.
[134] Ibid. caption to plate 12 and p. 181.
[135] D. P. McCracken and E. M. McCracken, *The Way to Kirstenbosch* (Cape Town, 1988), 17–18, 22.

section of them only'.[136] This line of criticism was elaborated by 'Hortulanus' who, in 1857, argued that the Garden was failing to serve the broader interests of the country. The writer suggested that the Garden should be expanded to 500 acres and used to instruct farmers and improving landowners in the selection of crops and use of indigenous species. Moreover, its purpose should not only be to receive and acclimatize exotic species, but also to collect and preserve the many indigenous species, such as the silver-tree or *witte-boom* of Table Mountain, that were 'fast disappearing from their native wilds'.[137] An 1856 parliamentary select committee bluntly concluded that the Botanical Gardens, while attractive as 'a place of pleasurable resort', were a 'failure' in scientific terms; in order to remedy this deficiency it was necessary to appoint a botanist or curator, on a similar basis to that of the Museum.[138]

These criticisms of the purely ornamental role of the Botantical Gardens, formulated in the idiom of scientific imperative and conservationist necessity, were duly acted upon.[139] In 1858 (the same year in which the foundation stone of the Library was laid) Dr Ludwig Pappe, a Hamburg medical doctor who emigrated to the Cape in 1831, was designated as the Cape's first 'colonial botanist'. Pappe was a close associate of the visiting botanist William Harvey. He also had strong connections to Kew and he was an ardent champion of the Cape's unique floral kingdom.[140] The creation of the post of colonial botanist had been urged by Sir William Hooker, director of Kew Gardens. In this way the Cape was further drawn in to the expanding network of global science and trade stimulated by Sir Joseph Banks's vision of Kew as 'a "great botantical exchange house for the empire" '.[141]

On the initiative of Governor George Grey, Pappe (who had also been a member of the Literary Society and possibly the Scientific Institution[142]) was appointed to the new chair of botany at the South African College, thereby reinforcing the local institutional networks of power and patronage which are so clearly revealed through even a casual perusal of cross-membership of committees in the fields of education, science, and the arts. The initial collection of plants for the Garden was purchased by the government from the estate of Baron von Ludwig following his death in 1847. Von Ludwig lavished great expense in laying out extensive private gardens (known as Ludwigsburg) at his home in Kloof Street. With assistance from Pappe these were stocked with some 1,600 exotic species and made available for public viewing and recreation upon purchase of a ticket.[143] Von Ludwig was known as an enthusiastic supporter of science and

[136] *The Cape Monitor*, 24 Mar. 1858.

[137] 'A Cape Botanical Garden, As It Should Be', by 'Hortulanus', *Cape Monthly*, 2: 9 (1857), 173–6.

[138] *Report of the Select Committee on the Botanical Garden House of Assembly, 1856*, p. iii–iv.

[139] On the rise of conservationist ideology see esp. W. Beinart, *The Rise of Conservation in South Africa: Settlers, Livestock, and the Environment 1770–1950* (Oxford, 2003).

[140] Ibid. 74–5.

[141] R. Drayton, *Nature's Government*, p. xiii.

[142] Bregman, ' "Snug Little Coteries" ', 219.

[143] McCracken and McCracken, *The Way to Kirstenbosch*, 17–18, 92.

the arts, and was congratulated, in an almost forgivable gardening pun, for his cultivation of 'science in the uncongenial soil of South Africa'.[144]

As was the case in the origins of the Museum and the Library, it was the private collections of eighteenth- and nineteenth-century European antiquarians and amateur men of science like von Dessin and von Ludwig which formed the nucleus of Victorian civic culture. This was not in itself exceptional. In nineteenth-century England, as Alberti has shown, the development of provincial city museums commonly saw a phased transmutation of their holdings: collections first acquired by private individuals or *virtuosi* subsequently moved to the custody of learned societies and voluntary associations, and thereafter came under the ownership and control of municipal authorities, which saw fit to house expanded collections in dedicated accommodation designed to foster civic pride.[145] What marks out the story of such institutions at the Cape is that in this case the status of colonial town, imperial province, and proto-nation were all intertwined. The fact that key Victorian knowledge-based institutions were founded upon the collections of two German immigrants, von Dessin and von Ludwig, is also of interest: their input provided the elements of that shared foundation myth of European cultural mutuality which champions of Anglo-Dutch cooperation were eager to underline.

The Museum

Having fallen into abeyance following the removal of its collections to the lackadaisical custody of the South African College, the South African Museum was resuscitated at mid-century, this time as a result of the government's initiative. Acting on advice given by the influential English natural historian Richard Owen, Governor George Grey invited Edgar Leopold Layard to join the Colonial Secretary's Office and to engage in museum work during his spare time. In 1855 a government inquiry was set up with the intention of establishing a 'Public or National Museum'. The Cape Town lawyer and man of letters Charles Aken Fairbridge played an important role in this initiative, yet it was Grey's personal support and patronage that persuaded some 300 leading public figures to pledge the considerable sum of 1 guinea per year in order to establish the museum. For its part the government itself provided a small £300 annual grant. (The government also offered a grant of £150 to the Albany Museum in Grahamstown, which was established shortly after the South African Museum, and which saw itself as a provincial rival.)[146] The South African Museum was duly incorporated by

144 *The Cape of Good Hope Literary Gazette*, 2: 6 (1848), 206.

145 Alberti, 'Placing Nature', 309.

146 N. Fowler, 'A History of the Albany Museum, 1855–1958', unpublished TS. (1968), Albany Museum. The museum's hasty creation originated with the Grahamstown Medical-Chirurgical Society, established by five doctors under the leadership of Dr William Atherstone in July 1855. This soon changed its name and expanded its objectives as the Literary, Scientific and Medical Society, which in turn became the nucleus of the Albany Museum.

legislation modelled closely on the act that established the Australian Museum in Sydney. Its first two trustees were the colonial secretary Rawson W. Rawson and the botanist Ludwig Pappe. The astronomer royal, Thomas Maclear, was elected by the subscribers as a third trustee. Charles Fairbridge replaced Rawson in 1860.[147]

Despite being temporarily housed in premises above a bookseller in St George's Street,[148] the Museum was established on a much more secure basis than its predecessor: not only did it now have the services of an energetic official curator, it also enjoyed firm statutory recognition and benefited from a small government grant. But the Museum was still a fledgling institution and was by no means fully professionalized. Lacking the advantage of an endowment (like its equivalent in Sydney) and constrained by funding shortages, it was heavily dependent on private subscriptions. Moreover, its curator retained his responsibilities at the Colonial Office. The Museum's initial collection was based on the mouldering and depleted remnants of Andrew Smith's earlier venture. The already lamentable state of many specimens had deteriorated further since the Literary and Scientific Institution passed them over to the South African College, and there was confusion about the ownership of various items and sub-collections.[149]

The trustees were actively involved with the Museum and brought useful expertise as well as practical experience to their offices. Pappe and Maclear were both associated with the Literary and Scientific Institution and thus represented a direct continuity with Cape Town's pioneer scientists of the 1830s, including Smith, Herschel, and von Ludwig. Maclear enjoyed particular esteem as a long-standing representative of metropolitan science who had made the Cape his home. Edgar Layard was well connected in his own right. An amateur naturalist, who had previously worked in Ceylon, he was the younger brother of the renowned archaeologist of Mesopotamia, Austen Henry Layard, and the son of a senior East India Company official.[150]

Notwithstanding Layard's specialist interests in ornithology, he conceived of the Museum as an old-fashioned 'cabinet of curiosities', believing that it should offer something of interest to everyone.[151] His energetic, if at times indiscriminate, collecting policy may have been the result of a lack of professional training, but it was also the product of practical necessity. Animals, insects, geological specimens, coins, as well as ethnographic materials were all collected and displayed. This approach entailed the active involvement of ordinary collectors in the field. Layard relied heavily on the donations of private individuals (often farmers, travellers, and enthusiasts), rewarding their generosity with a personal mention in the Museum's annual reports or in his regular column in the *Cape Monthly*. A typical entry in the *Monthly* might record the receipt of a collection of

[147] Summers, *History of the South African Museum*, 22–5.
[148] The Museum moved to dedicated space in the South African Library building in 1859–60.
[149] Ibid. 25, 28–9, 30, 33, 38. [150] Ibid. 21–2. [151] Ibid. 28.

birds from the Free State given by David Arnot jnr.; 128 birds and animals from the explorer Andersson; several birds from F. Green in Damaraland; a meteorite from W. H. Auret in Victoria West; a brass spur sent by P. D. Martin; and a petrel from the southern ocean given by Dr Dyer of the ship *Royal Charter*.[152] In the public mind, the validation of exhibits and of donor-collectors was part of a mutually reinforcing circuit of donation and recognition.

In many ways Layard's regime epitomized the pre-professional world of museum curating. He was more concerned with adding to the various collections and mounting impressive displays than he was with advancing specialist research, which became more of a priority during the curatorship of Ronald Trimen in the 1870s.[153] Given the need to establish the Museum on a firm footing, and its dependence on active public support, Layard's populist instincts made sense. His 'Descriptive Catalogue of the Mammalia', compiled in 1860, was intended 'to amuse as well as instruct the visitor to the Museum'. Likewise, the pocket-sized *Catalogue of the South African Museum* which he produced in 1862 contained handy descriptions and anecdotes about individual specimens drawn either from 'noted authorities or else from personal observation'.[154]

Despite appearances, Layard's curatorial approach was not altogether lacking in method or system. Shortly after moving to the new Library building in 1860 he recorded the underlying classificatory principles of the collection: mammalia, birds, and reptiles were arranged according to the system of Cuvier; the conchological collection according to Lamarckian principles; the lepidoptera according to de Jean; and the minerals, originally arranged by Mr Calvert, according to the system of Birzelius. But the exhibit of fossilized plants collected by the colonial geologist Mr Wyley was 'simply *exhibited*', since Layard professed himself inexperienced in this area.[155] The study of ornithology was given a particular boost under Layard, and despite being unable to undertake extensive fieldwork himself, he published *The Birds of South Africa* in 1867, the first comprehensive descriptive work on the subject and a much quoted source-book.

The success of the Museum depended not only on the involvement of the Cape's citizenry, but also on establishing and maintaining networks with individuals and institutions further afield. In the course of a lengthy, albeit disrupted, association with the Museum Layard spent periods of time abroad, including a year as George Grey's private secretary when Grey became governor of New Zealand for the second time in 1861. Layard was thereby afforded the opportunity to visit museums in Auckland, Sydney, Melbourne, and Adelaide and he duly established

152 'Report of the South African Museum for the Month of September', by E. L. Layard, *CMM* 8: 46 (1860), 253–4.

153 Summers, *History of the South African Museum*, ch. 4. According to Summers, Trimen was less socially acceptable than Smith or Layard but enjoyed greater scientific esteem as a lepidopterist and entomologist and was elected to a fellowship of the Royal Society.

154 G.23-'61, *Report of the Trustees of the South African Museum for the Year 1860*, p. 1; *The Catalogue of the South African Museum, Part I* (Cape Town, 1862).

155 *CMM* 7: 41 (1860), 319.

relations between these institutions. An active exchange system was also maintained with Louis Agassiz in America.[156] The ease with which Layard moved in the higher reaches of colonial society, his extensive social and intellectual networks, and his extrovert personality were probably as important as his devotion to science in making a success of his tenure as founding curator of the Museum.

Ethnographic collecting was another field in which the Museum was involved and, in light of the colony's view of itself as an offshoot of Europe in Africa, a comment on these holdings is apposite. In the first months of his curatorship Layard took possession of various Inuit objects as well as a collection of African musical instruments, and weapons, including examples from Polynesia which derived, via the Dessinian collection, from the voyages of Captain Cook. Eight unclothed life-sized wax figures said to represent the 'four principal tribes' of South Africa (Kafir, Basuto, Hottentot, and Bushmen) were sent for repair. These were now dressed in skins by the surveyor-general and artist Charles Bell, who had accompanied Andrew Smith on his 1834–6 expedition.[157]

This material was presented in a somewhat haphazard fashion. The walls and upper galleries of the Museum were casually adorned with weaponry and musical instruments, and 'a collection of manufactured articles, ancient as well as modern, pottery, glass, and Kafir ornaments, &c' was posted near the door.[158] Typically, such articles were referred to as miscellaneous 'objects of curiosity', articles of 'existing Barbarous Races', or simply as 'native implements'. Although the Museum's founding charter was dedicated to the acquisition of a 'complete collection of objects illustrative of the habits and customs of the natives of South Africa', it held only approximately 400 ethnological items at the century's end, and exhibits were acquired in a random, miscellaneous fashion.[159] Indifference to indigenous artefacts rather than a preoccupation with the 'otherness' of Africans may be the more reliable indicator of colonial attitudes.

Patricia Davison's work suggests that, by the 1890s, the association between ethnographic manufactures and 'uncivilised races' was made more explicit. At the turn of the century visits by overseas experts like Henry Balfour of the Pitt Rivers Museum in Oxford and Alfred Haddon from Cambridge encouraged a more rigorous scientific approach to anthropology. Concerted efforts were henceforth

[156] Summers, *History of the South African Museum*, 26; K. H. Barnard, 'History of the South African Museum (unpublished TS in SAM archives), 38–9; G11-'63, *Report of the Trustees of the South African Museum for the year 1862*, 1. In 1864 it was reported that 300 species of American fish had been received from Prof. Agassiz of Cambridge, Mass., in exchange for specimens of birds and mammalia and 'some native arms and utensils', as well as 72 bird specimens from the Museum of South Australia in return for 310 of the same. See G.22-'65 *Report of the Trustees of the South African Museum for the Year 1864*, 2.

[157] Summers, *History of the South African Museum*, 28, 41–3; *Cape Monthly*, 7: 41 (1860), 319; Barnard, 'History of the South African Museum', 49–51.

[158] Summers, *History of the South African Museum*, 40, 42–3.

[159] P. Davison, 'Material Culture, Context and Meaning: A Critical Investigation of Museum Practice, with particular reference to the South African Museum', Ph.D thesis, University of Cape Town (1991), 11, 104–5.

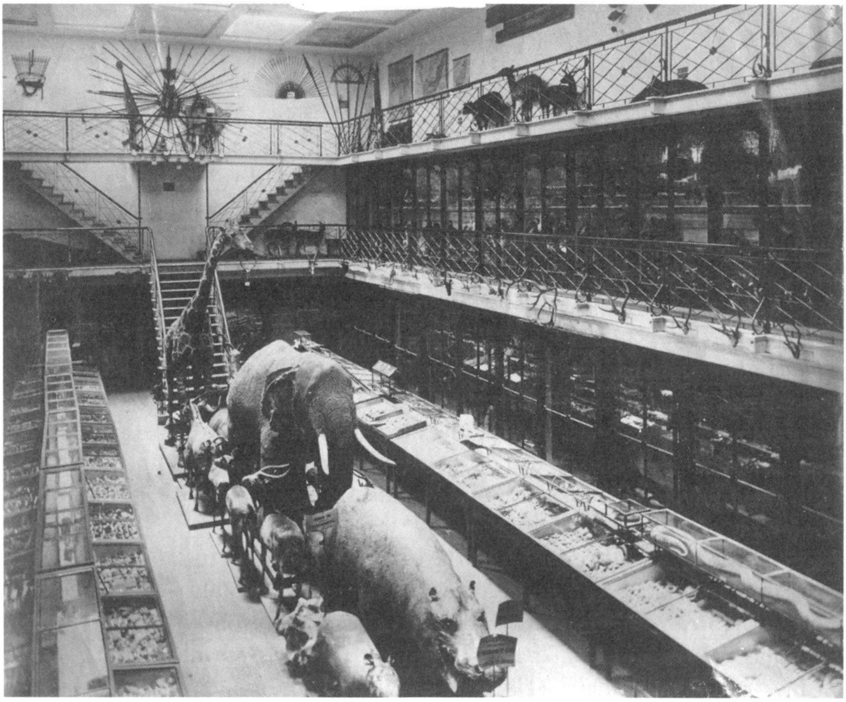

Fig 2. View of the interior of the South African Museum, *c.*1880. At this stage the museum shared premises with the South African Library. (Courtesy of Iziko South African Museum)

made towards mounting ethnographic exhibits in accordance with the systematic typologies and comparative classificatory mechanisms of the developing racial and anthropological paradigms.[160] The project set in motion by Louis Péringuey to cast real-life models of 'dying out' Bushmen 'types' was one outcome of this racializing impulse. Yet the available evidence suggests that the Museum at mid-century was a repository for a fairly random selection of donated ethnological objects, and that displays conveyed no overarching theory of European racial superiority. The juxtaposition of cultural artefacts with scientific specimens implied, problematically, that the curiosity value of native peoples was coextensive with the wonders of the natural world. It may thus be safe to assume that, at least at a subliminal level, the presentation of ethnological items would be read by a Victorian colonial audience as proof of the distance it had traversed from barbarism to civilization. But the Museum's curatorial gaze had not hardened, and we should not assume that exotic ethnological items bore the full freight of scientific racism before the start of the twentieth century.

[160] Davison, 'Material Culture, Context and Meaning', 105, 145.

MATURITY AND RESPECTABILITY

Rapid growth of the colonial economy made possible the expansion of public institutions outlined above. These, in turn, contributed to a sense that, in Walker's words, South Africa 'was truly on the march'.[161] From the late-1840s the rapid development of the merino sheep industry in the eastern Cape led to wool becoming the Colony's largest export, increasing more than tenfold in value from 1846 to 1866, and peaking in 1872 at £3.3 m.[162] Merchants became involved in the financing of the wool industry, whose requirements for new lines of credit and investment were met by a conspicuous expansion of merchant houses and banking institutions. Only twenty-five years after the creation of the Cape's first private bank in 1837, there were no fewer than twenty-eight operating in the Colony. Trade was given a further boost by the increase in sea traffic across the Atlantic following the discovery, in 1851, of gold in Australia.[163] Significantly, the prosperity achieved during the second half of the 1850s was also the 'heyday of the commercial bourgoisie's political power'.[164]

Optimism about the prospects of the Colony was further sustained—until the onset of drought and economic depression in the early 1860s—by a perception that Cape Town was no longer as isolated as it had been from either its European heartland or its African hinterland. The increase in trans-oceanic maritime traffic during the 1850s was matched by moves to enhance the capacity of Table Bay harbour. In 1860 Prince Alfred tipped the first rocks into a new breakwater and dock. This costly engineering project considerably enhanced the city's prestige and prosperity, though it proved politically controversial because it was achieved at the expense of the rival harbour of Port Elizabeth, which served the wool-producing regions of the eastern Cape. Communications with England were markedly improved by the introduction of a monthly steamship which managed the run to Devonport in six weeks: its arrival attracted much attention in the papers and was eagerly awaited by those expecting news and letters—and not least by those who placed bets on the precise time of its arrival.[165]

Internal communications were considerably improved by the introduction, in 1846, of a twice-weekly postal service linking the main population centres of the Colony. This replaced the haphazard and unpopular 'postholder' system of the early nineteenth century, which had hitherto obliged farmers to forward post in a relay fashion. Construction of the Colony's first railway line linking Cape Town with Wellington by way of Stellenbosch began in 1859, and was accompanied by excited public debate. A regular omnibus service to Wynberg was established

[161] E. A. Walker, *The South African College and the University of Cape Town 1829–1929* (Cape Town, 1929), 32; also id., *A History of South Africa* (London, 1928), ch. 10.

[162] Beinart, *Rise of Conservation*, 18.

[163] Houghton, *The South African Economy*, 10, 191.

[164] Meltzer, 'The Growth of Cape Town Commerce', 155.

[165] Hattersley, *Illustrated Social History*, 111.

in 1848, and the Colony's first telegraph lines, linking Cape Town with Simonstown, were erected in 1860.[166] Scientific and technical advances were laden with symbolism. As John Mackenzie points out: 'Railways, ships, telegraphs, and the universal post were... endowed with moral and redemptive powers' and viewed as 'alternating links in a chain of imperial development'.[167]

Road construction was another acknowledged register of social progress. Mention has already been made of the virtual isolation of the Cape Peninsula from the immediate interior and the poor state of its public roads and turnpikes. From the mid-1840s transportation became a major priority for John Montagu, the administrative reformer and 'road-making Colonial Secretary'.[168] Montagu engineered a road from Cape Town to Stellenbosch in 1848 which facilitated the introduction of a regular omnibus service at the end of that year. His determination to open up routes to the western Cape interior and the Karoo resulted in the construction of several mountain passes, involving large-scale investment and widespread use of convict labour—born of Montagu's experience as governor of Van Diemen's Land.[169] Beyond their obvious practical uses, the construction of roads was said to signify the dynamism and enterprise of the new British colonial order in contrast to the inertia that characterized Dutch rule.[170] Roads transformed understandings of the landscape and its scenery, allowing areas that were previously inaccessible and perceived as hostile or backward to be domesticated and surveyed with a freshly confident colonial vision.[171]

More than any other sphere, political and constitutional developments underwrote the Cape's self-conscious sense of maturity. The achievement of self-government in 1853 was a crowning event in a narrative of colonial self-government that began with the challenge to Somerset's autocratic rule. Key way-stations along the road included the report of the Commission of Inquiry (1826), the abolition of the oligarchic Burgher Senate (1828), slave emancipation (1834), the creation of a Cape Town municipality (1840), and opposition to the nominated Legislative Council (1825–53).[172] The successful popular agitation against the

[166] Ibid. 111, 207–8, 210; Worden *et al.*, *Cape Town*, 165–6.

[167] J. M. Mackenzie (ed.), *Imperialism and the Natural World* (Manchester, 1990), 'Introduction', p. 7.

[168] *CMM* 8 (May 1874), 289.

[169] Hattersley, *Illustrated Social History*, 107 ff.

[170] See e.g. *The Cape of Good Hope Literary Magazine*, 2: 5 (1848), 53: 'I shall not readily forget my first trip to Stellenbosch on my arrival in this Colony, when, seated in that antiquated wooden conveniency [the stage wagon] ... I was jolted along at the rate of 2.5 miles per hour, nearly pitched at every furlong into the lap of a fat Dutch vrouw, redolent of snuff, who was one of my compagnons de voyage ... Upon my inquiry of an Afrikander present, why such a description of vehicle was used, I was told, it was exactly suited to the country. My reply to this was,—"Then change the country; make roads, good roads, and you may then use carriages with springs, and travel quickly and comfortably, not at this snail's pace." My colloquialist only shrugged up his shoulders, and I understood him to say, that as his grand-papa and great-grand-papa had been satisfied with what existed, they were good enough for him.'

[171] J. Dubow, 'Rites of Passage: Travel and the Materiality of Vision at the Cape of Good Hope', in Barbara Bender and Margot Winer (eds.), *Contested Landscapes: Movement, Exile and Place* (Oxford and New York, 2001).

[172] Meltzer, 'The Growth of Cape Town Commerce', 135.

designation of the Cape as a penal colony (1848–9), which led directly to the acquisition of self-government, was another signal point on the journey.

This underestimated popular movement is interesting in the way it casts light on the development of colonial nationalist awareness and conceptions of popular sovereignty at mid-century. Here again, it is possible to detect ripples of the political ferment that raged throughout Europe and in other parts of the Empire at this time. Miles Taylor has shown how Britain used its Empire in order to deflect Chartist and Fenian unrest at home. Transportation was a key component of this containment strategy, but, in combination with efforts to reduce military and fiscal expenditure abroad, the consequence was an eruption of civil disobedience and demands for settler rights emanating from New South Wales and Van Diemen's Land to the Cape and Jamaica.[173]

The campaign against the proposed importation of convicts aboard the ship *Neptune* was led by John Fairbairn, doyen of the free press, and John Bardwell Ebden, a leading merchant and chairman of the Cape's first banking institution. What began as an unlikely, even parochial, battle by the anti-convict association rapidly took on heroic proportions, culminating in concerted acts of civil disobedience that paralysed the colonial government for a period. The *Cape Town Mail* asserted that the struggle had 'created a people in South Africa', while Fairbairn's *Commercial Advertiser* argued that, by their resistance, 'the people of the Cape of Good Hope have shown to the world what it is that constitutes a state'.[174] It is ironical that several of the convicts on the *Neptune*, like Irish rebel John Mitchel, were being deported for their involvement in the anti-colonial famine risings. But Mitchel could only observe the 'heroic phraseology' of the settlers from aboard ship.[175]

The anti-convict movement mobilized the citizenry of the Cape in an unprecedented display of unanimity that effectively bridged intra-white class and ethnic differences. New techniques of political propaganda, such as petitions, public meetings, pamphleteering, and newspaper campaigns, were deployed in order to facilitate the emergence of a collective public voice. Supporters of the anti-convict association pledged an economic boycott against anyone associated with the support or employment of convicts, especially government agencies. Substantial common cause was thereby found between the disparate and often conflicting constituencies of town and country, leading merchants and ordinary colonists, Dutch and British. In the view of André du Toit, the experience of anti-convict agitation even helped to precipitate a brief, albeit unfulfilled, 'liberal moment' in Afrikaner politics.[176]

[173] M. Taylor, 'The 1848 Revolutions and the British Empire', *Past and Present*, 166 (2000).

[174] Cited in A. F. Hattersley, *Convict Crisis and the Growth of Unity* (Pietermaritzburg, 1965), 80.

[175] Taylor, 'The 1848 Revolutions', 169.

[176] A. du Toit, 'The Cape Afrikaner's Failed Liberal Moment 1850–1870', in J. Butler *et al.*, *Democratic Liberalism in South Africa* (Cape Town and Middletown, 1987). But see du Plessis, 'Colonial Progress and Countryside Conservatism', 76–81, on van der Lingen's curious confrontation with the anti-convict committee.

Fig 3. *The Great Meeting Held in Front of the Commercial Hall, Cape Town, on the 4th of July, 1849*, in protest against plans to land convicts at the Cape. By Thomas Bowler. (Courtesy of Iziko William Fehr Collection, Cape Town)

Given the nature of coalition politics, divisions and differences in objectives soon became apparent: radicals and conservatives split into different factions, while easterner separatists took a firm loyalist line. Notwithstanding such tensions, a collective determination to represent the Cape as a respectable colony of settlement, untainted by convicts who threatened to undermine the moral fibre and worth of the country, was clearly evident. This quest, as Robert Ross has argued, is central to the understanding of nineteenth-century Cape history, and it was shared by all those—black as well as white—who sought recognition and advancement within the framework of the colonial state.[177]

In political terms the assertion of respectability was registered through demands for political status and civil rights, coupled with a rejection of 'the aristocratic forces that had long dominated the life of colonists'.[178] In cultural terms respectability was expressed by conscious efforts to carve out select public spaces and to differentiate reputable leisure activities from boisterous popular recreations like gambling, carousing, and dancing—forms of public sociability in which Cape Town's racially mixed working classes indulged with enthusiasm.[179] From a materialist point of view, notions of honour and probity can usefully be related

[177] R. Ross, 'Missions, Respectability and Civil Rights: The Cape Colony, 1828–1854', *JSAS* 25: 3 (1999).

[178] Hattersley, *Convict Crisis and the Growth of Unity*, 93.

[179] Worden *et al.*, *Cape Town*, 144–9, 156, 190 ff.

to the ways in which the nascent colonial bourgeoisie sought to defend their status as 'respectable men of property'.[180] The quest for honour and respectability in the anti-convict movement was also notably inflected by gender and familial relations, as McKenzie has demonstrated, with implications for personal morality, domesticity, business conduct, and marriage.[181]

The passions aroused by the anti-convict movement, and the echoes of Chartism that it evoked, fed directly into constitutional reform, in spite of efforts by Governor Smith to persuade the Colonial Office that British interests were under threat. Faced also by Boer rebellion in the Orange Free State and the Kat River uprising of disaffected Khoi in the eastern Cape—problems that were compounded by large reductions in his army—Smith was struggling to deal with a 'turbulent frontier pregnant with nationalism'.[182] Concession was the way out. A 'popular' constitution was framed in 1850 based on the draft prepared in 1848 by the attorney-general, William Porter, and in 1853 the Cape was granted representative government. The constitution proved a victory for popular opinion and a defeat for Smith and his conservative settler supporters: provision was made for an elected rather than a nominated second chamber, and a male non-racial franchise was established with a low £25 property qualification, sufficient to enable poorer whites as well as more affluent blacks to have a voice in government. The terms of the 1853 constitution were, at the time, probably the most democratic in the world, though ex-slaves and Khoi gained little from the dispensation.[183] Nevertheless, it took nearly 150 years for South Africa to improve on it.

SIR GEORGE GREY AND THE SOUTH AFRICAN LIBRARY

The achievement of representative government represented a signal victory for colonial nationalist opinion and a decisive rejection of what was seen as the arrogant imperial prerogative. Fittingly, Governor Smith's successors in the early 1850s evinced a more consensual and placatory style of administration. In 1854 Sir George Grey assumed the governorship, having recently served terms in charge of South Australia and New Zealand. As a civilian rather than a military ruler,

[180] Keegan, *Colonial South Africa*, 229.

[181] McKenzie, 'Gender and Honour', ch 8 and *passim*. See also McKenzie's more recent treatment of these themes in her *Scandal in the Colonies: Sydney and Cape Town, 1820–1850* (Melbourne, 2004).

[182] Taylor, 'The 1848 Revolutions', 168.

[183] Ross, *Status and Respectability in the Cape Colony, 1750–1870* (Cambridge, 1999) 4. Cf. Bank's discussion of Porter and Fairbairn in 'Losing Faith in the Civilizing Mission: The Premature Decline of Humanitarian Liberalism at the Cape, 1840–60' (in M. Daunton and R. Halpern (eds.) *Empire and Others: British Encounters with Indigenous Peoples, 1600–1850* (London, 1999)), where he sees the achievement of representative government in terms of a new 'conservative consensus'. The decline of emancipationist and humanitarian ideals was by no means coterminous with the decline of Cape liberalism.

Grey proved to be closely in tune with the aspirations of the Cape's increasingly confident colonial leadership during its formative experience of parliamentary government. The support he gave to public and intellectual institutions was intrinsic to his liberal creed. His personal contribution to the development of comparative ethnographic knowledge in South Africa, Australia, and New Zealand was part of his enlightened, progressive world-view, and helped to shape his approach to colonial governance. As a visionary imperial proconsul, with unyielding confidence in his own rectitude and expansive plans for the future—his was the first attempt to federate the colonies and republics of South Africa into a single whole—Grey was also a highly controversial figure. Anthony Trollope captured Grey's duality well when he dubbed him 'the most popular,—and probably the most high-handed' of colonial governors.[184]

Whereas many liberal commentators have seen Grey as the very best and most progressive of governors (Olive Schreiner called Grey 'God's Englishman', and thought of him as the greatest and most unique of Cape liberal figures[185]), radical critics have focused on his role in eroding Xhosa political and cultural independence through his manipulation of the 1857 cattle-killing episode. Keegan decries Grey as the 'archetypal utilitarian liberal of his age', a man who carried through the civilizing mission 'more thoroughly than his predecessors could, and without resort to large military expenditures'.[186] James Belich's wry comment on Grey's impact on New Zealand—that despite revisionist attacks, 'Grey looms by far the largest [of colonial governors] in both history and myth'—can equally be said of his South African legacy.[187]

Whatever the historical verdict—and the two views are by no means incompatible if one recalls Eric Stokes' evocation of the 'latent authoritarianism' in utilitarian liberal efforts to assimilate and to anglicize[188]—Grey was undoubtedly the most intellectually able and imaginative of Cape governors. Unlike many intellectual visionaries, he proved highly effective in bringing his ideas to fruition. Grey's practical support of the Library, Museum, and Botanical Gardens, has already been noted. An enthusiast for public works, notably the harbour development,

[184] A. Trollope, *South Africa*, repr. of 1878 edn., ed. J. H. Davidson (Cape Town, 1973), 61.

[185] Schreiner called Grey 'God's Englishman' in a letter to W. T. Stead in October 1896 (*Olive Schreiner Letters, Vol. 1*, ed. Richard Rive (Oxford, 1988), 292, and mentioned him along with Saul Solomon and William Porter as the greatest of English-speaking Cape liberals (see n. 200 below). Schreiner also dedicated her novel *Trooper Peter Halket* to Grey, citing his incorruptible sense of justice and broad humanity.

[186] Keegan, *Colonial South Africa and the Origins of the Racial Order*, 288. The traditional liberal view is stated by W. M. Macmillan, *Bantu, Boer and Briton: The Making of the South African Native Problem* (London, 1929), 292 ff. A powerful denunciation of Grey is offered by J. B. Peires, *The Dead Will Arise* (Johannesburg, 1989), ch. 2. For a more measured and rounded review of his career in Australia, New Zealand, and South Africa, see J. Rutherford, *Sir George Grey* (London, 1961).

[187] James Belich, *Making Peoples: A History of New Zealanders* (Auckland, 1996), 190.

[188] Stokes, *English Utilitarians and India*, p. xvi. As Stokes points out, utlitarian thinking after the death of Jeremy Bentham and James Mill was no longer a fixed programme or set of intellectual dogmas. 'It was rather a practical cast of mind, a strong aggressive logic with which a man of affairs could aproach specific political issues' (p. 249).

he also provided grants towards the building of hospitals and schools, including Zonnebloem College, designed to instil civilized values into the children of African chiefs.

Grey was one of a handful of statesmen whose intellectual ideas fed directly into their understanding of South African society. As an antiquarian collector with developed interests in philology, ethnology, folklore, and natural history, he counted leading Victorian thinkers among his friends and mentors, including Charles Lyell, Charles Babbage, J. A. Froude, Richard Whately, and Thomas Carlyle. In Australia and New Zealand Grey conducted significant research into aboriginal and Maori language and folklore. His belief in the underlying unity of mankind coexisted with an evangelical belief that the baneful effects of tribalism and superstition should be eradicated by extending Christian truths to all mankind. The historic, indeed providential, role of British imperialism was therefore to integrate the 'primitive' races—whether aboriginals, Maoris, or Africans—and to assimilate them within the expanding universe of western civilization.[189]

Such dedication to the imperial civilizing mission was not only articulated in terms of the need to convert Africans to European ways. It was equally directed towards white colonial society, in order to nourish the belief of colonists in their historic role as agents of progress and enlightenment. This dual conception of civilization in Africa is well reflected in Grey's close personal interest in the development of the South African Library, which remains one of his most important legacies. In the best tradition of gentlemen antiquarians, Grey accumulated a rich and diverse collection of philological and ethnographic materials, including illuminated medieval and Renaissance manuscripts, incunabula and Shakespearean folios, as well as fine early examples of Africana. In 1856 Grey engaged the services of the German philologist and ethnologist Wilhelm Bleek (who had been brought to Natal in 1855 by Bishop Colenso as a translator and Zulu linguist), and set him to work on cataloguing his own private library.[190] Soon after leaving the Cape in 1861 to reassume the governorship of New Zealand (where he immediately became founding president of the New Zealand Society, a forerunner of modern scientific research in that country), Grey donated the bulk of his collection to the South African Library. In his bequest he stated that he was giving his library to a 'new country' in the hope that it 'might prove to it a treasure of great value, to some extent helping to form the mind of each of its generations, as they come following on'.[191]

[189] Rutherford, *Sir George Grey*, 54, 224 and *passim*; Robert Thornton, 'Capture by Description: Ethnographic Writing in Nineteenth-century South Africa' (unpublished TS), ch. 6.

[190] Thornton, 'Capture by Description'. Bleek was a convinced disciple of Darwin and Huxley and did not baulk at distinguishing between primitive and advanced races. Thus his belief in the ultimate unity of human races must be set against the fact that his researches were predicated on the precise determination of racial difference. See Andrew Bank's recent reconsideration of Bleek's attitudes towards race in 'Evolution and Racial Theory: The Hidden Side of Wilhelm Bleek', *SAHJ* 43 (2000).

[191] D. H. Varley, 'The Grey Collection: Mirror of Western Culture', in C. Pama (ed.), *The South African Library*, 36; *Guide to the South African Library*, 2nd edn. (Cape Town, 1977).

The symbolic capital of the Grey collection worked at multiple levels. Like the von Dessin bequest which constituted the original nucleus of the South African Library, the Grey collection was intended to be seen as a 'mirror of Western Culture', reminding generations of white South African scholars and schoolchildren of their treasured European inheritance.[192] Yet its rich ethnographic holdings suggested that this unique collection amounted to more than a mere trophy or relic of that European past. It signified that the Cape was not only an outlying outpost of the British Empire but also a European bridgehead into Africa. Whether that message would be interpreted as a licence to impose European values, or as an invitation to recognize the worth of African alongside European cultures, lay somewhat beyond the gift of its endower.

At least as important as Grey's literary bequest were the provisions he made for the construction, at the bottom of the Botanical Gardens, of an impressive neo-classical building with a portico modelled on the Fitzwilliam Museum in Cambridge, designed to house the new South African Library and Museum. After years of temporary and changing accommodation they were at last to have a permanent home. Grey laid the foundation-stone in March 1858 and the building was opened in 1860 by the young Prince Alfred.[193]

Both occasions were marked by elaborate pomp and public ceremony. The 1858 event began with a procession beginning from the grounds of Government House led by a military band. Representatives of the Library, Museum, South African College, and Botanical Gardens were all present, as were members of the Legislative Council, the judiciary, and a number of foreign dignitaries. Masters of four masonic lodges dedicated the building, in turn, to Art, Science, and Literature.[194] Grey began his address by describing the location of the Library and Museum in 'a beautiful valley, embosomed in magnificent hills' within the Company Gardens begun by the founder of the colony himself, Jan van Riebeeck. Fittingly, it was spatially situated at the centre of the first colonial enterprise. The 'geographical position' of the Cape was also significant, lying 'half way between the rising kingdom of Australia, and the important countries of America'. Extending this sense of place—and in an apparent allusion to his grand scheme to unite the colonies and Boer republics into a single political federation—Grey invited his audience to 'look at our position with regard to the continent of South Africa' and to consider the spot from which he was talking as the 'school in which future generations of the inhabitants of this part of the continent of Africa must be educated'.

192 Varley, 'The Grey Collection: Mirror of Western Culture'. Once in New Zealand, Grey began a new collection which he donated to the Auckland Public Library. As Varley's article indicates, there has existed an element of rivalry between the bibliophiles of Cape Town and Auckland with respect to which was the finest in 'the southern hemisphere'.

193 Lewin Robinson, 'The History of the South African Library', 4.

194 Drawing on an MA thesis by A. Cooper, Worden *et al.*, *Cape Town*, 144, make the interesting link between freemasonry and the Dutch Patriot movement of the late eighteenth century. English masonry was closely linked with the rise of the nineteenth-century mercantile and administrative elite.

At the conclusion of his address the procession re-formed and walked back to Government House, where the masons and other gentlemen were thanked. As the watching crowds left the gardens, the ceremony concluded with a piquant post-script. A group of thirty convicts employed in construction of the new building sat down to a dinner of roast beef, mutton, and plum pudding. Thomas Maclear, the astronomer royal, presided at this sit-down dinner, and 'illustrated to the Kafirs the civilised mode of eating beef and mutton with knives and forks'. Thus was enacted the trope of civilization in Africa.[195]

Grey's febrile dreaming, and the faintly absurd nature of this *mise en scène* in the Botanical Gardens, appear to have aroused scepticism in some minds. The *Cape Argus* adopted a distinctly tongue-in-cheek approach to the pretensions of the governor, and editorialized accordingly:

> Another stone of that grand social edifice was placed yesterday, amidst a large concourse of spectators in laying the foundation stone of the new Library and Museum, at the bottom of the Botanic Garden. Another rivet, we suppose, was struck from the chain which still binds the Cape, though loosely, to barbarism and the past. We are an imposing people—at least the Governor told us so yesterday . . . [196]

When, in 1860, Prince Alfred toured the Cape, the library committee seized the opportunity to inaugurate the new building, even though it had not yet been completed. The royal presence made the occasion even more triumphal than the 1858 hors d'oeuvres. Admission tickets were restricted to subscribers and their families, with seating for 600 provided in the new library hall. Prince Alfred unveiled a large portrait of his mother, Queen Victoria, and ceremonially placed the first books on the new shelves—among them editions of the Bible, Charles Knight's *Shakespeare*, a *Pictorial History of England*, and a rare manuscript donated by Sir George Grey.

Grey's address, archly described by a local newspaper as 'one of the most impressive harangues that has ever been heard in South Africa', elaborated on his 1858 vision with even more confidence. He reiterated his plans for a university at one end of the garden that would educate 'all the youth of South Africa, of every race, who may desire to drink at the fount of knowledge . . .'. He likened the South African Library to the one in Alexandria, and compared Alfred to his Anglo-Saxon namesake, the good king who had driven back 'wandering hordes, aided in their civilisation, and gave peace to England, and established a university, and transmitted a lasting and permanent impulse to science and learning . . .'. In just the same way, the teenage prince had 'most materially aided in the spread of civilisation, Christianity, and learning throughout this continent'.[197]

[195] *Cape Argus*, 25 Mar. 1858; *Cape Times*, 25 Mar. 1858. Cf. Maclear's demonstration of the Observatory to Andries Waterboer in 1834 as recounted in Musselman, 'Worlds Displaced'.

[196] *Cape* Argus, 24 Mar. 1858 (editorial).

[197] Tyrrell-Glynn, 'History of the South African Public Library', 101–5; 'The Public Library, Cape Town', in *CMM* (Aug. 1861), 99.

Fig 4. Statue of Sir George Grey, sculpted by Calder Marshall and erected by public subscription in 1864. This picture, taken by J. W. Dugmore, was sent by Olive Schreiner to W. T. Stead, appearing in the *Review of Reviews* in 1897. Schreiner was depressed by the 'greed of wealth and power' in the aftermath of the Jameson Raid, and found a 'curious consolation' in Grey, whom she called 'God's Englishman'.

The spirit of governor Grey's address was reiterated by the *Cape Monthly Magazine*, which commended the Library as a vital register of civilization and advancement, viewing its inauguration—along with the commencement of grand public works in the harbour and the achievement of representative government—as a key marker of the Colony's progress. The article characterized Alfred's royal progress through the Cape, Natal, the Orange Free State, and 'numerous dependent and independent tribes' as a distinctive 'South African event, the first occasion in which the whole populations of this vast region from the Cape promontory to Natal, acknowledged the moral influence of one sentiment, and were ready to act in one cause'.[198] In a revealing inversion of the familial metaphor that bound colony to metropole, this process of maturation was symbolized by the royal progress of a young prince under the tutelage of Grey; the implicit message was that the newly mature colony and its governor were acting *in loco parentis* for the monarch herself.

The complex of institutions that were located in the gardens below Table Mountain and Signal Hill now included a museum and library, Government House, as well as land earmarked for the South African College. Close by was the new Cape Assembly (in the Goede Hoop Lodge), the Supreme Court, and St George's Cathedral. In time the South African Museum was to have its own imposing building in what was still known as the 'wilderness' section of the gardens,[199] the South African National Gallery likewise. Within a generation a new parliament would be built alongside Government Avenue, the main thoroughfare through the gardens. Grey's personal contribution to this constellation would be honoured by the erection of a statue of him standing directly in front of the portico of the Library, and paid for by public subscription.[200] Here indeed was a formidable expression of public life and culture, a visible architectural and spatial register of the collective achievements and ambitions of the aspirant middle-class colonial elite, which had gone a long way towards transforming a small Dutch way-station into an improving and civic-minded Victorian city. It was, in short, a place which had done a great deal, by the 1850s, towards realizing Joseph Priestley's enlightenment vision of an 'empire of reason'.[201]

[198] 'The Visit of His Royal Highness Prince Alfred', *CMM* 8: 46 (1860).

[199] The museum acquired its own building, also in the public gardens, in 1895.

[200] The sculptor was Calder Marshall. It was paid for by public subscription and unveiled in 1864. In 1896 Olive Schreiner sent W. T. Stead a photograph of this statue, saying that it provided her with 'a curious consolation' for her despair over Rhodes's role in the Jameson Raid and the 'greed of wealth and power' that persuaded English-speaking Cape parliamentarians to vote in favour of a 'Strop' Bill allowing for the flogging of black servants. She commented: 'Of the three large English-speaking men who have during the last fifty years appeared on the South African stage—William Pater [*sic*], Saul Solomon and Sir George Grey—I think the last was the greatest and the most unique. A "God's Englishman" if you like it. It is the thought that there have been such Englishmen that takes away one's despair for England's future.' *Review of Reviews*, 15 (1897), 120–1. Schreiner would no doubt have found it galling that a statue of Rhodes, gesturing towards Africa, points directly past that of Grey; yet it is fair to note that Rhodes and Grey shared certain characteristics in common, not least their visionary hopes for Cape expansionism.

[201] Cited in L. Davidoff and C. Hall, *Family Fortunes: Men and Women of the English Middle Class, 1780–1850* (London, 1987), 442. For an interesting discussion of the duality that characterized Priestley's view of science—a commitment to emancipation and liberation on the one hand, and a leaning towards discipline and authority on the other, see I. Kramnick, 'Eighteenth-century Science and Radical Social Theory: The Case of Joseph Priestley's Scientific Liberalism', *Journal of British Studies*, 25: 1 (1986).

2

'Of Special Colonial Interest': The *Cape Monthly Magazine* and the Circulation of Ideas

A further register of the confidence in colonial achievement that characterized the Cape at mid-century was the launch of the *Cape Monthly Magazine* in January 1857. Although not the first literary venture of this type, the *Monthly* was notably more ambitious and successful than any of its periodical predecessors. It lay at the very centre of the interlocking network of institutions that constituted the focus of the previous chapter—the South African Library, Museum, and the Botanical Gardens—and it did much to promote these by carrying reports and commentaries relating to their activities. Together with parliament, the press, and the Cape university, such examples of institutional public culture contributed significantly to the sense of colonial achievement and belonging that developed during the period of transition from representative (1853) to responsible government (1872).

In addition to publicizing the associational world of clubs, societies, and institutions, the *Cape Monthly* provided a much-needed outlet for the dissemination of new ideas and original research. The magazine reported, synthesized, and reacted to intellectual developments within the British world. Its editors' eyes were firmly fixed on developments in the imperial centres of London and Edinburgh, but their peripheral vision took in activities going on in rival imperial outposts of the southern hemisphere as well. Many pioneering studies of Africa and its peoples were first brought to public attention in the *Monthly*, which offers an unrivalled portal through which to view discussions of ideas and opinions ranging from the aesthetic and the literary, to philosophy, religion, and the natural sciences. Material on travel and on art, as well as botany, geology, paleontology, and anthropology, was all given due attention.

The *Cape Monthly* made a distinctive contribution to what might be thought of as a developing colonial conversation. Unlike newspapers, where opinion was largely broadcast in editorial columns or else via letters and was therefore conducted as a form of dialogue, the ongoing exchanges in the *Monthly* took on the form of a colloquy. Crucially, the magazine was reliant on active participation by its readers.

The periodical press existed largely outside of the developing imperial press system and was thus considerably less affected than newspapers by the constraints of new technology (the telegraph) or capital investment (permanent staff, presses, advertising copy). Nor was the periodical press subject to the same kind of direct influence that the rise of news agencies such as Reuters was to exercise from the 1870s.[1] By virtue of their smaller, self-selecting audiences, journals like the *Monthly* could be customized closely to the perceived interests of their more intimate circle of subscribers.

For all these reasons the problem of audience reception is somewhat less an issue for periodical literature than for newspapers. We know readers existed, because they addressed each other directly in debate and because many were (or were to become) prominent public figures. Moreover, constituencies of magazine subscribers and authors were far more likely to meet each other in the context of other social and intellectual activities: in the world of print culture, whose contribution to national identity is well established, theirs was arguably more of a real than a virtual or imagined community.

The demographic composition of this reading community nevertheless lacks precision, and analysis of the culture of reading in South Africa is at best rudimentary. Still, there are sufficient pointers to suggest that the size of the reading public was substantial and growing by mid-century. In 1858 there were no fewer than eight newspapers published in Cape Town (six in English, two in Dutch), with a combined circulation of around 3,500 and a considerably larger reach when one considers how many people may have read a single copy. Many inland and coastal towns had their own newspapers—the small rural centre of Graaff-Reinet, for example, sustained two English-language newspapers in the 1860s. According to the 1875 census, around half the population of Cape Town and its environs (total 57,000) was literate to some degree. There were five booksellers in the town at this time. Booksellers in Cape Town and the interior often enlarged their business by doubling up as informal circulating libraries. Literacy rates were significantly lower in rural areas, but even small towns boasted reading societies, libraries, or mechanics' institutes. Books to the value of £20,000 were imported into the Colony in 1870, £41,000 in 1875. Over 95,000 books were mailed inland that year.[2]

The high volume of books entering the Colony was facilitated by the Book Post Act of 1862, which allowed for free delivery of overseas literature to any post office in the Colony. Stamp duty on newspaper and other publications was abolished according to Ordinance 2 of 1848 by Sir Harry Smith (some years before this

[1] S. J. Potter, *News and the British World: The Emergence of an Imperial Press System 1876–1922* (Oxford, 2003).

[2] The best estimate of the reading public remains R. V. Turrell, 'A Cape Periodical. The *Cape Monthly Magazine* 1870–1875', BA (Hons) thesis, University of Cape Town (1974), 7–12. See also G.42–76, *Results of a Census of the Colony of the Cape of Good Hope, 1875*; Hattersley, *Illustrated Social History*, 244–5.

occurred in England). This move was most likely a concession to the rising middle-class political activism of that period and discussions about widening access to literature. It may also have been a response to trades-union organization amongst compositors and pressmen. The 1840s saw an efflorescence of activity in the world of journalism. Several new titles, including the satirical *Sam Sly's African Journal*, were published during this decade.[3] Unfortunately, the absence of stamp duty means that the circulation figures for the *Cape Monthly* can only be guessed at. In all probability, it never sold more than a few hundred copies a month. Yet, the mere fact of its longevity was a measure of success. Even if the *Monthly* was not read much beyond the social and family circles in which its contributors were based, this is in itself indicative of a sustainable critical mass of middle-class intellectuals and opinion-formers who exchanged ideas with one another and took their own opinions seriously.

The constant watchwords of the *Cape Monthly* were 'progress', 'improvement', and 'civilization'. These virtues were interpreted in a broad sense to include political, religious, social, and economic advancement as well as the cultivation of intellect. Thus, a sense of moral community was being called into being alongside incipient nationhood. Readers were frequently reminded that material achievement should not be the sole measure of progress and that an aesthetic and cultural sensibility was equally important. As the *Monthly*'s first editorial proclaimed:

> On all sides we hear of prosperity and progress in things material, though, with characteristic John-Bullism, we still grumble a bit because we do not move so fast as we would. But, while careering along the material road, need we forget the intellectual? A community that gives no signs of even a nascent literature of its own, is still far from civilization. Wealth may accumulate, but mind decay; though, indeed, the experience of the world has rather shown that a nation's intellectual vigor and its increase of wealth have been simultaneous. That such may be the case with the Cape Colony, all its true friends must ardently desire; and to them, therefore, the establishment of a local periodical like the present, must be a source of pleasure and satisfaction.[4]

Local achievement was a constant source of self-satisfaction and pride. Indeed, the very existence of the *Monthly* was offered as evidence of the Cape's newly established maturity and respectability. 'It was once erroneously remarked in our hearing', the editors remarked in amused tones of pompous self-regard, 'that three institutions existed in the Cape for which the community was not ripe, viz., the abolition of slavery, representative government, and the *Cape Monthly Magazine*...'[5] Yet, pride in colonial achievement never quite displaced underlying fears of provincial inferiority. England remained 'home', and assertions of the Cape's distinctiveness were more often defined in terms of regional particularity

[3] Hattersley, *Illustrated Social History*, 138, 140–1. The Stamp Act was finally abolished in England in 1855, and in 1861 tax on paper was ended. See e.g. R. M. C. Ensor, *Oxford History of England* (Oxford, 1938), 143.

[4] *CMM* 1 (Jan.1857), 1.

[5] *CMM* 11 (Jan. 1862), 53.

than in the discourse of nascent national assertion. The novelty of youth and the excitement of discovery made it possible to celebrate that which was different or unique and to make favourable comparisons with other parts of the Empire. But the desire to emulate British norms was always present and deference to the metropole was an ingrained reflex.

The first series of the *Cape Monthly Magazine*, jointly edited by Alfred Whaley Cole and Roderick Noble, ran from 1857 to 1862.[6] After an interregnum, which coincided with adverse economic conditions, the periodical was revived in 1870 under the sole editorship of Noble. Following Noble's death in 1875, the *Monthly* was continued by his brother John, who was also clerk of the Legislative Assembly.[7] All three editors were prominent and respected citizens who immigrated to the Cape around mid-century and contributed significantly to the Colony's developing anglophone civic and intellectual culture. In addition to their roles as English-speaking opinion-formers, they all enjoyed close personal and professional links with leading members of the Dutch-speaking community.[8] Brief biographical details must suffice to illustrate this point.

Roderick and John Noble were born in Inverness. Roderick (1829–75) was educated at Edinburgh University. His original intention had been to enter the ministry, but he became a journalist and a teacher instead. Noble emigrated to the Cape in 1850 and took up a position as a teacher of mathematics at Dr Antonine Changuion's Institute, a secondary school geared especially to educating the sons of the Dutch-speaking elite. He subsequently moved to the South African College as a lecturer in chemistry, and took on the professorship of physical science as well. Four years later Noble added the chair of English literature to his burgeoning portfolio. In addition to his teaching duties Noble gave public lectures on a range of literary and scientific topics, as well as writing for English and Dutch newspapers at the Cape. His energetic involvement in the civic life of the town ranged from participation in public meetings and functions to attendance at parliament, church synods, and the Commercial Exchange. It also included committee work at the South African Library, the Mechanics Institute, Meterological Commission, and the University Council. As a speaker of Dutch and an inspirational teacher at the South African College, Noble exerted a profound influence on a generation of late nineteenth-century leaders, including justices de Villiers and Kotzé, W. P. Schreiner, and J. H. 'Onze Jan' Hofmeyr. His premature death at 46 drew one of the largest crowds of mourners ever seen in the city, thereby testifying to his close association 'with the general life of the community'.[9]

[6] The collaboration between Cole and Noble ended in 1862. In 1867 Cole became a founding editor of the *South African Magazine* together with William Foster.

[7] John Noble, *South Africa, Past and Present: A Short History of the European Settlements at the Cape* (Cape Town and London, 1877).

[8] Unless otherwise indicated, the following biographical details are based on entries in the *DSAB*.

[9] 'In Memoriam: Roderick Noble', *CMM* 12 (Jan. 1876); R. V. Turrell, 'A Cape Periodical', 25–30; A. M. Lewin Robinson, *Selected Articles from the Cape Monthly Magazine (New Series 1870–76)* (Cape Town, 1978), 2–3; *Cape Argus*, 23 Jan. 1875.

John Noble, who succeeded his brother as editor of the *Cape Monthly*, came to the Cape in 1857 and began working as a journalist on Saul Solomon's *Cape Argus*. He developed a strong interest in early Cape history, in particular the role of the Huguenots, and married Susanna Francina Leibbrandt, sister of the first full-time Cape archivist, H. C. V. Leibbrandt. After John Fairbairn's death in 1864, Noble took over the editorship of the *South African Mercantile Advertiser and Mail*, and the following year he became clerk of the House of Assembly, a post he held until his retirement in 1897. He was highly respected for his knowledge of parliamentary procedure and developed a good relationship with members of the Afrikaner Bond. Noble developed a reputation as an editor and writer of literary and historical works, including his *Descriptive Handbook of the Cape Colony* (1875) and *South Africa Past and Present* (1877).

The third member of the editorial triumvirate was Alfred Whaley Cole (1823–96). Born to a middle-class Bath family and educated at London University, he first arrived at the Cape in 1841, en route to New Zealand. But his ship sank in Table Bay and he opted to spend five years in the Colony instead, before returning to London in 1846 to train as a barrister. An acquaintance of Dickens and Thackeray, as well as the illustrators George Cruikshank and William McConnell, Cole combined his legal interests with work as a journalist and novelist. He returned to the Cape in 1856, whereupon he became an advocate and parliamentarian in addition to founding the *Cape Monthly Magazine*. Cole eventually rose to become a senior judge, was an acknowledged authority on Roman-Dutch law, and lectured on the subject at the South African College. He held the Cape Dutch gentry in high esteem, considering them to be as well educated and cultivated as their fellow English colonists; in support of this conviction he married twice into the Cloete family.[10]

The *Monthly* was initially priced at 15*s*. per annum (rising to £1 in 1859) for subscribers (postage was extra). It appeared between grey covers bearing a picture of Table Mountain and the Bay, and was printed on Saul Solomon's steam press.[11] Like several British periodicals of the age—*Blackwood's* or *Macmillan's Magazine*, for example—the *Cape Monthly* aimed to be a 'miscellany', 'combining amusement with information, and affording equal space to literature, science, the fine arts, and commercial and statistical intelligence'.[12] In this regard the *Monthly* combined the seriousness of purpose characteristic of the highbrow British quarterlies like the *Edinburgh* and *Quarterly Reviews*, with the more popular approach of the

[10] A. W. Cole, *Reminiscences of My Life and of the Cape Bench and Bar* (Cape Town, 1896), 62; also *DSAB*.

[11] No circulation figures are available but the potential reading public was substantial: according to the 1875 census, 177,000 persons out of a total Cape population of 721,000 were able to read and write, at least in a rudimentary manner. Turrell's figures for the value of books imported into the Colony—£20,440 in 1870 and £41,909 in 1875—are perhaps a more meaningful indicator of the reading public. So, too, is the figure of 95,653 books mailed inland from the Cape in 1875. See G.42–76, *Results of a Census of the Colony of the Cape of Good Hope, 1875*, 9; Turrell, 'A Cape Periodical', 10.

[12] *CMM* 1 (Jan.1857), 2.

monthly magazines, though it steered clear of the waspish satire and lampoons characteristic of such publications. Unlike the British periodical press, which was segmented according to taste and political viewpoint, subscribers to the *Cape Monthly* were assured of the journal's non-partisan approach to politics.[13] While this probably reflected a wish to reach as broad a readership as possible, it was entirely consonant with the Noble brothers' broad-minded civic liberalism and the consistently inclusive view they adopted towards anglicized and educated Afrikaners.

Mindful of their inability to compete directly with British journals (which were widely available), and consciously following in the tradition of earlier literary periodicals like the *South African Journal*, the editors of the *Monthly* sought to 'refine and elevate the every-day life of colonial readers' by covering topics of local importance and seeking to 'chronicle those indications of improved civilization which are happily becoming each year more frequent'.[14] Thus, although a varying proportion of its content was imported from abroad, the *Monthly* was no mere imitation of British periodicals. Emphasis was placed on local material, including antiquarian colonial history, accounts of travel and recent exploration, reports of scientific discoveries, biographical sketches of Cape notables, photographic portraits, literary reviews, fiction and poetry, meterological tables, news and events, and so on. In addition, there were regular features on the progress of commerce, agricultural societies, the South African Library, and the Museum.

The *Monthly*'s stress on local content increased when it resumed publication in 1870.[15] The first issue of this series reminded potential subscribers that it was the only periodical in the country which specialized 'in Literary and Scientific as well as Social interest'. Stressing continuity and striving to build on the success of the first run, Roderick Noble stated his main object as being 'to render the Magazine one of special Colonial interest and that, in a certain sense of the term, it should be racy of the soil'. Throughout South Africa, he continued,

> there are historical facts to be collected, reminiscences to be recorded, legends and traditions, both of the Natives and European settlers, to be gleaned and preserved for the use of the future historian, as well as for the gratification of the present reader. There are literary workers and scientific observers scattered over the length and breadth of the country, whose studies and pursuits can be rendered more satisfactory to themselves and beneficial to the public by being thus brought together as into a common focus in the pages of a Colonial Magazine—which will be placed unreservedly at their disposal.[16]

The *Monthly* functioned as a vital brokerage for the exchange of ideas, and did much to facilitate the emergence of a colonial intelligentsia. In the customary manner

[13] *CMM* 5 (June 1859), 373–4.

[14] *CMM* 11 (Jan. 1862), 53.

[15] Its format was virtually unchanged from the first series, and the subscription price was set slightly lower at 12*s*. with 3*s*. postage.

[16] *CMM* 1 (July 1870), 1–2.

of periodical literature, articles were often left unsigned or bore a pseudonym, which makes authorial identification difficult. It is nonetheless clear that one group of regular contributors were public figures of intellectual distinction like Langham Dale, A. W. Cole, James Adamson, W. H. Ross, J. H. de Villiers, John Noble, William Bisset Berry, John Robinson, F. W. Reitz, and T. B. Glanville.[17] The *Monthly* also provided an outlet for the discussion of original scientific and literary research by writers with specialist knowledge. The more outstanding and lasting contributions in this category include essays by the pioneer ethnologist and comparative philologist W. H. I. Bleek, the colonial historian G. M. Theal, and the botanist Harry Bolus, as well as geologists like W. G. Atherstone, R. N. Rubidge, and G. W. Stow. Other contributors of note include H. W. Piers, Roland Trimen, Alfred 'Gogga' Brown, Henry Callaway, F. W. Kolbe, E. L. Layard, John Shaw, Thomas Maclear, Charles Brownlee, R. W. Murray, J. M. Orpen, E. J. Dunn, and Thomas Baines. Reviews of new books on South African topics provided another important focus for discussion, as did reports of speeches by leading politicians and dignitaries with a bearing on the cultural and intellectual life of the country.

In addition to its function as a clearing house for new ideas, the *Cape Monthly* did a great deal to assist the development of cultural and scientific institutions. The South African Library, the Museum, and the Botanical Gardens were treated as a venerable trinity and their respective progress was afforded regular and detailed attention. Campaigns to encourage the formation of new institutions were also encouraged, including the creation of a permanent colonial art gallery.[18] Improving agencies with more utilitarian objectives were also afforded attention, such as the formation of agricultural societies.[19]

The need for tertiary education was given extensive coverage in the *Monthly*'s columns. Particular attention was devoted to the Cape of Good Hope University, whose formation in 1873 was welcomed as a milestone in colonial development and linked directly to the award of responsible government in the same year.[20] Following in the tradition of the Library address, the *Monthly* reported the annual chancellor's speech to the University, and carried articles by the likes of William Porter and Langham Dale on matters such as the acceptance of women students, the sorts of degrees that should be offered (for example, medicine), and whether the university should continue to exist purely as a federal examining body or else aspire to become a teaching university.[21]

17 Turrell, 'A Cape Periodical', 34.

18 *CMM* 5 (Feb. 1859), 106; 2 (Mar. 1871), 181–3; 2 (Apr. 1871), 237–46; 2 (May 1871), 318–19.

19 The *Monthly*'s coverage of agriculture was, however, sporadic and the journal never succeeded in exploiting the possibilities that existed in this sphere of applied science. But see e.g. *CMM* 5 (Mar. 1859), 158–60; 10 (Oct. 1861), 248; 11 (Apr. 1862), 246; 3 (Sept. 1871), 144 ff.; 3 (Oct. 1871), 244 ff.

20 See e.g. CMM 6 (June 1873), 'Universities'; 4 (Sept. 1873), 218; 8 (Feb. 1874), 'The University and Colleges of the Cape of Good Hope'.

21 See e.g. *CMM* 13: 74 (1876); 8 (Nov. 1876); 14 (Jan. 1877).

AN ART GALLERY

Cultivation of the public's aesthetic sensibility and taste was a constant concern for the *Monthly*, which viewed literature and art as a vital counterpart to material prosperity. Vigorous promotion of exhibitions and a public art gallery was one practical expression of this interest, discussions of the aesthetic virtues of the Cape's landscape and scenery (often linked to the Colony's desirability as a place of permanent settlement) were another. Both of these concerns had a bearing on colonial identity formation.

The campaign to establish an infrastructural base for the fine arts progressed along similar lines to the initiatives behind the Library, Museum, and Botanical Gardens. In 1851 a public exhibition of paintings and sculpture was organized, the first of its type. Nearly 3,000 people paid 1*s*. each to see the works on display. Medals were awarded to Thomas Bowler for a watercolour of Table Bay, to Charles Bell for his heroic history painting *Landing of van Riebeeck, 1652*, and to the architect and secretary of the organizing committee, Gilbert MacDougall, who exhibited a model of his design for the Cape's prospective new parliamentary buildings. Another, smaller exhibition was held the following year, but a withering series of newspaper articles on the pictures by a critic calling himself 'Mastic Varnish'—most likely Thomas Bowler—had a bruising effect on public enthusiasm.[22]

In the absence of a dedicated public gallery, exhibitions relied on the willingness of private owners to put their works on loan. The committee responsible for curating exhibitions was led by prominent citizens, businessmen, and dignitaries, whose knowledge of art was distinctly limited. In 1858 a further exhibition was organized by public notables in the Legislative Council Chambers.[23] In the *Cape Monthly* 'Z' reported that the event, at which some 600 specimens of art were shown—many good, many indifferent, and some, it must be admitted, unquestionably *bad*'—could not be repeated in perpetuity. A dedicated public institution like the Library was essential in order to house and build on the nucleus of thirty-two paintings left to the Colony by the 'patriotic' connoisseur von Dessin.[24]

[22] Governor Grey acted as patron and the committee included the colonial secretary Rawson W. Rawson, William Porter, John Fairbairn, E. L. Layard, and Thomas Bowler (who was also the largest single exhibitor). Bowler's protégé, the landscape painter and surveyor-general Abraham de Smidt, acted as secretary. See F. K. Kendall, *A Short History of the South African Fine Arts Association* (CapeTown, 1941), 6–8; Marjorie Bull, 'Introduction', *One Hundred Years Ago, South African National Gallery* (Cape Town, 1975); Joyce Murray (ed.), *In Mid-Victorian Cape Town: Letters from Miss Rutherfoord* (Cape Town, 1953), 23–4, 26–7.

[23] Kendall, *A Short History*, 8–9. Examples of work by Van Dyck, Turner, Landseer, Chantrey, and Prout were also exhibited. De Smidt (1829–1908) was born in Cape Town into a family of prominent Cape civil servants and landowners. He was educated at the South African College and became a government surveyor in 1848. A noted landscape artist, who painted both in South Africa and in Europe, he was a foundation member and trustee of the South African Fine Arts Association.

[24] *CMM* 5 (Feb. 1859), 103–6; R. R. Langham-Carter, 'The Founding of the Public Collection', in *South African National Gallery 1871–1971* (Cape Town, n.d.), 3.

Moves to establish a fine arts society stalled for some years, and no further exhibition appears to have been mounted until 1866. The format was similar to previous exhibitions, but it was noticeable that a higher proportion of work by local professional and amateur artists was on display.[25] In 1871 momentum to establish an art gallery revived, coinciding with the revival of the *Monthly*.[26] An article in March of that year proposed the establishment of an Art Association in order 'to form the nucleus of a national collection for all time to come'.[27] A further contribution, devoted to the theme of 'Colonial Art', focused on a recent exhibition of oil paintings by W. Hermann at Mr Barnard's photographic studio. This commended Hermann's ability to capture the brilliant foliage, light, and atmosphere of Cape scenery. The artist's representation of the Knysna landscape was said to be startling in its 'truthfulness' and could be interpreted as 'a bit of condensed patriotism': 'It is typical of that perfect freedom to which we are bred in this country, and which imparts an additional zest to every lovely aspect of nature, because it is instinct with true colonial liberty. We are not here cramped with the restraints of old world fences and game laws.'[28] The following issue of the *Monthly* carried an article (unsigned but almost certainly written by Abraham de Smidt) entitled 'An Art Gallery for South Africa', which enjoined the public to support the creation of a permanent gallery. The example of Melbourne, which boasted 'a handsome building devoted to a Library, Museum and Art Gallery', was cited as a model. Strikingly, de Smidt appealed not only to the higher sensibilities of art lovers and the 'ennobling influence' of art on human nature, but also took up the theme of public interest. The 'universal language' of art, he suggested, was intelligible to all social classes and constituted 'a powerful incentive to moral and intellectual improvement'.[29]

Events were moving swiftly. A month later it was reported that a meeting of gentlemen had unanimously resolved to issue a prospectus for the formation of a subscriber-based fine arts association, with a view to the creation of a permanent collection and gallery. 'We hail this movement as a step in the right direction', chimed the *Monthly*. 'Our library, Museum and Botanic Gardens all sprang from small beginnings.'[30] The South African Fine Arts Association was duly formed in August 1871. Richard Southey chaired a committee of fifteen, supported by subscribing members, and the governor, Henry Barkly, acted as president.[31]

The death later that year of one of the association's committee members, T. Butterworth Bayley, gave the gallery an unexpected boost. Bayley, who was born in India and served in the Bengal Civil Service, had moved to the Cape, where he became a progressive farmer and racehorse breeder, as well as a keen art

25 Bull, 'Introduction', *One Hundred Years Ago*.

26 This is not to suggest a causal connection—but the coincidence is striking.

27 *CMM* 2 (Mar. 1871), 'Concerning an Art Gallery', 182–3.

28 Ibid., 'Art at the Cape', 158–9.

29 *CMM* 2 (Apr. 1871), 237, 238–40, 242.

30 *CMM* 2 (May 1871), 318–9; *Rules of the S.A. Fine Arts Association, Cape Town 1871* (Cape Town, 1878).

31 Kendall, *A Short History*, 11.

collector. The catalyst was his bequest of more than forty South African and European paintings to the association, in addition to a sum of £500 towards the purchase of a suitable building to house them.[32] Abraham de Smidt, who had helped Bayley amass his collection, was appointed Bayley's trustee and became the driving force behind the Fine Arts Association. Shortly thereafter premises were acquired in Queen Victoria Street, adjacent to the Botanical Gardens. In 1875 the new South African Art Gallery opened to the public.[33] Thus, by the mid-1870s the basis of a national art collection had been established, though whether the new gallery was a civic institution or a national one—and thereby eligible for state subsidy—remained an open question.[34]

The catalogue of the Fine Arts Association's inaugural exhibition reveals that nineteenth-century English landscape pictures formed the bulk and major focus of the collection, supplemented by a substantial local contingent, including rising colonial artists like Bowler, Baines, Hermann, and de Smidt. The encouragment given by critics to these artists was tinged by an awareness that, as the *Penny Post* put it, 'colonial' art 'labours under a stigma'.[35] In order to correct this perception, the promoters of local artistic production chose not to set themselves up against European or English art. Acknowledgement and respect for local work rather than parity with overseas art was sought. The status of colonial art was regarded as being enhanced by virtue of being displayed alongside selected work by European masters, whose example would act as a spur to excellence.

The output of a number of Cape women artists was represented in some strength in 1871, including pictures by Miss Maclear, Mrs C. N. Thomas, Miss Denyssen, Lady Eyre, and Mrs Ebden. Judging by the catalogue descriptions, mountain passes, ravines, wild flowers, and views of Cape scenery predominated. Amateur women artists also figured prominently in subsequent art exhibitions. Later still, society figures such as Mrs Smyth and Lady Elizabeth Loch became leading members of the amateur South African Drawing Club (est. 1889). Painting of local landscape scenes was by this time an established and genteel leisure-time activity, in which the realms of public and private spheres were interconnected by means of excursions, picnics, tea parties, and exhibitions. Art was in the process of being domesticated, in more senses than one.[36]

[32] Bayley was born in India in 1810 to a wealthy Mancunian merchant family. In South Africa he served as a member of the Legislative Council, the Cape Roads Board, and as president of the Cape Agricultural Society. There are few indications as to how he became interested in art, but his friendship with de Smidt was key. See R. R. Langham-Carter, 'The Originator', *South African National Gallery*, 5. Bayley offered his funds on condition that a further £1,500 be found within 18 months after his death.

[33] Langham-Carter, 'The Founding of the Public Collection', 4; Kendall, 'A Short History', 11; M. Bull, *Abraham de Smidt 1829–1908: Artist and Surveyor-General of the Cape Colony* (Cape Town, 1981), 69, 76, 115. The building in Niewstraat, now Queen Victoria Street, was purchased from the old Dutch educational institution, 'Tot nut van't Algemeen'.

[34] *CMM* 14 (June 1877), 'The South African Fine Arts Association'. A parliamentary grant of £100 p.a. was secured in 1880–1.

[35] Cited in Bull, *Abraham de Smidt*, p. 69.

[36] *Catalogue of the First Exhibition of the South African Fine Arts Association, 1871*, South African National Gallery archives, Box 1. The catalogue lists artist's name, subject-matter, and owner's name. Typically, the pictures by women were owned by family members or themselves. Whether these

AESTHETIC VISIONS

Expressions of pride in the artistic progress that had been made did not mask those fears of colonial inferiority and disdain for provincial philistinism which often marked contributions to the *Cape Monthly*. In commenting on the 1858 fine arts exhibition, for example, 'Z' contrasted the 'rapid strides' that had been made in the Colony towards the achievement of political liberty with the desolate state of cultural achievement. Press censorship, absolute gubernatorial power, and the 'disgrace' of slavery had all been ended. Liberty had been won. But was the present generation capable of 'enjoying the blessings of freedom' through 'the cultivation of a sound taste for the beauties of nature or art'? Alas no, was the author's mournful rhetorical response. In his view—and we must assume gender in this case, as elsewhere—attempts to 'foster a love for the sublime and beautiful' in public institutions were distinctly lacking. Poetry was the only art form that maintained a presence at the Cape, but even its power was apt to 'flag in this dull town'. Yet, the force of poetry could be felt 'in the presence of nature's majestic works—in solitude and silence, under the shadow of the mighty mountain front which towers above us', and 'with the grand sweep of the Atlantic spread out at our feet, smooth as a mirror . . .'.

> But leave your book and your solitude and come again into the society of men, and what have you? Not a single object strikes your sight which can give rise to pleasant emotions. No gallery of pictures to which you can resort; no sculptures: but, instead, those hideous squares of tawdry brick and mortar; copious talk; abundance of personalities, on all sides dissatisfaction, ennui, self, mammon. We have but one relief—the cool shades of that magnificent [Government] avenue, for which may the *manes* of our Dutch ancestors be for ever blest.[37]

A similarly despairing note was struck at the same time in an article on 'South African Poetry', which contended that the 'moral atmosphere of colonial life' was not sufficiently 'nutritive or conducive to poetic thought', given the pervasive climate of 'hard money-grubbing' and 'sharp commercial speculation'. All that existed of genuine value was Pringle's 'exquisite' poem 'Afar in the Desert'. Like 'Z', this writer also looked to nature in order to stimulate the poetic imagination:

> In our rugged iron-bound coasts and surging seas,— in our abrupt, bold, towering mountains, deep ravines, and dense unpenetrable forests,—in our sounding cataract that haunts us like a passion,—in the lone luxuriant grassy slopes of emerald, dappled richly with their mimosa clumps, the lover of nature, and nature's aspects might revel in supreme delight.

works were watercolours or oils is not noted. Marjorie Bull, 'Introduction' to *One Hundred Years Ago*, comments of the 1875 exhibition that: 'Lady visitors predominated, gathering mostly in the room which contained an assortment of watercolours, photographs and contributions from local artists.' Cf. *Catalogue of the Fifth Exhibition of the South African Fine Arts Association, 1879*, in SANG archives box 1, in which watercolours of flowers by local women artists are a conspicuous feature. See also Bull, *Abraham de Smidt*, 99–101.

37 *CMM* 5 (Feb. 1859), 103.

And even the solitary wastes of the desert are not wholly unimpressive to the rightly-constituted poetic sense.[38]

The manner in which colonists made sense of the landscape of the Cape had important bearings on their feelings of belonging and the manner of their insertion into the African continent. There is no easy or simple answer to the questions posed with such clarity by the writer and critic John Coetzee: 'How are we to read the African landscape? Is it readable at all? Is it readable only through African eyes, writeable only in an African language?' Coetzee argues that the Edenic garden myth, with its connotations of innocence and earthly paradise, 'failed to take root in the garden colony of the Cape'. At the most simple level, he explains, Africa could not be envisioned as an earthly paradise because it was *not* part of the new world: it was '. . . the farthest extremity of the Old: it was Lapland of the south, peopled by natives whose way of life occasioned curiosity or disgust but never admiration'. Isolated from the European world and 'walled in by oceans and an unexpected northern wilderness', the Cape came to be conceived of by visitors as an 'anti-garden', a country evoking a future that promised 'less of the perfection of man in a recovered original innocence than of the degeneration of man into brute'.[39]

Coetzee extends this idea by looking at representations of people as well as landscape. Hottentots (or Khoi), he argues, were consistently represented by travellers during the Dutch colonial period in terms that evoke laziness, torpor, and ugly brutishness.[40] With the coming of the British in the early nineteenth century, the image of Khoi indolence was further reinforced by travellers and observers. It was extended, in tones of fresh moral outrage, to descriptions of Boers, whose manners and habits were similarly portrayed as squalid and slovenly. That Boer listlessness was made possible by their dependence on the labour of slaves and servants was regarded as especially reprehensible. The likelihood that Boers had regressed into a state of idleness—whereas Hottentot indolence was assumed to be original or inherent in their nature—gave rise to further moral condemnation on the grounds that their habits and behaviour constituted an affront to the civilizing ideal.[41]

With respect to understandings of the landscape, Coetzee employs the established eighteenth-century European distinction between the picturesque and the sublime.[42] In his view neither category suited the African environment. The sublime could not usefully be evoked, because the open spaces of the interior lacked the vertical elements that were so conspicuous a feature of European landscape

[38] *CMM*, 4 (Dec. 1858), 'South African Poetry', 342–3.

[39] J. M. Coetzee, *White Writing: On the Culture of Letters in South Africa* (Cape Town, 1988), 2–3, 62.

[40] Ibid. 17–18. [41] Ibid. 26–7, 28.

[42] The picturesque conventionally refers to aesthetic principles of domesticated landscape composition—typically, Claude Lorrain's ideal of three bands of recessional space: a detailed foreground yielding to a more simplified middle distance with a central icon (often a lake), thereafter receding into the infinite horizon. The sublime encompassed grand topographic features (often mountains) whose majesty, overwhelming size, and inhospitability is suggestive of the freedom that comes with uncontained space and movement.

(like the Alps). Notions of the picturesque offered more potential. But the harsh atmospheric conditions and the absence of changing light and shade, water and green foliage, made it difficult to transpose this category to African conditions without substantial modification.[43]

Coetzee's framing of this problem is incisive, and it has been influential. Yet, the evidence from the *Cape Monthly* is rather more mixed. Attempts to grapple with the meaning of the Cape and its hinterland reveal a wide range of responses, and the questions raised by Coetzee were repeatedly posed and reformulated over successive generations. In much the same way that Tim Bonyhady has challenged conventional formulations about the nineteenth-century colonial encounter with the Australian landscape (that its flora and fauna were topsy-turvy or misshapen, and that Europeans regarded the environment as a resource to be plundered rather than conserved),[44] so there is plenty of evidence of more empathetic and varied reactions to the Cape and its peoples on the part of colonial writers and visitors.

One of the most common tropes deployed by contributors to the *Monthly* was the view of Cape Town and its environs as seen through the eyes of a new arrival.[45] Such 'first impressions' were contrasted with the considered reflections of subsequent visitors or of colonial residents, sometimes through the medium of unsigned editorial comment. The *Monthly*'s dual role as the proud (though on occasion critical) voice of local opinion, as well as the intermediary between colony and metropole, helped to stimulate an ongoing dialogue. At one extreme were the opinions of those who regarded Cape Town as singularly unpleasant from the outset.[46] Charles Darwin, who spent nearly nearly three weeks in the south-western Cape in 1836 while the *Beagle* was docked in Simonstown, found little to admire. The landscape of the Cape was devoid of animals, people, and homesteads, and the sandstone mountains around Franschhoek 'gave to the country a bleak and desolate aspect'.[47] Darwin's views bear comparison with those of Captain James Cook, extracts from whose journals were published by the *Monthly* in 1858. Cook, the *Monthly* informed its readers, reported that:

> no other country that we saw during the voyage makes a more forlorn appearance or is in reality a more sterile desert. The land over the Cape, which constitutes the peninsula formed by Table Bay on the north and False Bay on the south, consists of high mountains, altogether naked and desolate: the land behind them to the east, which may be considered as the isthmus, is a plain of vast extent, consisting almost wholly of a light kind of sea sand, which produces nothing but heath, and is utterly incapable of cultivation.[48]

[43] Ibid. 39–40, 42–4, 52, 61–2.

[44] T. Bonyhady, *The Colonial Earth* (Melbourne, 2000).

[45] For a more theorized account than my own, see Jessica Dubow, 'Colonial Space, Colonial Identity: Perception and the South African Landscape', Ph.D thesis, University of London (2000), ch. 2.

[46] See e.g. *CMM* 4 (July 1872), 'An Englishman's First Impressions of Cape Town', and the editorial comment by Noble.

[47] Cited in W. S. Barnard, 'Darwin at the Cape', *South African Journal of Science*, 100 (2004), 245.

[48] *CMM* 3 (Jan. 1858), 'Captain Cook's Visits to the Cape', 7.

The *Monthly*'s point in publishing this negative description was to turn its message round by showing what had been achieved since that time or to indicate what remained to be achieved. Thus, Cook's remarks were presented as a 'well-drawn picture of the naturally unattractive and desolate character of the country' as it must have appeared at first to 'the enterprising men who first began to plant civilization in South Africa'. They 'convey appropriate praise to the triumphs of industry' in a land which Europeans 'had thought unworthy of culture' before the arrival of Van Riebeeck, and which remained 'dreary' a century after Dutch colonization.[49]

Depictions of Table Mountain presented their own peculiar descriptive difficulties. In the fictional historical romance 'The Blasted Bough', set in the 1670s, the 'wall of Table Mountain' rose 'vast and grey out of a sea of sombre foliage', and the whole scene of 'wild and unbroken nature had more of gloominess than of mirth about it'. There was an absence of human life, 'a dull monotony in the hues of colour, a general sadness of tone' that overpowered the sense of pleasure that a wide and varied landscape usually evoked.[50] Here again, the sense of threat associated with the mountain was more easily conveyed by the distancing mechanism of historical fiction, which implicitly contrasted the gloomy and glowering past to the more civilized and enlightened present.

A recent study indicates how perceptions of Table Mountain shifted through the nineteenth century. Whereas records of ascents during the first British occupation lay stress on conquest and mastery of a new imperial domain, by mid-century the wild mountain had been symbolically 'domesticated', as climbers embarked on day excursions and held tea-parties on the summit.[51] That perceptions of Table Mountain changed over time is undoubted, yet one should not assume that this shift was a unilinear process, nor treat observers as merely representative of sequential points of view. Exchanges in the *Monthly* reveal that contrary positions were argued at the same time, and also that similar ideas recurred over time. Moreover, writers were often sufficiently self-aware to hold more than one position and to reflect on their own evolving perceptions.

Consider, for example, 'Table Mountain. An Art Study', in which Thomas Fuller, parliamentarian, journalist, and leading light of the Cape fine arts movement, recorded his gradual familiarization with the mountain over a period of five years and his changing responses to its physical and emotional presence.[52] In so doing,

[49] Ibid. A similar strategy was deployed in the editor's response to 'An Englishman's First Impressions of Cape Town', *CMM* 4 (July 1872).

[50] *CMM* 9 (Jan. 1861), 'The Blasted Bough. A Legend of Table Valley', 9–10.

[51] Megan Voss, 'Wildness Domesticated: Nineteenth-century Perceptions of Table Mountain', in N. Worden (ed.), *Historical Approaches: Research Papers by History Major Students of the University of Cape Town, Vol. 1, 2002* (Cape Town, 2002). See also N. Vergunst, *Hoerikwaggo: Images of Table Mountain*, (South African National Gallery, 2001), for a striking pictorial-based interpretation of representations of Table Mountain.

[52] Thomas E. Fuller (1831–1910) was trained as a Baptist minister in England. He was brought out to the Cape by Saul Solomon to edit the *Cape Argus*, which he did from 1864 to 1873. Fuller later pursued a career as a parliamentarian and became a confidant of Cecil Rhodes. He was a strong advocate of responsible government and a supporter of the Cape African franchise. Fuller was also a leading figure in the Fine Arts Association and a trustee of the South African Library.

Fuller condensed into a relatively short time-span the process of cognitive acclimatization that characterized his colonial conversion. Fuller began by recalling the first time he saw the mountain from the deck of the steamer that brought him to the Cape. In the gloom of the night the mountain 'looked like some huge solitary mass of rock in the wild waste of waters'.[53] All through the night he paced the deck, 'contemplating steadfastly the weird-like grandeur of the bold rock' while thinking of the land he had left. Daylight brought striking new insights as he contemplated the scene before him. There was 'little of the tender beauty of quiet pastoral scenery with which an Englishman is familiar in his native land'. He had bid farewell to the 'green hills and valleys of Old England, the quiet country lanes, with the wild-briar and honey-suckle, the meadows rich in the tints of green and gold', and 'changed it for a country where the light was rich and brilliant, the atmosphere surpassingly bright and clear, and the scenery bold, spacious, and grand'.[54]

On closer examination of the mountain, Fuller was struck by the fact that it was more or less equal in height, breadth, and length. The traveller familiar with European scenery might miss the *aiguille* peaks which characterize the 'glory of the Alpine ranges'. But, to his relief, Fuller discovered that Table Mountain was indeed constructed 'after the true mountain pattern', even if it lacked the 'artistic finish' provided by high snow-capped peaks. There were compensations for the absence of these 'grander features'. Although the harsh light during the day was 'too glaring for true atmospheric perspective', the scene changed towards evening as the sun sank low: 'Broad bosses of hard sandstone, bold rocky escarpments, deep and nearly perpendicular ravines, appear as if by magic. In no landscape that we call to mind is the work of the shadow more strikingly illustrated.'[55]

Fuller continued his description of the structure and appearance of the mountain from its many different aspects, citing Ruskin and Turner for the purposes of considered intellectual triangulation. He wrote lyrically about the kloofs and ravines, as well as its vegetation. The foliage, he confessed, was somewhat stunted and dull as compared with those in Killarney, and the pines, 'though not of the most picturesque form', were nonetheless soft and rich from a distance. The silver trees, if not at their best from afar, were 'quite in keeping with the colour of the rock' and gave the sort of 'venerable appearance to the old mountain which grey locks do to the human figure'.[56] More familiar was the eastern side of the mountain, from Mowbray to Constantia, where the vegetation was denser with 'thick, broad belts and clusters of pine, brushwood, and oak, relieved by patches of white-headed silver-trees'. The view over the Atlantic, whose 'strange

[53] T. E. Fuller, 'Table Mountain. An art study', in R. Noble (ed.), *The Cape and its People* (Cape Town, 1869), 233. Noble's preface records that the book was intended 'to present within a moderate compass a fair representation of the ideas of some of our principal thinkers and writers in South Africa on questions of literary, scientific and social interest'. Most of the contributors were associated with the *Cape Monthly Magazine*, which had temporarily ceased publication.

[54] Ibid. 234–5.

[55] Ibid. 236–7.

[56] Ibid. 239–40.

mingling of the bounded and the boundless' mirrored life itself, was counterposed to the eastern side where the kloofs were more retired, symbolizing the 'best and deepest joys of solitude'. This side of the mountain, with its waterfalls, tree ferns, mosses, and half-tones of colour, awoke chords that were not touched by 'the broad, open landscape outside'. (Comparisons with Australian artists' enthusiasm for ferns and gulleys in the Victorian Dandenongs are apposite here.)[57] Fuller also paid painterly attention to the 'atmospheric effects' associated with the mountain's 'bold heights'. The action of wind, cloud, and light created 'ever-varying, ever-beautiful appearances'. It was by means of its constantly changing appearance that the mountain's ' "gloom and glory" become a possession'.[58]

Fuller's description contains elements of both the sublime and the picturesque. The process of taking symbolic possession of the mountain is facilitated not only by virtue of its variety and interest but also from seeing it both from a distance and, having ascended its heights, using it as a viewing platform from which to survey the landscape and the sea. European scenery and aesthetic conventions remain the reference points for Fuller's assessment of the Cape, and on certain criteria Table Mountain is the poorer by comparison, but the author remains sufficiently flexible to comprehend its distinctive and complex beauty and thereby to make it his own.

A similar process of coming to terms with the landscape of the Cape has been described in the case of one of the Cape's leading landscape artists and pictorial historians, Thomas Bowler (1812–69), who arrived in the Colony in 1834 as the servant of the astronomer royal, Thomas Maclear. Bowler soon freed himself from this position and succeeded in establishing himself independently as a professional artist. His rapid upward trajectory culminated in his ascent to become a respected member of colonial society. A faithful follower of the English watercolourist tradition of landscape painting, and an admirer of Turner in particular, Bowler was not an especially innovative painter, nor did he found a distinctive South African school of art; in many respects his view remains that of the outsider documenting exotic and picturesque colonial scenes for visitors to the Cape.[59]

The art historian Michael Godby has nevertheless demonstrated a transformation in Bowler's artistic sensibility and subject-matter from the 1840s to the 1860s, linking this to his gradual acquisition of a Cape identity. Bowler, Godby points out, began by painting picturesque scenes of Cape Town and the eastern frontier, many of which were purchased by Anglo-Indian visitors. These workmanlike watercolours and oils were marked by a sense of formal detachment and were principally designed to record topographical and natural facts. Exotic elements, such as 'Malay' figures, were frequently interposed onto the foreground,

[57] Bonyhady, *The Colonial Earth*, ch. 4.

[58] Fuller, 'Table Mountain', 238, 241, 244.

[59] E. and F. Bradlow, *Thomas Bowler of the Cape of Good Hope* (Cape Town, 1955), 105. See also F. Bradlow, *Thomas Bowler: His Life and Work* (Cape Town, 1967).

not only to satisfy artistic convention but also in order to render the pictures more memorable and saleable.[60] From the 1850s Bowler began to take a closer interest in the social activities of Cape Town by depicting scenes of economic activity and prosperity, drawing attention to key tropes of technological progress, and highlighting major public buildings and key cultural institutions like the Botanical Gardens and the Library.[61]

The 1849 convict crisis appears to have been the event which prompted Bowler to adopt a colonial persona. In his letters and in his drawings of the political events of the time, Bowler revealed a strong political attachment to the colonists' cause. His 'overtly patriotic' lithograph *The anti-Convict Meeting*, depicting the dramatic protest of some 7,000 people outside the Commerical Exchange in the teeth of winter storms, is a case in point. Whereas in 1846 Bowler declared that he would rather own an inch of ground in England to half the African continent, by the 1850s his descriptions of the Cape landscape came to be expressed in 'patently emotional language'. Writing in the *Cape Monthly* in 1857 as 'An Artist', Bowler celebrated the Cape's distinctive light, beautiful mountain ranges, fine scenery, and 'extraordinary rich skies to which those who inhabit Northern latitudes are strangers . . .'. In his late paintings of the 1860s Bowler's adoption of a low viewing point, together with his altered use of colour, light, and space, allowed him to present the landscape 'in terms of mood rather than fact'. Instead of remaining a distant observer, viewers were thus enticed into the picture itself.[62]

Also contemporary with Bowler, and illustrative of similar themes, is the remarkable collection of letters sent in 1861–2 and published anonymously in the *Cape Monthly* a decade later as 'Life at the Cape by A Lady'. The author was almost certainly Louisa Grace Ross, who arrived at the Cape as an infant around the time of the convict crisis.[63] Like other women diarists—most notably Lady Anne Barnard during the first British occupation—Louisa Ross felt free to

[60] M. Godby, 'Settlers and Travellers: Different Ideas of "Home" in the Representation of the South African Landscape by Thomas Bowler (1812–1869) and Thomas Baines (1820–1875)', in T. Barringer and D. Fordham (eds.), *Art and the British Empire* (Manchester, forthcoming).

[61] M. Godby, 'The Colonial Experience of Space: Thomas Bowler's Representation of the Figure in the Landscape in Lithographs and Other Views of the Cape', UCT Centre for African Studies seminar paper, (1992), 6–10.

[62] Godby, 'Settlers and Travellers'. See also *CMM* (May 1857), 'Art at the Cape', by 'An Artist'.

[63] *Life at the Cape a Hundred Years Ago. By a Lady* (Cape Town, 1963). This volume reprints the letters that were published in *The Cape Monthly Magazine* from 1870 to 1871 and includes illustrations by Thomas Bowler. Authorship of the letters was deliberately concealed, but it is highly probable that Louisa Grace Ross was responsible for them. For further details and discussion of her identity, see A. C. Lloyd, 'Some Secrets of South African Literature', *South African Bookman*, 4 (1911); postscript to *Life at the Cape*; E. Bradlow, *Who Was the Lady*? (Cape Town, 1963). Louisa's husband, Dr W. H. Ross, was a well-known doctor who was closely linked with the *Cape Monthly*. He eventually became surgeon-superintendent on Robben Island in 1884, where Louisa, who died a year later, is buried. The success of these letters prompted publication of further extracts from Grahamstown and Natal. These are believed to have been written by Mrs Glanville and Sir John Robinson, respectively.

write about social and domestic matters rather than dwell on the weighty public issues that male writers were inclined to address in portentous tones.[64] The tantalizing mixture of detail and concealment undoubtedly aroused readers' interest in the anonymous letters, but their popularity has more to do with their wit, insight, and zest.

The letters of 'A Lady' are uniquely revealing of Cape society in the 1860s and early 1870s. They merit more sympathetic treatment than they are sometimes accorded by historians. Louisa Ross wrote candidly about her domestic life, the social rounds of the day, and her love of riding and 'open-air' life. She rapidly immersed herself in the local cultural mix of the Cape and was genuinely taken with the indigenous flowers and heaths (fynbos) that grew on the mountain. Many of these—proteas, *avond bloemetjies*, *Africanders*—she called by their vernacular names, and she referred similiarly to the food that she ate—*bobotie, sosaartje, harders, meibos, biltong.*

Aside from occasional comments which the modern reader might find jarring (like her apparent defence of Cape slavery as a benign institution and casual remarks about the appearance and laziness of African labourers), Ross was remarkably lacking in social prejudice and celebrated the relative absence of class and ethnic distinctions in colonial society: 'it requires a cleverer head than mine to detect the existence of the upper ten thousand from the gentry and middle-classes.'[65] Her curiosity led her to attend the wedding of her Malay washerwoman's daughter as well as a ball at Government House; both were treated with a similar mixture of appreciative respect and gentle humour. Ross was particularly critical of English snobbery towards Afrikaners:

> Oh! It makes me ill to hear pert subalterns' wives talking so sulkily and ungratefully of this Colony . . . To see women of this stamp, who have perhaps never moved in Europe in other than third-rate middle-class society, turn up their chiselled noses at good-natured, and by no means vulgar Africanders, *because* their husbands or brothers are engaged in business, is one of the saddest proofs of insular pride and power of human conceit.[66]

Most of all, Ross delighted in the natural beauty of the Cape. Her very first letter revelled in the appearance of the town and and its scenic attractions: 'The sky here is intensely blue, and Table Mountain is of marvellous beauty; so, instead of wishing myself back in England, I honestly yearned that day for the society of all my old friends, so as to let them share the enjoyment we could not but feel, amid scenes and an atmosphere so intensely fresh and exhilarating.'[67] Like several other

[64] On this theme, but in a slightly earlier period, see N. Erlank, 'Writing Women In(to) Early Nineteenth Century Cape Town', *Kronos*, 23 (1996). Erlank builds her argument round the distinction between public and private spheres in middle-class society. That the letters of 'A Lady' were published in the public domain suggests that the boundaries between spheres was by no means impermeable.

[65] *Life at the* Cape, 81.

[66] Ibid. 34

[67] Ibid. 1.

commentators on the Cape, Louisa Ross's first point of reference for Table Bay was the Bay of Naples. It might not present quite so perfect a picture, but 'its brilliant hues *pulsating* in cloudland' were worthy of description by Ruskin. The surrounding hills were 'really Swiss mountains', and Devil's Peak, 'a perfect poem in itself', was characterized by constantly shifting light, colour and cloud: 'At one moment its jagged crests are swallowed up in vapour, then a whirl of wind, and the fleecy masses are torn into shreds and sent howling over the precipices in tortuous eddies. Anon the whole mountain becomes deep purple as the sun is obscured, only to break out again into wealth of colour as the passing colour has melted into thin air.'[68] Ross was a keen rider, and she took every opportunity to explore the Cape peninsula on horseback. The most favoured bridle-paths for equestrians followed the eastern flank of the mountain, which was wetter and more densely wooded than the Atlantic side. On account of their sylvan character, the villages stretching from Mowbray through Newlands and Rondebosch to Wynberg and Constantia became the favoured places of residence for the nineteenth-century middle classes, having been made socially respectable by military officials and Anglo-Indian visitors in the 1830s and 1840s. Anthony Trollope gave voice to these preferences when he commended the 'Elysian' aspect of verdant Wynberg and its surrounding suburbs over the villas in Sea Point with their open Atlantic views.[69]

Louisa Ross was less conventional in her tastes. She found Wynberg somewhat dull, and preferred to ride her horse over Kloof-Road to Camp's Bay, 'a most lovely and sequestered little bay' which 'ought to be the future Brighton of the Cape'.[70] On a longer excursion to the western Cape wine and wheat towns of Stellenbosch, Worcester, and Paarl, she delighted in her encounters with countryfolk. The route from Wellington to Worcester provided 'a glorious succession of mountain vistas and Claudian landscapes', while Bain's Kloof was 'the perfection of mountain scenery . . . it seems to me perfectly wonderful that no artist of merit has ever thought of studying these hill-side effects instead of drudging at Wales and Switzerland'.[71]

For readers of the *Cape Monthly*, Louisa Ross's open-minded and enthusiastic response to the Cape was a welcome validation of the colonial society of which they were part. Although her points of comparison were, like other travellers and writers of the time, European, she was more inclined to experience the Cape's indigenous flavours in their own terms. In this regard she differed from those who sought to transform the Cape's attractions into something better and more recognizably European by acting upon and transforming nature—building promenades, constructing roads, and so on. Nor did she conform to the common tendency to express admiration for the Cape only in

[68] Ibid. 1, 5.
[69] A. Trollope, *South Africa*, repr. of 1878 edn., ed. J. H. Davidson (Cape Town, 1973), 87.
[70] *Life at the* Cape, 15. [71] Ibid. 89–90.

guarded and qualified terms, as typified by the following extract from a poem printed in the *Monthly*:

> Our Cape its beauties hath, 'tis true:
> Our Table Mountain's always grand,
> Our sun is bright, our sky is blue;
> The Maker's bounteous hand,
> From which all beauty hath its birth,
> Made this far corner of His earth.
>
> Yet must a Briton love his home
> The more for absence, as I ween,
> And greatly do I long to roam
> Through daisied meadows green,
> Perchance made dulcet by the swell
> Of distant chiming village bell.[72]

The feminine perspective on the Cape's charms offered by 'A Lady' was parallelled by the professional endorsement given by her husband, William H. Ross, a noted medical doctor, liberal man of letters, and supporter of the fine arts. In an essay entitled 'Our Climate', which formed part of a volume on *The Cape and its People*, edited by Roderick Noble (and whose contributors included several of the *Cape Monthly*'s regular writers), William Ross expounded in detail the salubrious nature of the Cape's temperate climate. In the reassuring tones of a trusted doctor, Ross advised readers that the Cape was an attractive proposition for those with delicate constitutions, as well as for travellers and sportsmen.[73]

One of the objectives of Noble's collection of essays was to promote the Cape's potential for tourism and immigration. This was a matter of some concern during the economic depression of the 1860s, especially following the departure of the valued Anglo-Indian visitors. As Harriet Deacon has shown, portrayals of the Cape as a health resort at this time were closely linked to the politics of colonial identity formation. A healthy, temperate climate was indicative of civilized status and mature national character. Yet, these conditions did not exist in and of themselves. The pleasant state of *rus in urbe* on which they relied depended first on conquering and then domesticating the environment. The Cape's status as a healthy colony could therefore be seen as a metaphorical expression of the extent to which social, moral, and institutional order had been secured in a new area of settlement with great future potential.[74]

[72] *CMM* 11 (Mar. 1862), 'The Faded Photograph' by 'C.L.'.

[73] W. H. Ross, 'Our Climate: In Its Relation to Health and Disease', in Noble (ed.), *The Cape and its People*.

[74] H. Deacon, 'The Image of the Cape as a Health Resort in the Nineteenth Century or, The prehistory of "Braaivleis, Sunny Skies and Chevrolet"', paper presented to Pathologies of Travel conference, Oxford Brookes University, 1996.

VENTURING TO THE INTERIOR

By the 1850s and 1860s the centre of Cape Town had linked up with outlying districts like Rondebosch and Wynberg which, benefiting from a regular omnibus service, were becoming fashionable suburbs rather than self-contained rural villages. The construction of reliable roads likewise rendered western Cape hamlets like Paarl, Wellington, and Worcester easy destinations for travellers and day-tourists like the Rosses. Mountains, which had earlier been represented as posing formidable and fearful obstacles to travel, were more and more admired as picturesque settings for welcoming rural hamlets.

The development of roads, a key signifier of progress, encouraged altered visions of the country. Roads tamed and reclaimed the wilderness, functioned as arteries of civilization and prosperity, and altered the very pace of life. They also transformed the ways in which the country and its inhabitants were perceived. As well as offering an obvious narrative thread, travel by road provided opportunities for ethnographic comment and philosophical reflection. Consider, for example, an article on 'Roads' which appeared in James FitzPatrick's *Cape of Good Hope Literary Magazine* in 1848. Here the author recalls his first visit to Stellenbosch, and draws a direct parallel between the mode of conveyance and the slothfulness of the Dutch-speaking population. Travelling in an 'antiquated wooden conveniency', he was

> jolted along at the rate of 2.5 miles per hour, nearly pitched at every furlong into the lap of a fat Dutch vrouw, redolent of snuff, who was one of my compagnons de voyage . . . Upon my inquiry of an Afrikander present, why such a description of vehicle was used, I was told, it was exactly suited to the country. My reply to this was,—'Then change the country; make roads, good roads, and you may then use carriages with springs, and travel quickly and comfortably, not at this snail's pace.' My colloquialist only shrugged up his shoulders, and I understood him to say, that as his grand-papa and great-grand-papa had been satisfied with what existed, they were good enough for him.

The contrast was with the present. For now, with the construction of a hard road, the delights of Stellenbosch were only a day's journey by coach from Cape Town. Indeed, it was perfectly possible for 'parties of pleasure' to 'start on a pic-nic to Eerste River in the morning' and return by evening.[75]

As the countryside around Cape Town was domesticated and rendered fit for middle-class leisure pursuits, so the interior of the country became more accessible. The success of the merino wool industry from the 1840s, and the emergence of a powerful and improving agricultural gentry in regions like Albany and the Cape midlands, encouraged reappraisals of land previously considered unfit for habitation. Nodes of British settlement inspired efforts to exert spatial and

[75] *The Cape of Good Hope Literary Magazine*, 2: 5, (1848), 53, 59–60.

ideological control over a remade landscape through mechanisms such as architecture, irrigation, and tree-planting.[76]

The discovery of diamonds spurred new interest in distant, inhospitable regions, and added greatly to the allure of the northern Cape. The 'Fields', as they soon came to be known, fell uncertainly between several contested colonial boundaries, beyond even the Great Karoo, that vast region of semi-arid veld extending southwards from the Orange River. The Karoo was a byword for degradation, isolation, and retrogression in the early nineteenth century and well after: terms like 'back-country' and 'backwoodsman' signified moral and intellectual backwardness as much as they connoted remoteness.

The urge to date such perceptual shifts with precision should be avoided: the power of images is related to their capacity to generalize and endure, as well as to be reconstituted for new purposes—including acting as a foil for the generation of counter-stereotypes. Close reading of the *Monthly* provides unmistakable evidence that a broad change of opinion was under way by the early 1870s as regards perceptions of the Cape interior. This is clearly borne out in the genre of travel accounts, and mirrors diverging views of 'the Boer' as well.

In 1862 an article on the 'Post-Cart' discussed the comparative merits of travelling by different means in a country 'almost destitute of the means of locomotion'.[77] Travelling on horseback, stopping every two hours or so to allow the animal to roll on the ground, was the simplest method. While pleasant enough in an 'interesting country' where there was plenty to occupy the attention and an abundance of comfortable hostelries, riding was a very different proposition in the Cape, 'where the scenery is monotonous and oppressive' and where one had to cover fifty to sixty miles after daybreak in order to secure a bed for the night. Travel by ox-wagon could not be seriously entertained: while not unpleasant if time was of no consequence, at an average speed of 2 miles an hour it was 'too slow even for the Cape'. The recommended form of conveyance was therefore the mailcart which, as the only regular and rapid means of transit between distant centres, symbolized the imposition of spatial order and the establishment of reliable networks of communication. Proceeding at an average 10 miles per hour 'through daylight and dark, toiling up the rises, plunging down the descents, "springing them" over the flats', the mailcart was a constant if 'restless, moving speck on the wide, still landscape'.[78]

Characterizations of the Karoo as a wilderness were still widely in evidence by the late 1870s. 'M's account of travelling 'By Post-Cart to Beaufort West' decried the bare and barren landscape, which looked 'still more desolate by the

[76] For details of the economic shifts in landownership and the emergence of a Cape gentry, see R. Ross, *Beyond the Pale: Essays on the History of Colonial South Africa* (Johannesburg, 1993), chs. 1, 2, and C. Crais, *The Making of the Colonial Order* (Johannesburg, 1992). On the development of scientific and environmental awareness in the Cape interior and midlands, see W. Beinart, *The Rise of Conservation in South Africa: Settlers, Livestock, and the Environment 1770–1950* (Oxford, 2003).

[77] *CMM* 11 (Jan. 1862), 20. [78] Ibid. 21, 22.

sickly light of the stars—a long dreary waste of mile upon mile without a tree to relieve the monotonous sameness . . .'.[79] This article evoked a quick response from 'Beta', a self-declared 'dear friend' of the Karoo, who celebrated its many improvements—of which the coming of the railway was a prime example. The point of this piece was to show that the Karoo was 'fully keeping pace with the improving times'.[80] Yet, even if there are suggestions of aesthetic appreciation of the Karoo (in references to mimosa, for example), it is striking that Beta's defence was mostly couched on the safer grounds of material advancement and progress. The broad expanses of South African veld were evidently not yet ripe for general praise.

Indications as to how such a landscape could be encompassed within a romantic world of solitude and spaciousness are more evident in 'Colonial Roads, Routes, and Modes of Travel. A Sketch for Home Readers', published in 1874. Here the writer links the development of roads since the 1840s with a considerably revised view of the interior. He assures travellers who had hitherto taken for granted 'the very common imaginary idea of the monotonously sterile character of the country', that they would be pleasantly surprised at the varied and often 'magnificent, landscapes which are to be met with along some of these routes. Cape scenery generally is very strange, but it possesses peculiar charms from its exquisite vegetation and brilliant atmosphere; and it has a fair proportion of grand and majestic as well as richly beautiful and picturesque points.'[81] This grandeur was most apparent in the case of mountains—which compared favourably to anything in Wales, the west of Ireland, or Scotland. The hills, lakes, and woodlands of George and Knysna (the subject of Hermann's admired paintings) bore comparison with Cumberland and Westmoreland; the 'luxuriant slopes and glades of the frontier districts' possessed all the 'soft and tranquil characteristics of Arcadian life described by Pringle in his "Evening Rambles"'. Indeed, 'Even the wide extending pastoral plains of the Karoo' had 'a certain attraction for some'—not only as valuable sheep country, or for its wild game—'but from the freedom and airiness, and that strange sense of solitude which is realised as the eye gazes over the unbroken perspective of blue sky and fading distance to Pringle's evocation of a "blank horizon round and round, / Spread void of living sight and sound"'.[82]

[79] *CMM* 16 (Mar. 1878), 159.

[80] *CMM* 16 (May 1878), 'The Rail, Through the Karroo', by 'Beta', 316, 320.

[81] *CMM* 8 (May 1874), 'Colonial Roads, Routes, and Modes of Travel. A Sketch for Home Readers', 290. See also *CMM* 1 (Aug. 1870), 81–2: 'Letters from the Interior', by Thomas Baines. Here the artist-explorer countered the association of the Cape 'with treeless wastes and barren Saharas, sweltering under an African sun . . . Perhaps no Colony presents more attractive varieties of scenery, climate, and products. The dense forests of the southern coast-range; the grassy downs of Albany; the well-wooded, romantic vales of the Kaffrarian Amatolas; the unique water-scenery of the Keurboom, the Ghio, the Keiskama, the Kowie, and the Buffalo;—what eye needs a richer feast of natural beauty? The dry Karoo tracts, which constitute the valuable sheep-walks of Beaufort and the Northern districts, are not so pleasing to the European eye; but these furnish our staple exports, and supply our tables with mutton and beef . . .'.

[82] *CMM* 8 (May 1874), 'Colonial Roads, Routes, and Modes of Travel', 290.

BOTANICAL SCIENCE

The natural counterpart to literary representations of the Cape were scientific and ethnographic accounts of the land and its inhabitants. Here too, the *Monthly* played a distinctive role. Developments in botany, one of the most long-established natural sciences in South Africa,[83] were afforded close attention.

Even before the arrival of the Dutch in the mid-seventeenth century, drawings and descriptions of the Cape's indigenous flowers were well known in Europe. Visits by the Swedish collectors Carl Thunberg and Anders Sparrman in the 1770s began the pioneering phase of taxonomic classification whereby Cape flora were integrated into the Linnaean system. Late eighteenth-century accounts of travel and plant collecting provided by the Scottish horticulturalists W. Paterson and Francis Masson, as well as by the soldier-explorer Robert Gordon, did much to extend European awareness and understanding of the Cape's unique botanical signficance. So, too, did the collecting activities and botanical observations of explorers like W.H.C. Lichtenstein, John Barrow, and William Burchell in the first two decades of the nineteenth century. There was indeed a considerable vogue in Europe for cultivating Cape heaths or *fynbos* at this time.[84]

During the first two centuries of Dutch and English colonization the aesthetic and scientific curiosity of Europeans succeeded in bringing South African flora to wider notice. Of greatest importance was the imperative to classify, propagate, and absorb knowledge of South African plant species within the global taxonomic schemes of Swedish, British, and German science. Yet it was only during the second half of the nineteenth century—and particularly in the 1880s and 1890s—that botanical science began to establish a permanent and broad institutional basis for itself within South Africa.[85] One consequence was that South African plants came to be regarded not only as 'marvels or curiosities imported from a distant land', but as the indigenous plants amongst which ordinary colonists lived, and which required appropriate contextual understanding and protection.[86] However, it was only in the early twentieth century that suburban middle-class gardeners' 'penchant for the exotic' began to be replaced by a developing taste for 'African plants'.[87]

Important developments were set in train with the appointment of Dr Ludwig Pappe as colonial botanist, as well as the creation of botanical gardens at Cape

[83] Beinart, *Rise of Conservation*, 64.

[84] Vernon S. Forbes, 'Some Scientific Matters in Early Writings on the Cape', in A. C. Brown (ed), *A History of Scientific Endeavour in South Africa* (Cape Town, 1977), 41–3; E. Percy Phillips, 'A Brief Historical Sketch of the Development of Botanical Science in South Africa and the Contribution of South Africa to Botany', *South African Journal of Science*, 27 (1930), 45; Beinart, *The Rise of Conservation*, ch. 1.

[85] Phillips, 'A Brief Historical Sketch', 48 ff; L. van Sittert, 'From "mere weeds" and "bosjes" to a Cape Floral Kingdom: The Reimagining of Indigenous Flora at the Cape, *c.*1890–1939', *Kronos*, 28 (2002).

[86] R. H. Compton, *Kirstenbosch, Garden for a Nation* (Cape Town, 1965), 31.

[87] Van Sittert, 'From "mere weeds" ', 104–5.

Town, Grahamstown, and Durban between 1848 and 1851. These institutions were well placed to coordinate the collecting activities of a growing band of eager and knowledgeable amateur plant lovers, who dispatched plant specimens to the new centres from their residences in towns and villages throughout the country. Many amateur collectors were women (who also figure prominently as painters of wild flowers and plants from the 1850s onwards).[88] Also important during this phase of intellectual and artistic acclimatization was the publication from the 1860s of research conducted by the first generation of South African-based botanists, including Mary Barber, Peter MacOwan, Ludwig Pappe, and Harry Bolus. The beginnings of settler conservationist discourse, stressing the need to prevent deforestation and environmental degradation, are similarly evident at this time, notably in the writings of John Croumbie Brown.[89]

Bolus and Pappe both published in the *Cape Monthly*, which offered a much-needed outlet for scientific reports, in addition to championing the Cape's Botanical Gardens' economic role in the Colony and its place in the recreational and civic life of the town.[90] In 1848 Pappe argued that a section of the Botanic Gardens be reserved for indigenous plants, 'not only to appreciate the richness of our vegetation, but to make them known throughout the world'.[91] In 1860 he reviewed William Harvey's *Thesaurus Capensis* and took the opportunity to remind readers of the extraordinary range and diversity of South African flora. So, too, did J. McGibbon in his celebration of the distinctive beauty of Table Mountain's indigenous vegetation, published in Noble's *Cape and its People*.[92]

In the 1870s Harry Bolus contributed articles and book reviews which brought to the attention of the *Monthly*'s readers the important work conducted by the German botanist Ernst Meyer on the regional distribution of South African plants, and the significance of J. F. Drege's collection of approximately 8,000 species of South African plants in the 1820s and 1830s.[93] Bolus's claim that the vast number of different plant forms found in the country far exceeded that of all other known regions of the world was to culminate, in the 1890s and 1900s, in a campaign to preserve plant diversity; as a key figure in the promotion and academic professionalization of botany, he skilfully linked the country's unique botanical heritage to wider calls to awaken patriotic South African sentiment.

[88] Lady Margaret Herschel was a notable painter of Cape wild flowers. So was Mrs Arabella Roupell, whose *Specimens of the Flora of South Africa, by a Lady* (London, 1849) was published for an exclusive group of English subscribers, including Queen Victoria.

[89] See R. Grove, 'Scotland in South Africa: John Croumbie Brown and the Roots of Settler Environmentalism', in T. Griffiths and L. Robin (eds.), *Ecology and Empire: Environmental History of Settler Societies* (Edinburgh, 1997); Beinart, *The Rise of Conservation*, ch. 3.

[90] *CMM* 2 (Sept. 1857), 'A Cape Botanical Garden, As it Should Be', by Hortulanus.

[91] Pappe to William Hooker, 1848, cited in Beinart, *Rise of Conservation*, 75.

[92] *CMM* 7 (Jan. 1860), 54–5, 'Dr Harvey's Botany of South Africa'; 8 (Oct. 1860), 'Botanical Geography; and Distribution of South African Plants', by the Colonial Botanist. See also J. McGibbon, 'The Botany of Table Mountain', in Noble (ed.), *The Cape and its People*.

[93] *CMM* 4 (Sept. 1873), 193–8.

EARTH HISTORY AND EVOLUTIONISM

The study of geology and earth history was, arguably, of even greater significance as a colonial science. In addition to its obvious connections to the burgeoning mineral extractive industries, geology provided a platform for cognate studies in palaeontology, anthropology, and pre-colonial history. Indeed, it formed an essential part of the intellectual substructure that would later coalesce and become known as African Studies.

In Britain and Europe geology had become firmly established as a leading intellectual discipline by the beginning of the nineteenth century, its popularity drawing on a huge corpus of work conducted by amateur fossil hunters as well as a growing corps of professional practitioners connected to institutions like the Geological Society of London (est. 1807). Changing attitudes towards the earth and its formation provided the conceptual and material basis for new and deeply contested understandings of life, historical time, and the universe. In the language of literature, art, and political thought, geological metaphors were widely deployed in order to represent competing 'perceptions of historic time and human agency within it'.[94]

Underlying many of these geological debates, especially after mid-century, was the increasingly vexed question of evolution. At the Cape, as elsewhere in the colonial world, intellectuals tended to address evolution in a cautious and circumlocutory manner.[95] While a few contributors to the *Monthly* were minded to accept Darwinian theory, at least in private, the predominant view ranged between hostility and agnosticism, and the situation was confused by a number of rearguard attempts to reconcile evolutionist thinking within a Christian framework.[96] This unease reflected developments in Britain and the English-speaking colonial world, where natural theology proved highly adaptable and tenacious in resisting—or incorporating—geological and other scientific evidence within a revised Christian

[94] J. Burrow, 'Images of Time: From Carlylean Vulcanism to Sedimentary Gradualism', in S. Collini *et al.* (eds.), *History, Religion, and Culture: British Intellectual History 1750–1950* (Cambridge, 2000), 199. The literature on historical geology in Britain is vast. See e.g. R. Porter, *The Making of Geology: Earth Science in Britain, 1660–1815* (Cambridge, 1977); N. A. Rupke, *The Great Chain of History: William Buckland and the English School of Geology (1814–1849)* (Oxford, 1983); E. Buffetaut, *A Short History of Vertebrate Palaeontology* (Beckenham, 1987); D. Cadbury, *The Dinosaur Hunters* (London, 2000).

[95] See R. Numbers and J. Stenhouse (eds.), *Disseminating Darwin: The Role of Place, Race, Religion, and Gender* (Cambridge, 1999), which considers local responses to Darwinian thought. This volume has chapters on Australia, Canada, and New Zealand, but South Africa is a notable omission.

[96] Turrell, *A Cape Periodical*, ch. 3, provides a good survey and discussion of Darwinian debates. In summarizing the changing treatment of natural theology in British periodicals from 1800 to 1880 Robert Young usefully suggests that the role of religion changed 'from that of providing the *context* for the debate to that of acting as the point of view in a *conflict*'. See R. M. Young, 'Natural Theology, Victorian Periodicals and the Fragmentation of a Common Context', in C. Chant and J. Fauvel (eds.), *Darwin to Einstein: Historical Studies on Science and Belief* (Harlow, 1980), 76.

schema. Nevertheless, the hold of biblical authority was visibly weakening from the 1860s onward, and evolutionist ideas were becoming the implicit framework in which natural history was understood.

As editor of the *Monthly*, Roderick Noble gave prominence to discussions bearing on theology, geology, and human origins. Noble was himself fascinated by natural theology, that slippery field of philosophical enquiry which sought to reconcile evolutionist ideas within a broad Christian framework that assumed regularity of design and rank in nature. In his first venture into South African public debate, a lecture given in 1854 to the Cape Town Mechanics' Institute, he considered the relationship of geology to sacred scripture. Here Noble attempted to reconcile biblical truths with contemporary scientific evidence of the earth's antiquity. In his 1868 address to the subscribers of the South African Library, Noble was more prepared to acknowledge the influence of evolution and of a highly qualified version of Darwinism, but he claimed that this did not undermine underlying scriptural truths: 'I think it only enlarges our views of the scale and stage on which Omnipotence has wrought.'[97]

The accretion of evidence in support of evolutionist theory, and growing recognition of mankind's deep antiquity, gradually eroded adherence to creationist ideas. Yet the notion that mankind may have been the result of an act of separate and divine creation remained a popular fallback position for contributors to the *Monthly* throughout the 1870s. Questions relating to the mechanism of evolutionary progress and observations about gaps in the fossil record were often raised by way of objection to evolutionist theories. As Gillian Beer has so compellingly demonstrated in the field of British literature, it was one thing to accept the evolutionist hypothesis as regards the natural world, quite another to apply it to humankind.[98]

The publication of Darwin's *Descent of Man* (1871) appears to have exerted a significantly greater impact on the Cape's reading public than *The Origin of Species* (1859).[99] But even if it became generally accepted that the history of mankind had to be measured in geological rather than scriptural time, the Spencerian concept of the 'survival of the fittest' and the notion of human descent from primates remained highly contentious propositions. While many of the *Cape Monthly*'s contributors were eager to engage with the latest scientific thinking on evolution during the 1870s, 'mechanical' explanations devoid of divine design and intimations of atheism were staunchly resisted.[100]

The *Descent of Man* was accorded a respectful review in 1871. In the same year the governor of the Cape, Sir Henry Barkly, tackled Darwinian theory in an address to subscribers of the South African Library. Ten years earlier, as governor

[97] Address by R. Noble to subscribers to the Public Library, 23 May 1868, *Proceedings at the 39th Anniversary Meeting* (Cape Town, 1868), 26.

[98] G. Beer, *Darwin's Plots*, 2nd edn. (Cambridge, 2000).

[99] *CMM* (Nov. 1876), 'Origin and Antiquity of Man', 312.

[100] See, e.g. W. Porter, 'On Scholastic and Philosophic Studies', *CMM* (Aug. 1876).

of Victoria, Australia, and in the context of the destabilizing social effects of the gold-rushes, Barkly had expressed anxiety that Darwinian ideas might contribute to political discord.[101] In the comparatively more stable social environment of the Cape (which was only just embarking on its own mineral revolution) Barkly proved willing to accept evolution in the plant and animal kingdom. However, he expressed horror at the idea of mankind's simian origins.[102]

Other leading figures in Cape intellectual and political life, such as William Porter (attorney-general of the Cape) and Langham Dale (superintendent of education), addressed evolution as an unproven if compelling hypothesis. This strategy of containment is typified by Dale's ponderous address to the 1874 meeting of South African Library subscribers, where he surveyed the claims and implications of evolutionist theory. Dale avoided taking any decisive stand on the salient issues but made direct references to recent scientific discoveries at the Cape, including the discovery of stone implements and Wilhelm Bleek's quest to reveal the original unity of language in his studies of Bushman society.[103] William Bisset Berry, a Scottish-trained physician based in Queenstown, countered with a scientifically well-informed and coherent defence of evolutionism (a controversial viewpoint that had earlier impelled him to retreat to the town of Burgersdorp).[104] But Dale's status as an influential educationist and all-round

[101] B. W. Butcher, 'Darwin Down Under: Science, Religion, and Evolution in Australia', in Numbers and Stenhouse (eds.), *Disseminating Darwinism*, 45–6, 56.

[102] *CMM* (June, 1871), 'The Descent of Man' (unsigned); also 'Notes by the Editor'. See also Barkly's address in *Proceedings at the 42nd Anniversary Meeting of the Subscribers to the South African Public Library 27th May, 1871* (Cape Town, 1871). Here Barkly surveyed the progress of science in the Colony, arguing (rather like Jan Smuts and other scientific nationalists 50 years later) that the country offered unique possibilities for scientific discovery: 'In regard to scientific enquiry, on the other hand, those who live in a region which has not been very long inhabited by civilized races, and which possesses remarkable peculiarities of soil and climate, are far more advantageously situated for prosecuting researches into many branches of natural history than the inhabitants of densely-peopled and highly civilized countries, which in the lapse of centuries have become vast cemeteries, where the relics of the past are commingled and confused, and where every division of the Fauna and Flora, extinct or existing, has been so long and closely explored that little novelty can be now expected. In few countries is this contrast with the old world more strongly marked than in South Africa.'

[103] Langham Dale, 'Anthropology—A Review of Modern Theories, *CMM* 8 (June 1874). See also his 'Anthropology', *CMM* (Feb. 1876). Langham Dale (1826–98) was a frequent contributor to the *Cape Monthly* on a issues ranging from philosophy, theology, literature, and classics to education. He wrote under the pseudonym 'Delta'. Dale came out to the Cape in 1848, at the invitation of Sir John Herschel, to take up the professorship of English and Classics at the South African College. He became superintendent-general of education in 1859, campaigned for the establishment of a university, and became the first vice-chancellor of the University of the Cape of Good Hope in 1873. He is considered to be one of the pioneers of archaeology through his interest in collecting stone age tools. For Porter's address to the Cape University in his capacity as chancellor, see 'On Scholastic and Philosophic Studies', *CMM* (Aug. 1876). For biographical details on Porter, see Ch. 3.

[104] *CMM* 9 (Aug. and Sept. 1874), 'Dr Dale on Evolution of Life. A Criticism in Two Chapters', by 'W. B. B.' (W. Bisset Berry). See also 9 (July, 1874), 'On the Origin of Language'; 9 (Oct. 1874), 'The Library Address', by Delta, and 'The Darwinian Theory (from a Theological Point of View)', by 'W. G.'; 9 (Nov. 1874), 'Selection-Artificial and Natural'. Berry (1839–1922) was born and educated

intellectual authority was difficult to counter, and Turrell is therefore probably correct to conclude that 'the stand taken by Dr. Dale right throughout the 70s of a willing suspension of belief was the order of the day'.[105]

The study of geology, and its allied fields of interest, were of reciprocal interest to metropolitan and colonial intellectuals. Britain's overseas dominions offered a rich repository of ideas for an imperial imagination that was eager to exploit mineral riches, receptive to the exotic appeal of new locations, and keen to imprint familiar European tropes onto the unknown. For leading metropolitan scientists (like Roderick Murchison, Henry de la Beche, and Richard Owen), geographical, geological, and palaeontological knowledge together provided the means to assert symbolic order over the British Dominions as well as to recover the 'lost worlds' of the distant and remote prehistoric past and to revisit 'the ancestral stages of their own culture'.[106]

Colonial intellectuals were animated by much the same concerns as their metropolitan peers, but their particular location in the chain of knowledge generation gave them a distinctive perspective. Ideas travelled easily through the Empire and penetrated to its furthermost recesses with remarkable speed.[107] As Christopher Bayly has recently observed, using the example of medical science, knowledge 'spread from Edinburgh to the more isolated parts of England at more or less the same pace as it spread from England to India and China and out into the hinterlands of Calcutta and Shanghai'.[108] It would be wrong to infer that scientific knowledge was merely derivative: it was often shaped by local issues, proved sensitive to context, and was capable from an early stage of adding distinctively to imperial or 'western' science.[109]

In a small community lacking a developed intellectual infrastructure or capacity for disciplinary specialization—but with much to discover and investigate—a spirit of catholic scientific interest could flourish. This was certainly the case in the study of South African geology, whose early practitioners can be characterized, with rare exceptions, as an enthusiastic group of autodidacts who revelled in their status as amateur or provincial men of science and did not feel the need to declare

in Aberdeen, where he qualified as a doctor. He became speaker of the House of Assembly in 1898 and also served as a member of council of the University of the Cape of Good Hope. According to the *DSAB*, controversy over his public support for Darwinian evolution resulted in his having to retreat to Burgersdorp for a few years from 1867.

105 Turrell, 'A Cape Periodical', 57. See also Turrell's helpful discussion of Darwinism (ch. 3).

106 R. A. Stafford, 'Scientific Exploration and Empire', in A. Porter (ed.), *The Oxford History of the British Empire*, Vol. 3 (Oxford, 1999), 317.

107 Several articles in I. Inkster and J. Morrell (eds.), *Metropolis and Province: Science in British Culture, 1780–1850* (London, 1983) are suggestive in this context.

108 C. A. Bayly, *The Birth of the Modern World 1780–1914: Global Connections and Comparisons* (Oxford, 2004), 321. See also ch. 8.

109 In making this point I have in mind the inadequacies in the so-called 'Basalla thesis', as expressed in his 'The Spread of Western Science', *Science*, 156 (1967). For a lively challenge to Basalla, focusing on J. F. W. Herschel's sojourn at the Cape in the mid-1830s, see E. Green Musselman, 'Swords into Ploughshares: John Herschel's Progressive View of Astronomical and Imperial Governance', *BJHS* 31 (1998).

themselves adherents of one or another fixed intellectual position. The pioneer generation of geologists and paleontologists who sent samples of their discoveries to the British Museum may not always have considered their efforts to have been adequately rewarded or recognized by metropolitan scientists; for the most part they were content to see themselves as empirical field-workers, and they derived considerable satisfaction through participating in the wider imperial network of knowledge generation.[110]

Interest in geology, seldom purely scientific, was often conducted in the context of travel, exploration, or prospecting. It offered a means of surveying the landscape and its inhabitants, of locating oneself in time and place, and of laying claim to the country and its resources. As the importance of local discoveries received public recognition, the distinctiveness of South African natural history became apparent. In due course individual achievements helped to foster local colonial identity and pride. Tom Griffiths has shown, in his important study of the Australian 'antiquarian imagination', that amateur scientists were inspired by 'a historical sensibility particularly attuned to the material evidence of the past, and possessing a powerful sense of place'.[111] This was no less the case in South Africa, where the work of the colonial *naturalist* can be said to have helped to shape the outlook and aspirations of the emergent colonial *nationalist*.

Pioneer geologists who contributed to the *Cape Monthly*, like Andrew Geddes Bain, W. G. Atherstone, John Shaw, R. N. Rubidge, George Stow, and Alfred 'Gogga' Brown,[112] were all involved in practical or professional activities that supported their wider scientific interests. Of these, Bain (1797–1864) was the most important and amongst the most colourful. He emigrated to the Cape from Scotland around 1816, married the daughter of a Dutch officer, settled in the rural centre of Graaff-Reinet, and embarked on an itinerant career as a saddler, hunter, ivory trader, explorer, soldier, and road-builder.[113] In addition to these varied activities Bain was also a frequent contributor to Cape newspapers and

[110] When they did not feel their work was taken sufficiently seriously or fully acknowledged, however, local scientists expressed irritation. For example, William Atherstone complained in 1871 that the saurian reptile bones unearthed by himself and Andrew Bain 'lie still in the vaults of the British Museum, unknown, unnamed, unexamined'. Alfred Brown also considered that his work was unrecognized by metropolitan scientific experts (see n. 112 below).

[111] T. Griffiths, *Hunters and Collectors* (Cambridge, 1996), 3.

[112] Alfred Brown (1834–1920) offers a useful contrast to Bain since his pioneering amateur work in geology, palaeontology, and archaeology largely went unrecognized for many years. Brown arrived at the Cape in 1858 and moved to Aliwal North. He was variously engaged as a teacher, clerk, postmaster, and librarian. Brown accumulated a huge collection of reptile fossils from the Stormberg region and sent them off to T. H. Huxley, who used this material without due acknowledgement or recompense. H. G. Seeley likewise treated him badly. Brown's achievements were finally recognized by Robert Broom. His story, as told from his journals, was finally made public by M. R. Drennan, in *Gogga Brown: The Life-story of Alfred Brown, South Africa's Hermit-Naturalist* (Cape Town, 1938). J. C. Smuts contributed a prefatory note. For Brown's contributions to the *CMM*, see *CMM* 3 (July 1871), 'The Monitor Albogularis', and 9 (Aug. 1874), 'The Dicynodon'.

[113] A. H. Lister, *Journals of Andrew Geddes Bain* (Cape Town, 1949), pp. xii–xiii.

periodicals. His dramatic sketch *Kaatje Kekkelbek; or, Life among the Hottentots* has been credited as the earliest example of creolized Dutch (Afrikaans) literature.[114]

This literary distinction was somewhat incidental to Bain's varied career, for it was as a road engineer and builder of mountain passes rather than a writer that he made his mark in the public imagination. Road-building afforded Bain the opportunity of extensive travel and necessitated close observation of rock formations. It also served as his entrée into geology. In the course of constructing the militarily important Queen's Road from Grahamstown to Fort Brown in 1837, Bain chanced upon a copy of Charles Lyell's *Principles of Geology*.[115] The discovery could not have been more fortuitous. Lyell's work was of seminal importance in discrediting biblical theories of creation, and it exerted a significant influence on Darwin's intellectual development (although Lyell himself was a reluctant convert to evolutionary theory). Bain pronounced himself 'smitten', and in what he described as 'a severe attack of *Lithomania*', duly resolved to become a geologist. He filled out his knowledge by purchasing William Buckland's 1836 Bridgewater Treatise on *Geology and Mineralogy* and Mantell's *Wonders of Geology*, and teamed up with M. Borcherds, the civil commissioner of Fort Beaufort, in a passionate hunt for fossils. Bain was rewarded in 1838 when he chiselled a skull with two canine tusks that at first resembled a tiger but, on inspection, looked like no known animal. He called it 'Bidental'.[116]

Borcherds and Bain soon accumulated a vast number of these specimens, as well as the fossilized skeleton of a saurian-like monster resembling a crocodile and various other unknown reptiles. This work aroused the interest of the Grahamstown medical doctor and polymath William Guybon Atherstone, who now added fossil collecting to his many scientific interests, and the two men soon became close collaborators. Bain's offer to loan his collection to 'a certain literary institution in Grahamstown' in order to form the nucleus of a museum met with a discouraging response, so he sent them instead to the Geological Society of London in 1844. He also submitted a paper to the *Transactions of the Geological Society* detailing the geological structure of a wide expanse of country in the eastern Cape and Karoo surrounding the area in which his reptiles had been found.[117]

Bain was rewarded for his efforts in 1845 when the Geological Society awarded him a grant from its Wollaston Fund in recognition of his fossil discoveries and extensive geological exploration of South Africa. Most important of all were the Bidental remains. These aroused the close attention of T. H. Huxley and Richard Owen (originator of the term 'dinosauria'), who came up with the name Dicynodon (two-canine-toothed reptile) to describe the affinities of the new

[114] M. Chapman, *South African Literatures* (London, 1996), 100–1.

[115] *The Eastern Province Monthly Magazine* (*EPMM*) 1 (Sept. 1856), 'Geology of South Africa', by A. G. Bain, 7.

[116] Ibid. 7–9.

[117] Ibid. 12–13; A. C. Brown, 'The Amateur Scientist', in Brown (ed.), *A History of Scientific Endeavour*, 473.

specimen to other reptiles.[118] The combination of Bain's pioneering geological survey work and his discovery of hitherto unknown fossil vertebrate specimens drew immediate international interest to the unique reptiles that had left their fossilized traces in the Karoo basin, and his findings were brought to the attention of authorities like Herschel, de la Beche, and Murchison. His work initiated the systematic study of vertebrate palaeontology in South Africa. In the longer term it become clear that the mammal-like reptiles of the Karoo represented an important transitional stage in the evolution of dinosaurs and living back-boned animals.[119]

By the mid-1850s Bain had consolidated his reputation and unveiled further discoveries. In his capacity as inspector of roads, and with further financial support from the Geological Society of London, he was able to tour the country in search of further examples of new reptile fossils. In 1851 Bain submitted the first comprehensive geological survey of South Africa to the Society, along with a collection of fossils. He also posited the remarkable idea, based on his observation of fossils, that the interior bowl of the Cape, centred on the Karoo beds, had once been a vast freshwater lake, possibly double the size of all the Canadian lakes combined.[120]

These findings were widely reported in the Grahamstown-based *Eastern Province Monthly Magazine*, which ran from 1856 to 1858 and existed in a state of friendly rivalry with the *Cape Monthly Magazine* (mirroring the political tensions between eastern and western parts of the Colony that are developed in the next chapter).[121] Bain's geological research, together with the findings of Richard N. Rubidge and William Atherstone, constituted the core of the *Eastern Province Monthly*'s scientific output and generated a lively internal debate. Richard Rubidge might have disagreed with Bain on his interpretation of the Zuurberg's

[118] In 1874 the British Museum directed Owen to prepare a catalogue of its holdings of fossils from South Africa sent by, among others, Bain, Atherstone, Rubidge, Sir George Grey, and Sir Henry Barkly. This beautifully illustrated volume was published as *Descriptive and Illustrated Catalogue of the Fossil Reptilia of South Africa in the Collection of the British Museum* (London, 1876).

[119] Bain, 'Geology of South Africa', 14–15; M. A. Cluver and T. H. Barry, 'Advances in South African Vertebrate Palaeontology', in Brown (ed.), *A History of Scientific Endeavour*, 321–2. Buffetaut, *Short History of Vertebrate Palaeontology*, 176–9. Also of significance was the discovery by J. M. Orpen, in 1853, of fossil bones at Harrismith. He sent these to Richard Owen, who in 1854 described them as *Massospondylus carinatus*— the first dinosaur identified outside of Europe.

[120] Bain, 'Geology of South Africa', 18–20. Also S. H. Haughton, 'The Father of South African Geology', *Africana Notes and News*, 16: 3 (1964); Rogers, 'The Pioneers in South Africa', 31. This observation was also linked to the idea of the gradual desiccation of the African subcontinent—a theory that at once suggested the depth of geological time and associated Africa with degradation and denudation. See Beinart, *Rise of Conservation*, ch. 2.

[121] Many of the institutions which Cape Town boasted had direct counterparts in Grahamstown which, as frontier town and self-consciously British settlement, compared itself to and competed with Cape Town. In 1850 a botanical garden was established; 1855 saw the establishment of St Andrew's College as well as the beginnings of the Albany Museum and the establishment of a Literary, Scientific, and Medical Society; and a year later the short-lived *Eastern Province Monthly Magazine* began publication. For details, see e.g. M.Gibbens, 'Two Decades in the Life of a City: Grahamstown 1862–1882', MA thesis, Rhodes University, (1982).

geological structure and its formation, but the three men nonetheless thought of their work as essentially complementary: they delighted in their buccaneering spirit and their doubly provincial identity, not only as colonials, but also as 'Eastern Cape men'.[122]

The discovery of diamonds gave geology and mineralogy fresh impetus. Observing the popular 'craving' for knowledge of the country's mineral potential, the acting curator of the South African Museum, Henry Piers, called for more structured government support of geological survey work.[123] He also acknowledged that the Museum's extensive mineralogical collection, which had hitherto barely been sorted, was too 'cosmopolitan and general' to satisfy the interest of those attracted by the country's diamondiferous regions.[124] John Shaw, the Colesberg medical doctor, geologist, and amateur botanist, sought to cast further light on the matter in a series of articles written for the *Cape Monthly* on these promising parts of the country, and the artist-explorer Thomas Baines reported from the Tati goldfields in present-day Zimbabwe.[125]

The prospects of mineral discoveries proved alluring to amateur geologists, but such men were equally driven by a sense of intellectual curiosity and adventure. This was certainly true of William Guybon Atherstone, whose reputation as a geological expert was enhanced by the distinction of having verified the first diamond found near Hopetown in 1867.[126] In 1870 Atherstone was commissioned by the Cape government to investigate reports of gold in the region of the Karoo between the Swartberg and the Nuweveld known as the Gouph.[127] The reports of his travels with Thomas Bain (son of Andrew) in the *Cape Monthly* took the form of a travelogue in which he commented freely on the towns and

[122] See, e.g. *EPMM* 1 (Mar. 1857), 'Gleanings'; 1 (Apr. 1857), 'Geology of South Africa, Part 1', by A. G. Bain; 1 (June and July 1857), 'Geology of Uitenhage', by W. G. Atherstone; 2 (Dec. 1857), 'Section of the Zuurberg', by R. N. Rubidge; 2 (Apr. and May 1858), 'Aspects of the Primeval Earth'; 2 (May 1858), 'On the Relationship of Certain Geological Formations in this Colony', by R. Rubidge. The *EPMM* also carried regular features on the physical geography of South Africa by H. Hall.

[123] *CMM* 3 (Oct. 1871), 'Suggestions for the Search of Minerals', by Henry W. Piers, 255.

[124] *CMM* 2 (June 1871), 'The South African Museum and its Defects', by H. W. Piers.

[125] See, e.g. *CMM* 1 (Sept. 1870), 'The Geology of the Diamondiferous Tracts of South Africa,' by Dr. John Shaw of Colesberg; 1 (Aug. 1870), 'Letters from the Interior: The Gold Mines and Matabililand', by Thomas Baines; 4 (Feb. 1872), 'At the Tati Gold-Fields', by Thomas Baines.

[126] W. G. Atherstone (1814–98) came to the Cape as an 1820 settler. He became interested in medicine and geology as a young man, serving as a military doctor in the 1834–5 frontier war. Atherstone spent time in the 1830s in London and Dublin, where he developed wide scientific and artistic interests. He was one of the first doctors outside of Europe or America to make use of an anaesthetic (1847) during a surgical operation. A prominent figure in Grahamstown, his scientific interests included geology, palaeontology, mineralogy, and botany. He was also active in eastern Cape politics and civic affairs, and gave evidence on behalf of the eastern Cape settlers to the 1836 Select Committee on Aborigines.

[127] These were reported in the *CMM* in 1871 and are reprinted in A. M. Lewin Robinson (ed.), *Selected Articles from the Cape Monthly Magazine (New Series 1870–76)* (Cape Town, 1978). Atherstone's report to the House of Assembly is contained in A. 30–'71, *Report of a Tour* . . . (1871). He concluded that the gold nugget allegedly found at Spreeuwfontein could not have derived from rocks in the area.

villages through which he passed, as well as on the land and its rural inhabitants. Atherstone's accounts were perfectly pitched and amply satisfied his editor's request for 'a gossiping article, with a golden thread of geology'.[128] In the absence of mineral riches, Atherstone declared that the true 'nuggets' of the Gouph were the extraordinary rock-formations he encountered and the fossils contained therein.

When Atherstone stopped his cart to pick up a fossilized stone, he was amused by his Khoi driver 'Hendrik', who excitedly thought they had discovered gold. Atherstone took particular delight in poking fun at the stereotypical figure of the 'Karoo Boer' who had little 'to occupy his mind or excite his attention'. In response to a tall, tanned man who disturbed them while they were hammering rocks to enquire whether they had struck gold, Bain admonished the interloper for his apparent lack of biblical faith, handed him a fossilized bush-louse, and declared that this was a relic of the Flood. 'That must certainly be a screw of the ark!' said the Boer, 'but magtig! [heavens!] *How rusty!*'[129]

The primeval remoteness of the Gouph led Atherstone to speculate facetiously about the action of Darwinian evolution on human development. On encountering another deeply tanned farmer, he joked that future Darwinian ethnologists might take this as evidence of 'the development of Boer into Hottentot by the mere solarisation of a dozen summers'.[130] Atherstone also wondered whether the 'wild Bushman', 'kept back for countless ages in his primeval state', possessed any sense of moral nature or shame. Was he, indeed, 'born in the garden where the fig-trees grew?' 'I know them and their language but from casual gossip of the fireside in their own wild country, not from books. Should ever Darwin or Professor Huxley venture in amongst these chattering races of the rocks, what wonderful discoveries, what inductions, what intellectual "nuggets" might they not pick up of this old Gouph world—the world before the Flood!'[131]

The ugly phrase 'chattering races of the rocks' was only one of many in a growing metaphorical repertoire suggesting a link between the antiquity of the earth and the racial status of the country's primeval human inhabitants. Indeed, Atherstone's intimations of polygenesis and his association of the geological record with what were later to become known as the 'stone age people' anticipates the scientific racism displayed by subsequent generations of paleontologists and physical anthropologists who sought the evolutionary 'missing link' in South Africa's prehistory. It certainly echoed the views of influential scientific contemporaries like Murchison, who easily linked the 'physical conservatism' of Africa and Australasia with 'accounts of the primitiveness of Africans, Australian aborigines, and other native peoples'.[132] But Atherstone's ideas on evolutionary development

[128] 'At the Gouph', by W. G. Atherstone, in Lewin Robinson (ed.), *Selected Articles*, 92.

[129] Ibid. 88, 90, 91. Cf. Andrew Bain's teasing of Boers about 'Noah's Wildebeest', in his 'Geology of South Africa', 9–10.

[130] Ibid. 96.

[131] 'Nuggets of the Gouph', by W. G. Atherstone, in ibid. 101.

[132] R. A. Stafford, 'Annexing the Landscapes of the Past: British Imperial Geology in the Nineteenth Century', in J. M. McKenzie (ed.), *Imperialism and the Natural World* (Manchester, 1990) 84.

remained unformed and unsystematic because he, like Bain, bore only light theoretical baggage and did not think of himself as a professional geologist. In common with other scientific amateurs, the mystery and wonder of nature was part of its allure, and he did not feel compelled to resolve its secrets himself.

On the question of creation Atherstone appears ambivalent. In 1857 he evidently took comfort in the fact that no evidence of human fossil remains had yet been discovered to complement the primeval traces of birds, reptiles, and ferns that he found in the Uitenhage district: this was surely testimony to the truth of the written word which 'assures us that man—the last, the noblest of His creatures—has been a tenant on this earth but for 6,000 years—a mere moment of geological time!'[133] By the 1870s Atherstone seems to have been less sure of his ground. He now suggested that South Africa, situated at the extreme of the 'old world', might have escaped the influence of the Flood which 'swept the northern races from the earth', along with their crimes and their civilization, leaving 'type specimens' of man and animals behind '*in their primeval savage state*'.[134]

The clear implication was that history had passed by Africa and its original peoples, a view that observers from Hegel to Hugh Trevor-Roper have casually endorsed. But Atherstone refrained from addressing the question of creation directly, wondering instead what the real experts—Huxley, Owen, and Darwin—might make of it all. He teased others about the Flood and ridiculed the suggestion made by his driver that footprint-like impressions in the rock might imply that Adam wore *veldschoen* sandals.[135] Such jocularity came naturally to an English-speaking man of science who was confident of his accomplishments and his superiority to crude Boers as well as Hottentots. But if Atherstone questioned fundamentally the notion of the deluge and of God's role in creation, he kept such doubts to himself.

HUMAN ANTIQUITY

The idea that human habitation in South Africa might be measured in geological time engendered a spirited debate in the *Cape Monthly* with respect to finds of coastal deposits and ancient stone implements.[136] Thomas H. Bowker, the frontier militiaman who unearthed palaeolithic implements along the Fish River in the eastern Cape during the 1850s, is credited with accumulating the first

133 *EPMM* (July 1857), 594, 'Geology of Uitenhague', by W. G. Atherstone.

134 'Nuggets of the Gouph', by Atherstone, 100.

135 'My Trip to the Diamond Fields', and 'On to the Vaal', by W. G. Atherstone, in Lewin Robinson (ed.), *Selected Articles*, 136, 144.

136 The timing of this debate closely followed that in Britain, where argument about the antiquity of man moved centre-stage in the mid-1860s, as a consequence of research conducted at Brixham cave in 1858, and the publication of work by Lyell and Huxley in 1863. The first mention of shell caves at the Cape was made by way of a 'Query' by 'Barnacle' in *CMM* (June 1858).

systematic collection of stone artefacts in South Africa;[137] other important collectors of antiquaries from the region included George McKay and Bowker's sister, the botanist and entomologist Mary Barber, who corresponded with prominent scientific authorities in Britain (including Darwin, Harvey, and Hooker) and became a convinced follower of Darwinian natural selection theory.[138] During the 1860s Langham Dale found and recognized a number of stone tools on the Cape flats near his home at Maitland. In a short article on the subject Dale introduced to readers of the *Monthly* the subject of archaeology and pointed to its importance in the understanding of the remote human past. He called it 'a new science, interposing between Geology and History'.[139]

Dale was keenly aware of similar finds in Britain and Europe. He followed closely the controversial debates about the antiquity of palaeoliths engendered by Boucher de Perthes's discoveries in the Somme Valley as well as the influential writings of John Lubbock, originator of the term 'palaeolithic'. Indeed, Lubbock cited Dale's finds in support of his own understanding of evolution.[140] Dale was one of the first collectors to link the stone tools and pottery found at the Cape with living Khoisan peoples, yet he was reluctant to draw any firm conclusions: 'The stone-tools found in South Africa,' he wrote in 1870, 'do not contribute any evidence towards the solution of the state of primitive man; at the same time, they do not tend to discredit the notion of a remotely distant existence of mankind on the European continent.' As yet, there was insufficient evidence with which to construct a theory concerning the implements and pottery so far found in South Africa, and nothing to prove that they were 'prehistoric'.[141] This non-committal contribution, notable for being the first article on archaeology published in the country, soon provoked a response from other enthusiasts, including one from a scientific loner, Alfred 'Gogga' Brown, who drew attention to ancient mounds, stone implements, and cave paintings in the vicinity of Aliwal North and Basutoland.[142]

[137] J. Desmond Clark, *The Prehistory of Southern Africa* (Harmondsworth, 1959), 24; A. J. H. Goodwin, 'A Commentary Upon South African Prehistory up to the Introduction of the New Terminology', *Bantu Studies*, 9: 4 (1935), 295. Goodwin notes that the first written reports of flaked implements in South Africa date back to traveller accounts of the late eighteenth century.

[138] A. Cohen, 'Mary Elizabeth Barber, the Bowkers, and South African Prehistory', *South African Archaeological Bulletin*, 54: 170 (1999).

[139] 'Stone Implements in South Africa', by Delta (Langham Dale), *CMM* (Oct. 1870). Dale's definition recalls Lubbock's characterization of the 'new science' of archaeology which 'forms, in fact, the link between geology and history.' See J. Lubbock, *Prehistoric Times as Illustrated by the Ancient Remains and the Manners and Customs of Modern Savages*, 3rd edn. (London, 1872), 2.

[140] Lubbock, *Prehistoric Times*, 107. B. B. Trigger, *A History of Archaeological Thought* (Cambridge, 1989), 114 ff., maintains that Lubbock's application of Darwinian evolutionist principles to nineteenth-century archaeology helped to underwrite imperial notions of racial and cultural superiority.

[141] 'Stone Implements in South Africa', by Delta, 236–7. See also A. J. H. Goodwin, 'Sir Langham Dale's Collection of Stone Implements', *South African Journal of Science*, 25 (1928), 419–26. In 'A Commentary on the History and Present Position', 295–6, Goodwin notes that Dale's finds were cited as evidence by John Lubbock in his struggle with Thomas Huxley for control of the Anthropological Institute of Great Britain and Ireland.

[142] Note from 'A.B.' in *CMM* (Dec. 1870), 366–7.

In 1871 Atherstone visited caves in Mossel Bay where mounds of shells were widely evident. Earlier travellers had speculated that these could be the work of birds or else be part of a hitherto raised beach. The fact that the shells were of edible molluscs, and that bones and charcoal were to be found amongst them, led him to conclude that they were the remains of an ancient kitchen midden.[143] This idea was taken up by Henry Piers at the South African Museum in respect to similar accumulations of shells found in the vicinity of False Bay and Cape Point. Were these mounds, occurring at heights up to 300 feet above sea-level, evidence of aboriginal habitation, or were they the results of recent geological change or the action of the weather? Roderick Noble favoured the theory of 'gradual upheaval' over that of human habitation. On the other hand, 'S. T.' argued that the nature and action of the sea would be to scoop out shells, not deposit them, noting also that the remains were all of edible species and that complete human skulls and animal bones were to be found mixed in amongst them. The obvious conclusion, therefore, was that they belonged to 'Hottentot' tribes who frequented the coast.[144]

Like several of his contemporaries, Piers was aghast at the idea that the shell caves of False Bay might have been the work of human hands. Nor could he accept that other evidence of stone implements were indeed what they seemed. While conceding that geology was proving that the earth was 'of inconceivable age', the notion that 'man was contemporary with animals' and must therefore have been living long before Adam was 'a staggering proposition'. It was more probable that such implements were initially formed by natural processes of weathering and that 'savage man' had subsequently found and improved upon them through chipping and sharpening.[145] This theory was rejected by a vistor to the Cape, Dr Comrie, who reviewed the available evidence, including examples of pottery which had been sent to him by P. D. Martin of Simonstown. Comrie's view, delivered in an address with the lapidary title 'The Stone-Age at the Cape of Good Hope', was that the chips and flakes must have been worked by palaeolithic Strandlopers or Khoi. He noted that the shell mounds were comparable to similar mounds found in other parts of the world where aboriginal races lived. However simple and crude, these implements were of great significance and they indicated 'that our present civilization is the result of simple and rude beginnings'.[146]

A striking aspect of this early-1870s debate in the columns of the *Monthly* is the way in which discussions about geological and human antiquity were beginning to converge: troubling ontological questions about earth history were compounded

[143] 'On the Way', by W. G. Atherstone, in Lewin Robinson (ed.), *Selected Articles*, 85. The first description of the Mossel Bay caves was provided by the Cape official and imperial explorer John Barrow in 1801.

[144] *CMM* (Aug. 1871), 'Notes by the Editor', 125; (Sept. 1871), 'On the Geology of the Cape Peninsula', by Henry W. Piers; 'Shell Caves', by 'S.T.'.

[145] *CMM* 4 (Sept. 1873), 191–2, 'Stone Implements' by W. H. Piers. See also *CMM* (Nov. 1871), 'Upheaval of the Coast', by 'F.'. It was only in the late 1850s and 1860s that the relation of human antiquity to the palaeontological record emerged in the forefront of European scientific debate.

[146] *CMM* (Nov. 1874), 'The Stone-Age at the Cape of Good Hope', by Dr Comrie, 290.

by speculation over the time-scale of human habitation and the origins of civilization itself. Contributors like Atherstone and Piers were aware that if human habitation went back further than 6,000 years, this would raise troubling issues about biblical creation as well as the nature of primitive man and his potential for civilized development. But the theory of ancient habitation, albeit directly mentioned by Comrie, remained largely unformed in the journal's discussions, both because of a lack of empirical knowledge (no evidence of fossilized human remains had yet been discovered), and also because competing theories of evolution had not yet freed themselves from the ooze of theological received wisdom.[147]

In 1880 the Cape government geologist, E. J. Dunn, surveyed the existing evidence of stone implements throughout South Africa, concluding both that many were of great antiquity and also that they were of Bushman and Hottentot origin. He also ventured the first systematic attempt to categorize different sorts of stone tools (including bored stones, hammers, grinders, spearheads, scrapers, axes, and so on), with observations about their likely uses.[148] From about this time the idea of natural selection would increasingly be deployed to link theories of evolution and progress to ideas of racial origin and notions of innate superiority. But although most of the elements were already available to effect this noxious synthesis—and were already evident in embryonic form[149]—they had not yet coalesced into a clear ideological and explanatory framework.

PHILOLOGY AND ROCK ART

The systematic use of evolutionist ideas with regard to ethnological research came from a very different source—via the discipline of comparative philology, whose underlying purpose was nothing less than to explain the nature of humanity and human consciousness. By treating language as a universal scientific key, it would be possible to unlock the secrets of human history and reveal the underlying relationships and affinities between the different races of the world. The assumptions of comparative philology rested heavily on evolutionist thinking: indeed, it has been seen as the first science to regard evolution 'as its very core'.[150]

[147] See, e.g. 'Anthropology—A Review of Modern Theories', by Langham Dale, in which he addresses recent discoveries of stone tools at the Cape in respect of theories of human evolution and cultural development.

[148] E. J. Dunn, 'On the Stone Implements of South Africa', *Transactions of the South African Philosophical Society*, 2 (1879–81).

[149] In a letter to her brother T. H. Bowker, in 1865, Mary Barber criticized the theory that stone implements at the Cape might have been made by a primitive race of whites: 'Depend upon it that if they had been white they would not have passed away the white races of the human family are the superior races, and in the great struggle for "dear life" that is always going on through out the world it is the "black fellow" that "go to the wall" for they are the weakest both in intelligence and common sense and cannot stand against the white races . . .' Cited in A. Cohen, 'Mary Elizabeth Barber: South Africa's First Lady Natural Historian', *Archives of Natural History*, 27: 2 (2000), 204, n. 48.

[150] E. J. Hobsbawm, *The Age of Revolution* (London, 1977 edn.), 346.

Strikingly, these core premises could just as easily reinforce notions of inherent racial difference as they could provide evidence for the underlying unity of humankind.

The foremost representative of the philological tradition in South Africa was the German immigrant and scholar Wilhelm Bleek, who frequently contributed articles to the *Cape Monthly* on linguistic and ethnological topics.[151] With his rigorous training in the methods of comparative philology, his extensive connections with the German and European scholarly establishment, and his important position as the custodian of Governor Grey's literary legacy at the South African Library, Bleek enjoyed an unrivalled intellectual status at the Cape in the second half of the nineteenth century.[152]

The discussions about flints and middens in the *Cape Monthly* coincided with the collaborative research conducted by Bleek and Lucy Lloyd into San language, mythology, and culture. This project received an important boost when, in 1874, the farmer, surveyor, parliamentarian, and native administrator Joseph M. Orpen revealed the existence of Bushman paintings in the Maluti mountains near Wodehouse. These were duly copied and published in red, black, and buff in the *Cape Monthly* (marking the first use of chromo-lithographic techniques in the country). In his account Orpen also made mention of paintings which the pioneer geologist and ethnographer George Stow (another regular contributor to the *Monthly*[153]) had found in the district of Queenstown. Orpen noted that 'many of the paintings of animals and men are surprisingly well done', and he suggested that they had an 'apparently mythological meaning' or else represented 'certain quasi-mythological rites'. Roderick Noble showed Orpen's drawings to Bleek, who immediately recognized their significance. In an afterword to Orpen's article, Bleek offered an interpretation of the paintings in which he argued that they

[151] For articles by Bleek in the *CMM* see e.g. *CMM* 3 (Jan. 1858), 'South African Philology'; 1 (Sept. 1870), 3 (Dec. 1871), and 9 (Dec. 1874), 'African Folk-Lore'; 6 (Mar. 1873), 'Dr Fritsch's Natives of South Africa'; 4 (Sept. 1873), 'Scientific Reasons for the Study of the Bushman Language'; 7 (Feb. 1874), 'On Resemblances in Bushman and Australian Mythology'; 9 (Sept. 1874), 'On Inquiries into Australian Folklore'. F. W. Kolbe was another prominent linguistic authority. See e.g. *CMM* 1 (Dec. 1870), 'The Ban-Ntu Prefixes', 3 (Aug. 1871), 'On the Origin of Language'; 3 (Nov. 1871), 'On the Affinity between the Indo-European and the African Bantu Languages'; 16 (Jan. 1878), 'Philology'.

[152] For illuminating discussions of Bleek, see R. J. Thornton, '"This Dying Out Race": W. H. I. Bleek's Approach to the Languages of Southern Africa', *Social Dynamics*, 9: 2 (1983); P. Skotnes (ed.), *Miscast: Negotiating the Presence of the Bushmen* (Cape Town, 1996), esp. articles by Janette Deacon, Michael Godby, Martin Hall, and J. D. Lewis-Williams. Also K. Schoeman (ed.), *A Debt of Gratitude: Lucy Lloyd and the 'Bushman Work' of G. W. Stow* (Cape Town, 1997).

[153] For Stow's contributions, see e.g. *CMM* 5 (Aug. 1872), 'Griqualand West'; 10 (Feb. 1875), 10 (Mar. 1875), and 10 (May 1875), 'On the Study of Geology'. Stow (1822–82) was born in Nuneaton and emigrated to the Cape in 1843, settling in the Albany district. A keen geologist, he worked variously as a teacher, soldier, and trader. Stow published several articles on South African fossils and geology in the journal of the London Geological Society and became a fellow of the Society in 1872. A. W. Rogers, 'The Pioneers of Geology', 68, considers Stow to have as much right to the patriarchal title 'father of South African geology' as Bain. For a discussion of the connections between Stow, Lloyd, and Bleek, see Schoeman, *A Debt of Gratitude*.

represented 'no mere daubing of figures for idle pastime', but were instead a 'truly artistic' indicator of the 'conception of the ideas which most deeply moved the Bushman mind, and filled it with religious feelings'.[154]

Bleek's fervent desire to see Stow's drawings of rock art was fulfilled shortly before his death in 1875, by which time the voice of stone age culture was speaking loudly through the rocks—whether in the form of stone implements, kitchen middens, or chromatic paintings. But how would they be heard? David Lewis-Williams, a leading authority on San rock art, maintains that Bleek's sympathetic understanding of the significance of these finds, and his hope that they would help to counter images of the Bushmen as retrograde primitives, did not prevail.[155] That Bleek was sympathetic towards the Bushman is undoubted—but he was himself studiedly ambiguous about the racial status of what he called 'this highly curious people'.[156]

By the end of the 1870s sufficient material had been collected to sustain two differing (albeit linked) interpretations: on the one hand, a romanticized view which described the diminutive 'artistic' Bushmen as the 'true aborigines' of the country, thereby meriting respect and preservation; on the other, a view that cast them as survivors from the distant human evolutionary past who were naturally doomed to extinction in the face of the 'stronger races'. At least until the end of that decade, however, the implications of such notions remained relatively fluid; in other words, these alternative readings were not yet cast in stone.[157] Pressure to clarify their implications was, however, growing more intense, as the mineral revolution took hold in South Africa and as the ideological ramifications of hard-edged imperialist and racial-supremacist rhetoric became apparent.

RACE: BLACKS AND BOERS

The scientific agnosticism that marked most discussion of Darwinian ideas was mirrored by an absence of a developed discourse of scientific racism. In discussions of the 'native question'—a somewhat neglected theme in the first series of

[154] *CMM* 9 (July 1874), 'A Glimpse into the Mythology of the Maluti Bushmen', by J. M. Orpen, 1; also 'Remarks by Dr. Bleek', 12–13. Orpen's rambling and self-justificatory volume, *Reminiscences of Life in South Africa* (Cape Town, 1964), 20–25, challenges Stow's ethnographic competence and posits a powerfully racialized account of South Africa's aboriginal peoples.

[155] J. D. Lewis-Williams, 'The Ideas Generally Entertained with Regard to the Bushmen and Their Mental Condition', in Skotnes (ed.), *Miscast*, 307–8.

[156] 'Remarks by Dr. Bleek,' 11. For a discussion of Bleek's contribution to racial theory see A. Bank, 'Evolution and Racial Theory: The Hidden Side of Wilhelm Bleek', *South African Historical Journal*, 43 (2000).

[157] Both interpretations are evident in the uncompleted MS of George Stow, which was eventually published as *The Native Races of South Africa* (London, 1905). For further discussion of these points, see my *Scientific Racism in Modern South Africa* (Cambridge, 1995), 67–9, and 'Earth History, Natural History, and Prehistory at the Cape, 1860–1875', *Comparative Studies in Society and History*, 46 (2004); C. C. Saunders, *The Making of the South African Past*, part 1 (Cape Town, 1988); D. M. Schreuder, 'The Imperial Historian as "Colonial Nationalist": George McCall Theal and the Making of South African History', in G. Martel (ed.), *Studies in British Imperial History* (London, 1986).

the *Monthly*, but one which appeared with greater regularity from the early 1870s—the full spectrum of white English-speaking views was represented. These ranged from humanitarian professions of hope to outright expressions of settler bigotry. But the general tone was considered and detached and, by contrast with discussions in the *Eastern Province Magazine*, there was little sense of urgency or anxiety.

Some evidence can be adduced to support the idea that biological racism was taking root by the 1870s. For the most part, however, argument was still couched within the context of the viability of the civilizing mission rather than the hard-edged coin of social Darwinist thinking.[158] Discussion centred on the moral capacities of indigenous peoples, the continuing power of chiefs, and the baneful influence of superstitious beliefs. The death in 1871 of Tiyo Soga, the pioneer Xhosa writer, Presbyterian minister, and mediator between European and African cultural traditions, offered an opportunity to voice the question of African advancement more generally. In an appreciative and extended obituary, John A. Chalmers referred to Soga as one of South Africa's 'best and truest and most illustrious sons'. His biography of Soga, published in 1877, offered further opportunity to debate the intellectual capabilities of Africans. In the view of one sympathetic reviewer, this served to demonstrate 'what the Kafir is intellectually capable [of], and this in the first stage of emerging from barbarism'. Another critic, by contrast, professed himself unimpressed with the character of 'our sable fellow men', and attacked Chalmers for his sentimentality and partiality.[159]

The *Monthly*'s second series devoted rather more attention to political and economic relations between black and white. Charles Brownlee, the well-known Ciskeian administrator and soon-to-be Cape secretary for native affairs, published an extended discussion of the 'Present State and Future Prospects of the Kafirs' in 1873, in which he assessed the slow progress of Christian 'elevation and enlightenment', while attempting to steer a judicious course between humanitarian perceptions of Africans as victims of European oppression, and settler views of natives as intrinsically evil.[160]

Towards the end of the decade the *Monthly* carried an increasing number of articles evincing markedly more illiberal attitudes towards Africans—in particular, bemoaning the failure of missionary ideals, calling for the adoption of firmer administrative measures (including 'class legislation' and the stimulation of

[158] A two-part discussion of 'The Influence of Race on National Development' published in 1862 was almost entirely devoted to the history of interactions between European races and, although the writer was convinced that the influence of race was powerful and permanent, the implications of this assumption were applied to the future of Dutch and English in South Africa, not black and white. *CMM* 11 (May and June, 1862), 'On the Influence of Race on National Development' (unsigned).

[159] *CMM* 4 (Jan. 1872), 'Tiyo Soga', by John A. Chalmers; 16 (Jan. 1878), 'New Books'; 16 (Apr. 1878), 'Some Recent Cape Works. A Criticism', by 'G.B.'.

[160] *CMM* 7 (Nov. and Dec. 1873), 'Present State and Future Prospects of the Kafirs', by C. Brownlee.

'industrious manners'), and questioning the wisdom of the common franchise.[161] These contributions seem to be representative of broader attitudinal shifts on the part of English-speakers at this time, as political events and economic processes brought the 'native question' into sharper focus and encouraged the formulation of systematic political solutions.

No better example of this shift can be found than in the 1878 presidential address delivered by the governor and high commissioner, Bartle Frere, to the newly constituted South African Philosophical Society. Frere's choice of topic—'The Native Races of South Africa'—provides a striking insight into his thinking just months before his military assault on the Zulu kingdom. He began with a survey of contemporary ethnographic knowledge, and proposed a rudimentary classification of the 'yellow', 'black', and 'brown' races of South Africa in order to assess their 'probable future' and relations with the 'great body of northern European races'. Much of his talk dwelt admiringly on Wilhelm Bleek's research on Bushman mythology and language. He also lavished praise on Governor Grey's 'truly royal' gift to the South African Library.[162]

Like Grey, the original progenitor of political confederation and a proconsul with whom he shared much in common, Frere was a man of broad intellectual interests. A newly elected fellow of the Royal Society, he was a versatile linguist who wrote for the *Quarterly Review*. Prior to his arrival in South Africa he had served terms as president of the Royal Asiatic and Royal Geographical societies. These accomplishments allowed Frere to adopt a tone of benign scholarly curiosity as he discussed the customs and beliefs of the Bushmen, whose demise he regarded as inevitable.

Frere's tone altered markedly when he moved on to consider the 'indestructibility of the natives', whose future, he hinted, would have to be decided 'by legislation and other means'.[163] Knowledge of these races and efforts to assess their capacity for 'improvement' raised practical issues of immediate concern. While Frere distanced himself from the short-sighted 'commercial' or 'practical' view whereby Africans were regarded as mere units of labour, akin to beasts of burden, he did not deny that they ought to work for colonists. On the contrary, he agreed that Africans should be 'subdued and converted . . . to habits of useful labour and civilization' and thereby made into worthy members of the community, just as the Roman empire had done with its own barbarian invaders. This utilitarian message was fully in keeping with Frere's reminder that ethnographic and linguistic studies had 'a twofold use—the one social, connected with the arts and uses of common everyday life—the

[161] See e.g. *CMM* 15 (Dec. 1877) and 16 (Jan. 1878), 'Our Relations and Responsibilities to the Native Races', by Victor Sampson; 16 (Mar. 1878), 'The Outcome of our Native Policy', by 'Y'; 16 (Apr. 1878), 'Our Future Native Policy', by Henry R. Woodroffe and 'Governor and Governed' (unsigned); 17 (Aug., Sept., and Oct. 1878), 'The Kafir War and its Lessons' (unsigned).

[162] B. Frere, 'The Native Races of South Africa', *Transactions of the South African Philosophical Society*, 1 (1877–8), pp. xii, xvi.

[163] Ibid., p. xxii.

other scientific'.[164] We may safely surmise that the *inquisitive* spirit was not all that far removed from the *acquisitive* urge.

Imperatives of empire constituted another important theme in Frere's address. At a time when he was actively planning the confederation of South Africa (and was in dispute with the Cape's premier, John Molteno, over the domain of imperial versus colonial authority, as we shall see in the following chapter), this was a highly sensitive issue. The presence of Africans, and the manner in which they were treated, was a 'badge of nationality, a sign of empire. It is not given to mere provinces to bear such burdens or to exercise such privileges', he warned. The South African colonies were fast growing into a 'great national confederation', and it followed from this that what was done in local contexts affected the British Empire as a whole. In tones rich with portent and hubris—and which would shortly rebound to his own discredit—Frere reminded his audience that the way in which Britain dealt with its hundreds of millions of subject peoples amounted to the 'title deeds of an imperial race'. Similarly, South Africa's treatment of 'her native races' would be closely watched by the men of science, religion, and politics 'who influence the progress and improvement of mankind'.[165]

Frere's choice of topic, as well as its timing, was carefully chosen. So, too, was his underlying message about the role of knowledge. Simply put, science was too important an activity to be left to scientists alone; likewise, the domain of native policy was too vital to the future of South Africa and the Empire at large to be left to the discretion of local politicians and parties. Thus, while gracing the new Philosophical Society and its scholarly objectives with his gubernatorial patronage, Frere signalled firmly his intention to use scientific knowledge in pursuit of broader imperial ends while reminding his colonial subjects of their subordinate status.[166]

The growing stridency of imperialist discourse from the 1870s encouraged intellectual as well as social prejudice against blacks. In respect to Afrikaners, responses were more ambivalent as indicated by the *Cape Monthly*'s portrayal of them. On the one hand, it stayed true to its founding position by demonstrating that improving Dutch- and English-speakers were partners in the promotion of civilization and the creation of a common colonial identity. On the other, however, it gave vent to those who disdained the stereotypical 'Boer' of the interior as the enemy of progress. As political pressure on the Boer republics became more acute, the latter position was articulated with increasing venom.

Evidence has already been adduced to show that images of Boer backwardness or lassitude were common currency from the beginning of the nineteenth century. The notion that Boers lacked any awareness of time or sense of industry was frequently contrasted with English dynamism and resolve. Boer self-sufficiency and apparent lack of material desire were in themselves an affront to the spirit

164 Ibid., pp. xvi xxiii, xxiv. 165 Ibid, pp. xxiii, xxiv.
166 For more on this and the implications for Cape politics, see Ch. 3.

of commerical enterprise and improvement. Travel accounts provided easy opportunities for directing casual slights towards Boers. Take, for example, the complaint of a visitor in 1861 who likened hoteliers' unresponsiveness to the needs of their guests and their time, with the way in which Boers tended their crops: 'If it comes up, well and good', if not, no matter.[167] William Atherstone likewise chastised a 'wealthy' Karoo Boer for 'listlessly playing with his tame baboon, an amusement frequently indulged in *for hours*, "tot tijd verkorting",—to shorten the time!'[168]

Another familiar trope was the fat Boer housewife or 'vrouw', accompanied by her ineffectual husband dressed in filthy moleskin trousers and felt hat. Charles Bell's satirical ink-and-wash caricature of 'The Boer' likewise exemplifies the haughty attitude expressed by many elite English-speakers towards Afrikaner farmers.[169] Accounts of Boers' dirty habits, their ignorance, and their indifference to education and lack of improving energy were sometimes offset by patronizing accounts of their open-hearted hospitality and rustic wisdom. Recounting a visit to a Boer homestead in Natal, a traveller reported how their party peered through the door and found, 'seated around a naked-looking room some very fat women, with pasty faces, our courteous friend with the sjambok, another rather like him, and some young people'. This brief inspection of the inside of a Boer's house provided sufficient material to arrive at the following conclusion:

> No wonder that civilization sleeps and a land drifts backward under the rule and residence of such a race. That petty field of ten acres is a bitter satire on the industry and energy of these stalwart lads. I longed to apply their father's lash to their broad shoulders, and drive their lazy limbs into really useful action. The only thing they show any enthusiasm about are guns, horses and wagons; and not one of them is applied to any useful purpose.[170]

Images of Boer listlessness and imperviousness to the invigorating stimulus of commerce, culture, and civilization were similarly evoked by Langham Dale in a three-part discussion of 'Our Agricultural Population'.[171] But Dale's purpose was, as ever, didactic rather than merely satirical. Employing the language of racial decline and degradation, in a manner that foreshadows twentieth-century concern about poor-whiteism, he castigated the government and the Dutch Reformed Church in particular for their neglect of education. It was 'an insult to our European origin, a blot on our common Christianity, a disgrace to our civilized

[167] *CMM* 9 (Feb. 1861), 'First Impressions II—Public Buildings of Cape Town' (unsigned), 66.

[168] W. G. Atherstone, 'At the Gouph', in Lewin Robinson (ed.), *Selected Articles*, 88.

[169] M. Godby, 'The Art of Charles Bell: An Appraisal', in P. Brooke Simons, *The Life and Work of Charles Bell* (Cape Town, 1998), 141.

[170] *CMM* 5 (Aug. 1872), 'Life at Natal', by 'A Lady', 109–10. The earlier success of Louisa Ross's articles prompted this series, and it was supposed that they were written by the same person and set at the same time (the mid-1860s). The consensus of opinion and internal evidence points, however, to John Robinson, the journalist and first Natalian premier, as the author of this series. See *Life at Natal a Hundred Years Ago* (Cape Town, 1972), 'Introduction'; See also the hint in the editorial introduction to Robinson's fictional work 'George Linton' first published in *CMM* 14 (Apr. 1877), 211.

[171] 'Our Agricultural Population', by 'Delta', *CMM* 6 (Mar.–May 1873).

community' that 'the native heathen races' received education and training, while we 'overlook our own stock and blood degenerating into the grossest ignorance and lowest habits of life, from the absence of elevating and consoling ministrations of religion, and from the want of appliances for elementary instruction'.[172]

Dale's extended polemic aroused an even more lurid account by 'X.Z.' of Boer degradation in the countryside.[173] However, it also evoked a reasoned and subtly argued defence written by a minister from an unnamed frontier parish, said to be 'fairly typical of the back country', who insisted that portrayals of sheep-farming life as a 'soulless, vegetable existence' were a 'disgusting caricature of its patriarchal simplicity'.[174]

The sense that there was much to be admired in Boer peasant life was powerfully developed by J. A. Froude, the well-known historian of Tudor England and enthusiastic advocate of empire, who toured South Africa twice in 1874–5, charged with advising Carnarvon on his imperial strategy. Froude inverted notions of Boer backwardness and found much to admire in their steadfast, simple lives and devotion to the Bible.[175] This counter-stereotype was hardly any more accurate as a description of Boer life, but it proved appealing to anti-imperialists during the South African War (1899–1902) who celebrated the Boer as an anti-modern variant of the European noble savage.

The Revd J. Murray was rather closer to the mark when he argued that there was 'no such thing as a Boer', and attacked widely held misconceptions about Afrikaners, including those propagated by Froude. The 'Teutonic agriculturalist of Malmesbury' differed in appearance, dialect, and modes of thought from the 'Huguenot vine-dresser of Wellington and French Hoek'. There were greater differences still between the 'uncultivated squatter of the Karroo; the more civilized, but still nomadic, grazier of Calvinia; the half Anglified colonist of the Eastern frontier; the backwoodsman of the Transvaal, or the hunter of the Limpopo'. To the question 'What mental culture do they possess?', Murray conceded that the 'civilization of Cape Town, which is nourished by the telegraph, the daily newspaper, the weekly steamer from England, and the Public Library is not theirs'. Yet the Boer possessed his own 'old world civilization', which 'keeps him from sinking into the mere animal'.[176] With more than a hint of irony, Murray proceeded to offer a detailed ethnographic description of that 'specially antique and specially

172 *CMM* 6 (May 1873), 'Our Agricultural Population', by 'Delta', 274. See also *CMM* 14 (Feb. 1877), 'Trans-Gariep Courtships', by 'M'.

173 *CMM* 6 (May 1873).

174 *CMM* 7 (July 1873), 'Our Pastoral Population', by 'D.R.', 49, 51–2.

175 *CMM* 14 (Apr. 1877), 'The "Quarterly Review" on English Policy in South Africa'; 'English Policy in South Africa', *Quarterly Review*, (Jan. 1877), 143. See also J. W. Burrow, *A Liberal Descent: Victorian Historians and the English Past* (Cambridge, 1981), 285: 'In the devout, old-fashioned patriarchal households of the Cape Dutch Froude clearly found a charm and serenity he found nowhere else in the colony . . . What the Boers had not lost, the English might recover, an idyll of self-sufficiency, stern morality and apple-cheeked children in new lands "where the race might for ages renew its mighty youth" '.

176 *CMM* 15 (Dec. 1877), 'Some Characteristics of our Fellow Colonists', by the Revd. Prof. J. Murray, 369–70, 372.

un-English' Boer, the deeply conservative so-called *Doppers*, of whom Paul Kruger was an example:

> If anthropologists and philologists claim our interest on behalf of a Bushman race which is nearly extinct, much more are we to study the peculiarities of those who are in great measure to be, if not the brain, at all events the backbone of the nation that is being formed in South Africa ... The body politic can more easily do without sugar than without salt. While we value the spread of light through the Public Library, the University, and similar agencies, let us estimate at its right value salt, even of the coarsest quality, wherever it is to be found.[177]

Murray may well have intended these remarks about *Dopperism* and Cape Town's self-regarding sense of cultural achievement as a rebuke to the complacency of the *Cape Monthly*. But the *Monthly* was never consistently anti-Dutch. Operating within the assumptions of a liberal anglophone world-view, it continued to seek to incorporate progressive-minded Afrikaners within a shared sense of Anglo-Dutch Cape colonial identity.

One way in which it did so was to foster the idea of a common colonial heritage through accounts of shipwrecks, exploration, reminiscences, and so on. The *Monthly*'s decision to publish extracts from the journal of Jan van Riebeeck from 1860 (following the lead of *De Zuid-Afrikaan*) represented a commitment to the construction of a shared colonial memory. Every 'settlement of European origin' ought to have its historian and, with the 'ignominious' exception of the Cape, most did. In van Riebeeck's sadly neglected journals it possessed 'an authentic contemporary history of the first settlement of the country, written by the founder of the settlement himself'.[178]

The *Monthly* also lent its support to the notion that, as 'D.U.M.' put it in 1872, 'a fusion of the different European elements at the Cape' was taking place: having obtained the privilege of self-government, it was conceivable that 'we shall form but one race and have but one language'.[179] At a moment when Afrikaans was being mobilized as a political weapon, the language question was becoming a key test of nationality. If there was to be one race and one language, which would it be? 'Y' argued that English would have to be adopted on account of its beneficial effects in advancing the cause of 'civilization and refinement'. He expressed concern

[177] *CMM* 15 (Dec. 1877), 'Some Characteristics of our Fellow Colonists', by the Revd. Prof. J. Murray, 379.

[178] *CMM* 8 (Sept. 1860), 'Cape Annals—Van Riebeeck's Journal', 150–1, 152. In complaining about the lack of historical knowledge of the colonial past, the work of Moodie was dismissed as partial and published '*for party purposes*'. See also 7 (Jan. 1860), 'The Cape a Century Ago', 27–8. The *Monthly*'s concern with colonial history was of particular interest to John Noble, who wrote colonial history and married the sister of the Colony's first archivist, H. C. V. Leibbrandt. George McCall Theal, the first professional historian and champion of settler interests, used the *Monthly* to publish early colonial documents selected from the newly constituted Cape archives (of which he became the part-time official 'keeper' in 1879).

[179] *CMM* 4 (July 1872), 'English and Cape Dutch', by 'D.U.M.', 52.

that Dutch grammatical forms and vocabulary were affecting the way in which English was spoken at the Cape, but conceded that it was a mark of 'a *living* language, that it assimilates to itself new and strange words, drawn from many opposite quarters'.[180]

In 1876 J. H. de Villiers addressed this theme in a speech to the South African Library. Colonial born and educated, de Villiers epitomized the Anglo-Dutch liberal elite whose emergence was so fulsomely endorsed by the *Monthly*. He began by asking whether the future language of South Africa was likely to be the pure High Dutch of Holland, a local Cape Dutch dialect, or English. Those who took for granted that English would prevail in a British settler colony should not assume that ordinary inhabitants would discard their mother tongue.[181] Yet, de Villiers was sceptical of the pretensions of the new Afrikaans language movement as represented by the *Afrikaanse Patriot* newspaper. While recognizing the value of the earthy simplicity of Cape Dutch, he expressed doubt that it could ever become a literary language capable of competing with English. He was also strongly disapproving of backward-looking attempts to retain 'the customs of our forefathers', whether these were worth keeping or not. Citing the examples of America, Canada, and Australia, de Villiers insisted that it was perfectly possible to foster 'the patriotism of South Africans' through the use of English.[182]

De Villiers's closely argued article reflected the assumption of many at the time that English language and culture were in the ascendant and that Dutch was a declining asset in political and public life, whatever its cultural value in the private or social sphere.[183] In this manner he enunciated many of the core values and assumptions of the *Cape Monthly*: English (or British) imperial culture was synonymous with progress and advancement, but the march of civilization did not necessarily entail supplanting the cultural and historical traditions of ordinary Dutch-speakers. In order to fashion a shared sense of colonial patriotism, born of mutual respect and pride, the historical contribution of both English and Dutch elements ought to be recognized.

CONCLUSION

The evidence produced in this chapter points to the emergence of a cohort of inquisitive intellectuals, predominantly resident in the subcontinent, who addressed each other through a linked set of locally based institutions and evidently took

[180] *CMM* 5 (Sept. 1872), 'Cape English', by 'Y', 182–3. The author claimed, somewhat dubiously, that words like 'kloof', 'krantz', and 'vley' all had English equivalents.

[181] *CMM* 12 (June 1876), 'The Future Language of South Africa', by J. H. de Villiers, 359.

[182] Ibid. 362–3, 364–5, 368.

[183] V. Bickford-Smith, *Ethnic Pride and Racial Prejudice in Victorian Cape Town* (Cambridge, 1995), 34, 61.

pride in doing so. The networks, associations, and conversations which they developed helped to shape a distinctive 'epistemic culture',[184] in which the pursuit of colonial knowledge and the projection of a heightened colonial presence were kindred activities.

If the intellectual desire to comprehend South Africa, its land and its peoples, helped to generate a sense of collective settler identity and ownership, this was further achieved by increasingly rendering the country's indigenous peoples as appropriate subjects of scientific enquiry. In so doing, settler capacity to subjugate or control indigenous peoples was enlarged. The passages quoted from Bartle Frere are revealing in this regard. Yet, we should be careful not to collapse knowledge into power. The relationship between the two could be instrumental, but often it was not. Much of the material presented in this chapter indicates that colonial knowledge was closely bound up with colonial self-discovery and understanding. The urge to know about others was in the first instance an impulse to understand and to constitute a sense of individual or collective self.

Colonialism, we should remember, involved rather more than the remorseless process of displacing indigenous peoples from their lands or forcing newly dispossessed people to work as labourers on white-owned farms and mines. A significant minority of settlers who desired to make the Cape their permanent home sought to engage intellectually with the land and its peoples, not only for the purposes of governance and control, but also to lay political and aesthetic claim to the country, to conceive of it as a unity, and to nurture a shared sense of white identity and ownership. These were explicit as well as implicit concerns of the *Cape Monthly Magazine*, which viewed intellectual and social advancement as no less important than material progress. Rather like the Whig *Edinburgh Review* of the early nineteenth century, it strove to foster a moral as well as a commercial community and to give a distinct public voice to the rising middle-class intelligentsia.[185]

Unlike imported magazines and periodicals, the *Cape Monthly* was able to fill an important niche by serving as a vent for public opinion and an outlet for local authors. Until the 1870s it satisfied this constituency well, in part because scientific knowledge was still so unspecialized in disciplinary terms. In our discussion of earth history, for example, we have seen how newly developing associations between geography, geology, palaeontology, and anthropology brought together fragmentary knowledge about landscape, topography, rock formation, stone tools, and stone age cultures. Concerns with antiquity (geological time and human

[184] The term is adopted, out of context, from Karin Knorr Cetina's *Epistemic Cultures: How the Sciences Make Knowledge* (Boston, 1999).

[185] M. D. Fox, 'The Reverend Sydney Smith and the Culture of Controversy in the Early Nineteenth Century', Ph.D thesis, University of Leeds (2005), ch. 3; Wahrman, *Imagining the Middle Class*, 237, 303.

habitation) merged with studies of human and material culture (language, stone implements, and rock art). Evolutionist ideas underpinned many of these associated discoveries. A growing awareness of racial difference and of gathering conflagration gave these abstract ideas striking topicality.

If the first and second series of the *Cape Monthly* helped to develop intellectual knowledge and to support those institutions which were devoted to the furthering of such knowledge, and if it was able to do so in part because the boundaries of scientific, literary, and artistic expression remained porous, these conditions were changing. By the 1870s it was becoming increasingly apparent that a generalist journal could no longer act as an effective conduit and clearing-house for literary as well as scientific research, and that each should be accorded separate institutional status. Thus, with the active cooperation of the *Monthly*'s editor (John Noble), and its publisher and printer (J. C. Juta and Saul Solomon), two separate societies were duly created: the South African Fine Arts Association (1871) and the South African Philosophical Society (1877).[186] The formation of the Philosophical Society marked the inauguration of scientific specialization and professionalization in South Africa.[187] Through publication of its *Proceedings*, the Society rapidly established itself as the country's premier general scientific institution. It gained a royal charter in 1907, and remains in existence today as the Royal Society of South Africa.[188]

That the Philosophical Society styled itself 'South African' indicated its ambitions to be the foremost scientific institution in the subcontinent, and this leads on to the theme of the next chapter. Indeed, of the eighty gentleman subscribers who initially joined, a handful were corresponding members from Natal and the Transvaal.[189] Yet the strong continuities between the Society and the *Cape*

[186] A. J. H. Goodwin, 'Preface', *Index to the Transactions of the South African Philosophical Society (1878–1909) and the Transactions of the Royal Society of South Africa (1909–1955)* compiled by A. M. Talbot (Cape Town, 1956), p. iii. See also Goodwin, 'A Brief History of the Royal Society of South Africa', TS [1954] in BA 478 Royal Society of South Africa archive, University of Cape Town, pp. 1–2.

[187] This was indicated by the Philosophical Society's holding of regular meetings to discuss new scientific discoveries, the formation of separate sections to represent particular interests, and the publication of the Society's *Transactions* in series form from 1878.

[188] A. V. Hall, 'A history of the Royal Society of South Africa', in A. C. Brown (ed.), *A History of Scientific Endeavour in South Africa* (Cape Town, 1977).

[189] Although academic qualifications were not a prerequisite for joining (members were drawn together by 'community of interests and personal acceptability', and a high subscription rate of £2 p.a. helped to guarantee exclusivity and prestige), 'promoting original research' was always a prime objective. In its inaugural year, nine ordinary meetings were held at which 18 scientific papers were read. The explorer H. M. Stanley also addressed a special meeting on the topic of his recent journey to the mouth of the Congo river. For details of the Philosophical Society, see Goodwin, 'Preface', pp. iii–iv; also *Rules of the South African Philosophical Society . . . and list of members* (Cape Town, 1877) in BA 478, Royal Society of South Africa archive, Box A; *Transactions of the South African Philosophical Society*, I (1877–8). The Society's first president was the royal astronomer, David Gill. Its inaugural council included Harry Bolus, Langham Dale, J. G. Gamble, J. X. Merriman, J. Sivewright, and Ronald Trimen.

Monthly Magazine, its close association with the University of the Cape of Good Hope, and the involvement in its affairs of influential personages drawn from the colony's educational, administrative, and political elite all attest to the enduring salience of the Cape's intellectual and social networks. The *Monthly* may have lost its pre-eminent position as the voice of intellectual ideas, but it had amply succeeded in its ambition to stimulate public opinion and consolidate colonial institutions.

3

Colonialism, Imperialism, Constitutionalism

'SOUTH AFRICA'

The last quarter of the nineteenth century is perhaps the most intensively studied in South African history, justifiably so given its importance for the making of modern South Africa. It is also the most argued over. Amongst the most dramatic and contentious events of the era one must include Carnarvon's failed attempts at confederation; the discovery of diamonds and gold; the final wars of conquest against the Pedi, Xhosa, and Zulu; the annexation of the Transvaal and the associated rise of Afrikaner nationalism; the Jameson Raid, and the crises leading to the outbreak of war. Debates about the nature of imperialism and capitalism continue to frame analysis of these events and processes, and the focus of historical attention has naturally shifted towards the South African interior, site of the mineral discoveries. It has thus become a commonplace of South African history writing that Kimberley and Johannesburg hold the key to historical developments in modern South Africa. Yet, as the spring-tide of historical revisionism recedes—some twenty-five years since the emergence of the so-called radical critique, and more than a century after the events in question took place—older problems re-emerge like glaring hulks on the historiographical seabed.

Most glaringly, perhaps, we encounter the skeletal form of South Africa itself which, in the period before Union, existed only in the imagination as a conception or ideal. This chapter begins with an account of how the idea of South Africa and its defining 'problems' emerged as part of a historically informed dialogue between observers writing about the country from within and without. In order to forestall the teleological hurtle towards the early twentieth-century creation of a unitary state, the point of approach proceeds forwards from the 1870s rather than backwards from 1910. This requires a spatial shift as well: away from the industrial regions of the country that have become the principal points of departure for historians trying to understand the rise of modern South Africa, and back to the Cape which, as we have seen, was the largest, longest established, wealthiest, and by far the most populous element of the subcontinent, as well as the region's most developed political entity and home to some of its most important and enduring institutions.[1]

[1] I suggest this as a rebalancing process, rather than a permanent historiographical shift, and am conscious that this book has little to say about the endogenous contribution of the Boer republics to South Africanist sentiment.

The historicization of the country and the production of competing narratives around colonization and conquest were often conducted in relation to the interplay between imperial and settler impulses. A key area of contestation focused on the Cape's political institutions and its constitutional prerogative. This was in turn inflected by ongoing struggles for regional primacy between the eastern and western sectors of the Colony. Thus, the second part of the chapter focuses on official institutions like parliament and the law (complementing the discussion of voluntary and civic associations in earlier chapters). The acquisition of responsible government in 1873, which conveyed significant new powers to the Cape legislature, and confirmed a sense of colonial maturity, is one reason for this emphasis on formal rather than informal civic institutions. Another is that constitutionalism became a major site of struggle in a set of public debates that were of central importance to discussions of nation and nationality, citizenship and rights.

The body of jurisprudence which developed through Roman-Dutch law was one of the distinctive intellectual achievements of late nineteenth- and twentieth-century South Africa. Cape law, as it was often referred to, combined pre-Napoleonic continental legal systems with English common law in a distinctly hybrid manner. Cast like a steel alloy in the die of Anglo-Dutch partnership, a firm yet flexible legal lattice was extruded out of the Cape and subsequently helped to give form to the emergent South African nation. The evolution of Roman-Dutch law therefore provides an important way of understanding how the incipient colonial nationalism of the mid-nineteenth century transmuted into the broad South Africanism of the post-reconstruction era.

In order to appreciate its appeal and, indeed, to comprehend the development of a contemporary historical conceptualization of South Africa, it is necessary to revisit one of the most salient political debates of the late nineteenth century, namely, the clash between 'imperialism' and 'colonialism'. The distinction between these terms, though widely commented upon at the time, has been occluded in most recent scholarship. By reinstating the importance of their differences, an alternative reading of Cape liberalism also suggests itself, one that transcends conventional wisdom about its nature and purpose.

Prior to its becoming a political entity in 1910, 'South Africa' was little more than a figurative expression. Notwithstanding growing economic integration, the region was hugely diverse in political, cultural, and constitutional terms. Until the late-1870s, when a network of railways and a telegraph system was rapidly constructed to link the principal towns and cities, communications were poor, slow, and intermittent. There was neither the means nor the will to bring into effect the confederal aspirations of dreamers like George Grey. In this respect, as in so many others, the mineral revolution provided the catalyst. One of its major effects was to shift the old east–west axis linking Cape Town to Grahamstown counter-clockwise towards Kimberley and the Witwatersrand, a process that saw traditional political and economic rivalries between the eastern and western Cape overshadowed by tensions between the British colonies of the Cape and Natal, and the Boer republics of the Transvaal and Orange Free State. It was in this rapidly reconfigured geopolitical context that South Africa came to be comprehended and conceived.

The term 'South Africa' was current from the 1830s and earlier (as we have seen in the case of the *South African Journal* or the *Commercial Advertiser*), but until the early twentieth century it referred principally to a region extending northwards from the Cape peninsula to the Zambezi. Although the term was used with greater frequency in political discourse from the 1870s, as late as the 1890s only the constituent elements of the subcontinent—African territories and societies, British colonies, and Boer republics—had any definite constitutional meaning. Even then the make-up and boundaries of these states and societies were often vague. In *A Federal South Africa* (1894), P. A. Molteno wrote that: 'When we speak of South Africa, we speak of the country bounded by the sea on all sides except the north, where the boundaries may roughly be said to be the Cunene towards the west and the Zambesi towards the east.'[2] Similarly, G. M. Theal's single-volume *South Africa*, published in the 'Story of the Nations' series in 1894, had to be augmented by an explanatory subtitle: *The Cape Colony, Natal, Orange Free State, South African Republic and all the other territories South of the Zambesi.* Where the epithet South Africa was in use—as, for example, to describe late nineteenth-century political movements like the South African Native Congress, the South African League, or journals and newspapers like *De Zuid-Afrikaan* and *De Patriot*—this was a question of aspiration, a claim on citizenship rights, or else an effort to promote autochthonous forms of political or ethnic affiliation.

From the perspective of outsiders, the region languished in relative oblivion until the 1870s. Robinson and Gallagher speak of the 'south African colonies' as the most unpopular of all the white dominions at this time, and pointedly render 'south' in lower case to underline the unformed political character of the region.[3] J. H. Davidson remarks that the term 'South Africa' was so unfamiliar in England, and ignorance of the Cape so marked, that the *Illustrated London News* preferred to speak in the plural of the 'Cape colonies'. Yet, it was precisely during this period that the idea of South Africa began to attract fresh interest; in so doing, the term acquired new meaning. The Cape's assumption of responsible government in 1872, the discovery of diamonds, and the publication by the British parliament of its 'South African Bill' in 1877 providing for unification or federation of the region's states and colonies drew new attention to the idea of an integrated subcontinent. A detailed index of the Cape Town English press tellingly lists

[2] See e.g. P. A. Molteno, *A Federal South Africa* (London, 1896), 39: 'When we speak of South Africa, we speak of the country bounded by the sea on all sides except the north, where the boundaries may roughly be said to be the Cunene towards the west and the Zambesi towards the east.' Also C. G. Goodfellow, *Great Britain and South African Confederation 1870–1881* (Cape Town, 1966), 67.

[3] R. Robinson and J. Gallagher, *Africa and the Victorians* (London, 1961), *passim.*

only two entries under the subject 'South Africa' for 1871, as against more than forty for the year 1876.[4]

According to Davidson, 'the English newspapers discovered South Africa' in the year 1877—'during a lull in the Eastern Question'.[5] This discovery, following closely on the annexation of the Transvaal, was observed with a mixture of bemusement and satisfaction by Cape colonial newspapers like the *Standard and Mail*, the *Cape Argus*, and the *Capetown Daily News*. Awareness of growing outside interest encouraged colonists to conceive of their country in broader terms, though as yet the idea of South Africa remained inchoate and unreal. As the *Capetown Daily News* drolly editorialized:

> The popularity of South Africa with the English press is the oddest thing within the experience of man. We are no richer than we were a year ago, no wiser, and certainly in no better position to encourage expectation of a brilliant future, and yet the newspaper world to whom at so recent a period we were scarcely known, insists upon making our affairs the centre of colonial attraction. What have we done that we should be thus honoured? . . . It seems but the other day that we read of ourselves as known only to fame by the production of wine that no one could drink, and diamonds which no one would wear. The very Cinderella of colonies we were the least known, the least liked and the most despised of any portion of the British Empire. We certainly have no desire to remain in Cinderella's kitchen stage, nor does our humility by any means resemble that of Uriah Heep. We don't like to be despised. As South Africans we desire with all our hearts not merely to keep up to the level of British dependencies, but to win our way to esteem for enterprise, for courage, and for patriotism.[6]

The British annexation of the Transvaal, coupled with Colonial Office pressures for confederation, did much to crystallize the problem of South Africa. Hitherto, the outward movement of the British frontier in southern Africa had for the most part been 'merely acquiesced in by a reluctant imperial government'.[7] Now, a series of expansionist initiatives signalled the emergence of an increasingly interventionist imperial power anxious to secure and consolidate its position overseas. This shift was registered by local actors and observers, who were concerned that an invigorated imperial presence posed a significant threat to their hard-won autonomy. It is in the interplay of a developing imperial and colonial conversation between outside observers and locally based commentators and intellectuals that the process of imagining South Africa can usefully be discussed.

[4] R. P. Coates (comp.), *The Cape Town English Press Index for the year 1871* (South African Library, 1980).

[5] Davidson, 'Editor's Introduction' to A. Trollope, *South Africa*, repr. of 1878 edn. (Cape Town, 1973), 1. Davidson is perhaps exaggerating the singular importance of the year 1877, though his point is well taken. His interest in this particular year is influenced by the fact that Anthony Trollope visited South Africa in this year.

[6] *Capetown Daily News*, 31 Jan. 1876 (editorial). See also *Cape Argus*, 22 Jan. 1876 (editorial); 19 Feb. 1876 (editorial); *Standard and Mail*, 6 July 1871 (editorial).

[7] R. Cope, *Ploughshare of War: The Origins of the Anglo-Zulu War of 1879* (Pietermaritzburg, 1999), 80.

FROUDE, TROLLOPE, AND THEIR INTERLOCUTORS

One way of charting this dialogue is by giving attention to the wave of travellers and commentators who visited South Africa during the 1870s, spurred by growing metropolitan curiosity towards a neglected part of the imperial chain. A notable contemporary visitor was Henry Rider Haggard, who participated in the annexation of the Transvaal and memorialized his experiences in *Cetywayo and his White Neighbours* (1882). Another was the jingoistic adventure writer R. M. Ballantyne, who produced two books based on his experiences, with a strongly pro-British slant: *Settler and the Savage: A Tale of Peace and War in South Africa* (1877) and the peculiar travelogue *Six Months at the Cape or Letters to Periwinkle from South Africa* (1879).

These books were complemented by the more serious reflections of the celebrated English historian James Anthony Froude, and the renowned novelist and traveller Anthony Trollope, both of whom visited the Cape during the 1870s. The presence of these 'eminent' Victorians aroused considerable public controversy, and the ideas that they expressed had lasting impact. Rereading their works today, the imprint of their phrases and insights on subsequent generations of historians is all too apparent.

Froude's arrival in South Africa was fortuitous. Impelled to travel at a time of deep personal crisis, and in pursuit of his cherished dream of an imperial federation of Anglo-Saxon peoples, he decided 'to study the working of an English colony'. Australia was Froude's preferred destination, but the election of the Disraeli government in 1874 caused him to modify his plans. News of the controversial imprisonment of the Hlubi leader Langalibalele, and the competing claims to ownership of the diamond fields, were of growing concern to the new Tory administration. Federation was again being discussed as a means of dealing with these problems on a comprehensive basis. Froude's close acquaintance with several senior members of Disraeli's inner circle (most notably, Salisbury and Carnarvon) resulted in his invitation to travel to South Africa at government expense in order to assess the political situation at first hand.[8]

Although routinely mentioned by South African historians in connection with Carnarvon's confederation plans and his intelligence-gathering activities, Froude's ideas have not been closely studied in the South African context. This is surprising, for Froude was more than the colonial secretary's 'eyes and ears'.[9] Opinionated, contrary, and motivated by a very particular view of England's place in history and of the future of the Empire, he succeeded in provoking vigorous

[8] J. A. Froude, 'Leaves from a South African Journal', in *Short Studies on Great Subjects*, vol. 3 (London, 1898), 477; W. H. Dunn, *James Anthony Froude: A Biography 1857–1894* (Oxford, 1963), 390.

[9] Probably the most detailed description of Froude's activities is Goodfellow, *Great Britain and South African Confederation*, ch. 4. See also Dunn, *James Anthony Froude*, 392.

debate about the nature of South African society and its place in the Empire, which resonated long after the period of his visit.

From the 1860s Froude had used his position as editor of *Fraser's Magazine* to argue for the strengthening of relationships between Britain and her white settler colonies, sharing many of the same attitudes and assumptions of contemporary advocates of empire like Charles Dilke and J. R. Seeley.[10] A Tory Radical and a keen disciple of Carlyle, Froude's cast of mind was determinedly anti-philanthropic and unashamedly racist towards Africans. This stance enabled him to find common cause with English-speaking eastern Cape separatists and Transvaal Boer republicans, who were in other respects so opposed to each other.[11] As a conservative, Froude recoiled from the philosophy of laissez-faire individualism, abhorred the squalid meanness of industrializing England's overpopulated cities and suburbs, and feared a deterioration in the condition of the English race.[12] Instead, he found inspiration in the age of the Tudors and took nostalgic comfort in the idea of a vigorous and conservative-minded 'peasant proprietary', ruled over by a paternalist state that was firmly committed to Protestant values.[13]

Like many of the new imperialists (Milner, for example), Froude was strongly influenced by contemporary German romantic thinking on the virtues of an organic society. He was especially attracted to the ideal of a pre-eminent 'Anglo-Saxon' race which could reinvigorate its 'mighty youth' in the fresh air and open spaces of the Empire.[14] In his most extended treatment of this subject, *Oceana*, Froude envisioned a modern version of James Harrington's seventeenth-century utopian dream of 'a perfect commonwealth, half real, half ideal', composed of 'the Scotch, English and Anglo-Irish nations'. Bound together by common economic interests, shielded and united by naval power, and 'animated by a common spirit', Froude looked towards the day when Britain, together with Canada, South Africa, Australia, and New Zealand, 'would ride securely in self-supporting stations in the four quarters of the globe'.[15]

In the course of an extended four-month tour, Froude made his way through the Cape, Natal, the two Boer republics, and the diamond fields. His reputation as one of England's foremost writers, and his association with the imperial government, attracted close scrutiny. Few believed his claim to be little more than a gentleman traveller. Doubts about his real intentions and status only served to increase speculation about the true purpose of his visit.[16] Froude expressed measured support for the general policy of confederation, but declined to make

[10] J. R. Seeley, *The Expansion of England* (London, 1883); C. W. Dilke, *Problems of Greater Britain* (London, 1868).

[11] Thomas W. Thompson, *James Anthony Froude on Nation and Empire: A Study in Victorian Racialism* (New York and London, 1987).

[12] J. A. Froude, *Oceana or England and her Colonies* (London, 1886), 8–9, 386.

[13] J. W. Burrow, *A Liberal Descent: Victorian Historians and the English Past* (Cambridge, 1981), 240, 252, 266, 281–2.

[14] Burrow, *A Liberal Descent*, 285; Thompson, *James Anthony Froude*.

[15] Froude, *Oceana*, 1, 14–15.

[16] Dunn, *James Anthony Froude*, 412.

any firm proposals. He appears to have been broadly in accord with Carnarvon on key issues like the need to tackle the 'native problem' and the importance of defining a clear strategic policy founded on the Simonstown naval base. But many of Froude's observations and opinions were at odds with official views. It is unclear, for instance, how committed he was to imperial expansion and settlement in the subcontinent. He inclined to the view that Britain should choose between imposing direct rule on the Indian model or, preferably, restrict itself to holding the Cape peninsula as a military and naval station along the lines of Gibraltar, while leaving the rest of the country to itself.[17]

Froude's ambivalent endorsement of imperialism as he found it working in practice was further complicated by his increasingly strong philo-Boer sentiments. Even before setting foot on South African soil he expressed sympathy with Dutch settlers, a somewhat unconventional view for a 'new imperialist', albeit fully in keeping with his original, not to say idiosyncratic, turn of mind. Froude was convinced that the diamond fields properly belonged to the Orange Free State and should never have been annexed by the Cape. He felt that confederation could never come about until the understandable feelings of resentment on the part of the Dutch were assuaged.[18] And he spoke in favour of 'a South African native policy on "Transvaal lines" ', which would include the use of the apprenticeship system and forced labour.[19]

Yet, for those historians who subscribe to the view that confederation was primarily intended to take advantage of the diamond fields, manage the flow of African labour, and facilitate capitalist expansion,[20] it is well to remember that Froude, at least, expressed little interest in economic issues. He was wholly unsympathetic towards the 'adventurers' who sought to exploit the country's mineral wealth. Speaking at Bloemfontein, he reminded his audience that fortune-seekers such as these were not the men 'who in the English and Dutch fleets dyed the English Channel with the blood of the Spaniards of the Armada. Then, as always, the soldiers of liberty who risked their lives in defence of the country were the hardy yeomanry and peasantry and fishermen who were fighting for home and firesides.' In South Africa their equivalents were plainly the Boer farmers. Boers were 'the only true colonists. They alone make South Africa their home. They alone are cultivating the soil.'[21]

[17] Froude, 'Leaves', 528; *Oceana*, 36, 59–60, 229.

[18] Dunn, *James Anthony Froude*, 416. Notably, Froude blamed the inconsistencies and weaknesses of a vacillating imperial government for the Langalibalele imbroglio. See Dunn, *James Anthony Froude*, 401.

[19] Lewsen, *John X. Merriman*, 54–5.

[20] e.g. J. Guy, 'The Destruction and Reconstruction of Zulu Society', in S. Marks and R. Rathbone (eds.), *Industrialisation and Social Change in South Africa* (Harlow, 1982), 172. On Confederation see also Goodfellow, *Great Britain and South African Confederation*; R. Cope, 'C. W. de Kiewiet, the Imperial Factor, and South African "native" Policy', *JSAS* 15: 3 (1989); A. Atmore and S. Marks, 'The Imperial Factor in South Africa in the Nineteenth Century: Towards a Reassessment', *Journal of Imperial and Commonwealth History*, 3 (1974). The most recent extended discussion is in R. Cope, *Ploughshare of War*.

[21] Quoted in Dunn, *James Anthony Froude*, 414.

Writing for a British audience in the *Quarterly Review* of 1877, Froude lamented the common English view of the Boer as 'little better than a savage'. However unprogressive in outlook, the Boer was a 'strict Calvinist' and an instinctive republican (though untainted by any 'modern sense' of liberty). 'Without tenderness, without enthusiasm, and with the narrowest intellectual horizon, he has a stubborn practicality well suited for the work which he has chosen as the pioneer of African civilization.'[22] This unusual perspective on the civilizing idea was given a further twist through Froude's use of a geological metaphor to root Boers in the land, make a virtue of their timeless existence, and grant them privileged status as the original Europeans in Africa:

> You know what it is to find in a late geological formation the remains embedded of earlier structures. I have seen in the families of Boers in South Africa representatives of what their ancestors were in Holland two centuries ago, and what, for that matter, our own people were in Scotland and England at the same time. I saw young women who might have stepped down out of the canvas of Van Eyck. I saw young men who might have sat to Teniers. There was the big family Bible with the family registers for half a dozen generations.[23]

It is conceivable that Froude was deliberately pitching his remarks so as to flatter Afrikaner republican sentiment and thereby prepare the ground for confederation. However, the evidence of his wider writings, both private and in public, suggests that his regard for the Boers was genuine and his belief that the colonial government should be condemned for reneging on its agreements to recognize the independence of the Boer republics sincere. Time and again Froude expressed a sense of admiration, bordering on veneration, for the conservative attitudes and devout Protestant beliefs of Boer patriarchs and their simple yeoman lives. His efforts to set contemporary South African problems in their historical context, as described in *Oceana* and in his letters to Carnarvon, record a story of imperial blundering and betrayal that might easily have been written by later Afrikaner nationalists.[24]

Whereas Froude was enthusiastically received in many parts of the country, he received a hostile reception by opponents of confederation in Cape Town, most especially on the occasion of his second visit in 1875 when he was cold-shouldered by the governor, Henry Barkly, and severely criticized by the *Cape Argus*, which resented his 'Carlylese proclivities' and his lack of sympathy with 'free institutions' and 'constitutional government'.[25] There existed concentrated pockets of support for confederation in the country, especially amongst groupings in Natal and the separationist eastern Cape who saw economic and political rewards in imperial expansion coupled with the prospect of greater powers of local self-government.

[22] 'English Policy in South Africa' (unsigned but undoubtedly by Froude), *Quarterly Review*, 143: 285 (1877), 109. See also 147: 294 (1877), 'The South African Problem'.

[23] *CMM*, 14 (Apr. 1877), 'The "Quarterly Review" on English Policy in South Africa', (unsigned), 257, 258; cf. Dunn, *James Anthony Froude*, 407.

[24] This was, indeed, a charge laid against Froude by the outright imperialist W. Greswell. See below.

[25] *Cape Argus*, 22 Jan. 1876 (editorial).

But the established liberal-minded colonial elite who dominated Cape parliamentary politics was mostly opposed to further territorial annexation or consolidation. Fears about the financial burden that the Cape might have to bear in a confederated South Africa, together with concerns about the Zulu threat in Natal, and the implications of having to assume greater responsibility for colonial defence as a whole, loomed large in their calculations. Confederation also risked disturbing the delicate political balance in the Colony where long-standing rivalries between 'westerners' and 'easterners' were re-emerging.[26]

Without the cooperation of the Cape no scheme of confederation could succeed. But Molteno, the first prime minister of the Cape, was implacably opposed to Carnarvon's plans unless confederation was initiated from, and centred on, the Cape itself. Molteno and his supporters were intensely jealous of the Cape's recently acquired status as a colony enjoying 'responsible government', and they were conscious of the need to maintain the support of Cape Afrikaners who mistrusted imperial designs on the Afrikaner republics. Influential liberal politicians like Saul Solomon and John X. Merriman[27] feared the consequences for African political rights in the Cape if confederation was imposed on a unitary basis. In a manner that was strikingly reminiscent of the 1848 anti-convict agitation, a mixture of political pragmatism, high-minded principle, and a newly confident sense of Cape colonial patriotism came to a head in the battle to defend the constitutional independence of the Cape against unwarranted imperial interference. These impulses lay at the heart of Cape liberalism.

The issue of constitutional freedoms emerged as a pre-eminent concern in Froude's conflict with Saul Solomon, the leading colonial parliamentarian, owner of the *Argus* newspaper, and prime guardian of the principle of non-racialism, whom Froude referred to as 'a diminutive John Mill'.[28] Froude and Solomon clashed over Carnarvon's plans, but their private correspondence makes it clear that the principle of confederation was not the real reason for their dispute.[29] The issue

26 E. A. Walker, *A History of South Africa* (London, 1935), 364; Cope, *Ploughshare of War*, 94–5.

27 Merriman was a convert to responsible government, having opposed it until 1872. His previously illiberal attitudes began to change at the same time. See Phyllis Lewsen, *John X. Merriman: Paradoxical South African Statesman* (Cape Town, 1982), 25–7.

28 W. H. Dunn, *James Anthony Froude*, 396. Solomon was physically disabled and extremely short, a fact that was almost invariably commented on by observers. Froude prefaced his description of Solomon as a diminutive John Mill by referring to him as 'a dwarf about four feet high with a big head, and the consequentiality of a leading vestryman'.

29 W. E. G. Solomon, *Saul Solomon: 'The member for Cape Town'* (Cape Town, 1948), 181, 183. In *Oceana*, 63, Froude called Solomon 'one of the best men I ever knew'. In a letter to Solomon in 1879 written to set the record straight, Froude insisted that he had never subscribed to the Colonial Office's confederation plans and denied that he had any direct hand in Carnarvon's notorious dispatch on the matter or that his objectives were in any way 'sinister'. Froude assured Solomon that his primary motivation was to secure justice for the Free State over the issue of the diamond fields—and that this was also a precondition for any overall political solution. In his 1875 speech to parliament condemning Carnarvon's plans, Solomon had himself declared that he had been in favour of a 'United States of South Africa' for over twenty years. His major objection was thus not to the principle of confederation, but rather to the way in which it had been proposed. See Solomon, *Saul Solomon*, 191, 199–204.

that separated them turned on the *means* by which confederation was to be secured—as well as its timing. This was not a trivial difference, since it bore on a vital aspect of the liberal elite's view of their own and their colony's self-respect and dignity. In the view of Solomon and his followers, the need to defend the fragile constitutional integrity of the Cape parliament against unwarranted imperial interference was of prime importance. Like John Molteno, who entertained the idea of a federal solution when the time was right (and so long as this entailed incorporation of the rest of southern Africa by the Cape), Solomon was insistent that confederation could only be initiated from within the country.[30]

John X. Merriman, by now a leading figure in the pantheon of Cape liberal parliamentarians, similarly articulated the 'colonial' viewpoint. At a public meeting in 1875 to mark the opening of the railway-line to Uitenhage, he publicly castigated Froude's conduct as 'Imperial agitation promoted by a political emissary from Home', only to be pelted with buns by enraged admirers of Froude. No less provocative was Merriman's assertion that he was 'a Colonist first, then an Englishman'.[31] Froude, on the other hand, had little regard for constitutional and parliamentary privileges, especially in the colonial context. He expressed lofty disdain for the 'Cape politicians [who] strut about with their Constitution as a schoolboy newly promoted to a tail coat, and . . . imagine that they have the privileges of perfect independence, while we are to defend their coasts and keep troops to protect them in case of Kaffir insurrection'. Declining official government hospitality, he preferred to spend his time with naval officers in Simonstown who 'hate above all colonial politicians as heartily as I do'.[32]

The historian of ideas John Burrow points out that Froude often used the word 'constitutional' in a pejorative sense, believing that government which was exercised through a balanced representation of interests was no guarantee of moral order or responsibility. The irony, as Burrow points out, was that 'the attachment of the new English settlements across the Atlantic to ancient constitutionalism was to be perhaps more deeply-seated and lasting than that of England itself'.[33] This is not so surprising when one considers that respect for parliamentary sovereignty was closely bound up with recognition of and respect for the Colony's status and achievements. In the microcosm of Cape colonial politics, Froude, the Carlylean Tory, had run up against an articulate and stubborn strand of local Whiggism whose patriotism was expressed in terms of the indivisible rights of freeborn Englishmen, whether exercised at home or in the colonial arena. It is

[30] W. E. G. Solomon, *Saul Solomon: 'The member for Cape Town'* (Cape Town, 1948), 184, 189, 190. On Molteno's attitudes to confederation see Goodfellow, *Great Britain and South African Confederation*, 108, 133.

[31] P. Lewsen, *Selections from the Correspondence of J. X. Merriman 1870–1890* (Cape Town, 1960), 16; Lewsen, *John X. Merriman*, 56, 57.

[32] Dunn, *James Anthony Froude*, 435, 441. See also Goodfellow, *Great Britain and South African Confederation*, 92.

[33] Burrow, *A Liberal Descent*, 233, 236–7. Burrow is here thinking of America, but the point might be extended to colonies on the other side of the Atlantic.

surely in this sense that we should understand Solomon's 1875 declaration that he was 'a Cape colonist and an Englishman'. On the occasion of parliamentary nomination day, a year before, he professed himself to be an 'Afrikander', before adding that he gloried in being an Englishman and wished to remain under the jurisdiction of the British Crown.[34]

Froude's controversial intervention in South African affairs was followed two years later by the visit of the acclaimed novelist and traveller Anthony Trollope, whose intention to 'do' South Africa signalled a desire to satisfy his own and the British public's newfound curiosity about a forgotten corner of Empire. Like Froude, Trollope was interested in the English-speaking diaspora; he had already travelled in and written about the West Indies, North America, Australia, and New Zealand. Both writers were responding to contemporary political events, and they shared similar views on matters like race. But, whereas Froude's writings were geared towards exerting direct influence on the course of imperial development, Trollope adopted the approach of the wry literary observer. Their political instincts were also subtly different, Froude's radical Toryism existing in delicate counterpoise to Trollope's self-denomination as an 'advanced ... conservative Liberal'.[35]

An extraordinarily prolific writer, Trollope completed a novel the day before arriving in Table Bay in mid-1877 and, with scarcely a pause after disembarking, began to write his treatise *South Africa*. Some five months later, on the homeward-bound voyage, he finished the draft of this monumental two-volume work. More than a century on it retains a sense of energy and excitement, its impressionistic style and vivid narrative descriptions revealing much about the attitudes and state of mind of English-speaking colonists and officials at the time. The disjointed structure of the book and errors of detail are a consequence of the speed with which it was written. *South Africa* is nevertheless judged by a recent biographer to be his 'most discerning travel work; his thinking ... more subtle, more progressive, more amenable to change, and certainly less prejudiced in regard to the native races'.[36] Its impact was considerable in South Africa as well as England, where it attracted enthusiastic reviews. Four editions were issued within a year, followed by a single-volume abridged version in 1879 including a new chapter touching on the Zulu War. In Davidson's estimation *South Africa* was the most widely read book since Livingstone's *Missionary Travels and Researches*, and it was used as a standard work of reference by a generation of editorialists, parliamentarians, and polemicists.[37] For this reason alone it is of considerable historical significance.

Trollope's tour through South Africa was well trailed in local newspapers. He was treated as a celebrity wherever he ventured, and buttonholed by proud locals eager to display their town's distinctive achievements. Amongst the many

[34] Solomon, *Saul Solomon*, 193; E. Drus, 'The Political Career of Saul Solomon Member of the Cape Legislative Assembly from 1854 to 1883', MA thesis, University of Cape Town, (1939), 11.

[35] N. J. Hall, *Trollope: A Biography* (Oxford, 1991), 112. [36] Ibid. 433.

[37] Davidson, 'Editor's Introduction', 18; Hall, *Trollope*, 433–4.

senior personages from across the political spectrum who Trollope sought out and spent time with were Bartle Frere, J. C. Molteno, Thomas Burgers, Saul Solomon, Thomas Fuller, Henry Bulwer, Chief Langalibalele (whose refusal to yield to Shepstone in 1873 was a proximate cause of the Anglo-Zulu War), and Bishop Colenso. The approach of this distinguished outsider amounted to a national audit of a country in the making, and provided a unique opportunity to test colonial self-perception against outside opinion. Local responses to his presence were often tinged by a degree of apprehension, and even resentment, for memories of Froude's controversial tour were still fresh in the public mind. When Trollope's book was published it was extensively reviewed in South Africa.

Several writers found fault with particular claims and seized on inaccuracies, but on the whole it was graciously received and treated as a validation of colonial achievement.[38] Trollope's acknowledgement of the extensive use he had made of local publications was a source of considerable satisfaction. One reviewer called it a 'capital book'—while reserving the right to correct errors of fact and interpretation at painstaking length, and with evident delight.[39] Another commented in the *Cape Monthly*: 'It is only when active and energetic travellers come amongst us, and in a few weeks rattle gaily through the length and breadth of the land, and find their way up to Natal and the Transvaal, Free State and the Diamond-fields, that we wake up to the true magnitude and importance of our possessions.'[40] By the late-1870s, it seems, colonial opinion still deferred to external opinion. Nevertheless, it was sufficiently well developed to engage in a dialogue and to respond to judgements and criticisms from a position of growing confidence and specialist expertise, if not actual parity. The age had at last arrived where the traveller's gaze could be held—*and returned*—by a colonial audience.

As an experienced commentator on the comparative worth of Victorian civic institutions, Trollope duly toured and ticked off Cape Town's Botanical Gardens, Museum, Library, cathedral, and parliament, stopping too—out of professional interest—to inspect the post office and to give advice that he knew (from past experience of other colonies) would only be ignored.[41] His eye was by no means uncritical, but he was generally appreciative of what he saw. He was much taken by the scenic beauty of the Cape, comparing the majesty of its mountains to the Alps and the Pyrenees and enthusing about its tourist potential. Although rather disappointed with Cape Town itself, he expressed strong approval of its suburbs, somewhat predictably preferring the 'Elysian' aspect of Wynberg to the villas of Sea Point with their views of the Atlantic.

[38] For discussions of the response to *South Africa* in Britain and South Africa see Davidson, 'Editor's Introduction', 11–13, Hall, *Trollope*, 433–4.

[39] *CMM* 16 (June 1878), 'Trollope's South Africa', reviewed by 'SRN', 362. Amongst the books Trollope acknowledged using were Noble's *South Africa—Past and Present*, Theal's *Compendium of South African History and Geography*, Maclean's *Compendium of Kafir Laws and Customs*, Wilmot and Chase's *History of the Cape of Good Hope*, and Silver's *Handbook to South Africa*.

[40] *CMM* 16 (Jan. 1878), 'New Books', 61.

[41] Trollope, *South Africa*, 79–83.

It was as a social commentator that Trollope was most percipient. To the author's own surprise he was forced to revise his own preconceptions of the Boer as 'a European who had retrograded from civilization, and had become savage, barbarous and unkindly'. Instead, Trollope insisted: 'The Dutch Boer is what he is, not because he is Dutch or because he is a Boer, but because circumstances have isolated him.' For all their roughness and their crude manners, Trollope delighted in Boer hospitality and courtesy, and he joined Froude in going against conventional British opinion through his admiration for Boer old-world piety and gentlemanly conduct.[42] Pursuing a somewhat contorted line of argument, Trollope came out in favour of Shepstone's controversial annexation of the Transvaal while expressing admiration for the patriotism of the ousted Transvaal president, Thomas Burgers, and defending him from some of the grosser charges made against his administration. Despite Trollope's initial support for confederation, he came to regard Carnarvon's policies as impracticable and unsuited to South African political and social realities.

Trollope was quick to appreciate the critical importance of the diamond fields and, a decade before the discovery of the Witwatersrand reef, he was convinced that the gold-diggings in and around Lydenburg would have a huge impact on the future of the Transvaal. Industrial work rhythms, Trollope predicted, would do far more to 'civilize' Africans than could be achieved by any amount of missionaries 'teaching a great many hymns'. Trollope's disdain for Exeter Hall philanthropy was equalled only by his disapproval of those who insisted that Africans should be ruled by 'a rod of iron'. But he was an enthusiast neither for missionary education nor for the extension to blacks of franchise rights, considering Africans unfit for political power in the foreseeable future.[43]

Trollope's view of Africans veered between casual Carlylean contempt for their racial inferiority and a vague hope that they might in time prove amenable to civilizing processes under white tutelage. He was most impressed by 'the Zulu', for whose natural grace and picturesque primitiveness he expressed a degree of romantic appreciation. In the 1879 edition Trollope condemned Sir Bartle Frere for his precipitate actions against Cetshwayo, a 'peaceable neighbour' of British Natal. As he caustically remarked: 'I have no fears myself that Natal will be overrun by hostile Zulus; but much fear that Zululand should be overrun by hostile Britons.'[44]

Such comments may not have had the impact on British thinking that Rider Haggard soon came to exert through his influential book *Cetywayo and his White Neighbours*.[45] Yet Trollope's views about 'the Zulu', whose cause he vigorously championed, nonetheless contributed to an imperial view that increasingly constructed the South African problem in terms of racial enmity. As Russell Martin has observed, the Anglo-Zulu War 'marked a new and decisive phase in the history

[42] Ibid. 258–9, 325–6, 452. [43] Ibid. 39, 453, 456–7.

[44] Ibid. 231 ff., 466, 471. On Trollope's support for the Zulu cause see also Hall, *Trollope*, 434.

[45] C. Hamilton, *Terrific Majesty: The Powers of Shaka Zulu and the Limits of Historical Invention* (Cambridge, Mass., 1998), 113–16.

of British awareness of the Zulu. From being a relatively little known African people, the Zulu achieved a distinct and recognisable identity. They became in 1879 a nation whom the British had to reckon with, militarily, politically and imaginatively.'[46] Colonial fears or fantasies that 'Cetshwayo was forging a pan-African conspiracy to "drive the white man into the sea" '[47] made it all the more imperative to conceive of the 'native problem' from an overall South African point of view.

Although drawn to the drama of the Anglo-Zulu conflict, Trollope's vision was more appropriately fixed on the central importance of the 'native question' as a whole, since he recognized that it was this that marked South Africa out from the other white colonial dominions. Hence his much-quoted aphorism: 'South Africa is a country of black men,—and not of white men. It has been so; it is so; and it will continue to be so. In this respect it is altogether unlike Australia, unlike the Canadas, and unlike New Zealand.'[48] The point of this observation was not, as is sometimes inferred, to deny that 'the white man has to be master and the black man servant'; on the contrary, Trollope fully endorsed the idea of white ascendancy.[49] Rather, he was pointing to South Africa's exceptional colonial status and warning that, unlike other British colonies of settlement, the indigenous races of South Africa enjoyed overwhelming numerical superiority and were not likely to die out in the face of colonization—a perspective that stands in striking contrast to Trollope's view of other regions of the colonial world. For many nineteenth-century writers the demographic (and eugenic) reality of South Africa's 'vigorous natives' was one of the country's most distinctive and troubling features. The preponderance of blacks over whites meant that South Africa might never become part of the Empire in the same way as the other dominions. And it implied that conditions might not be conducive to South Africa becoming a self-governing federal state.[50] This pessimistic view of the country's suitability as an imperial realm of settlement bears comparison with that of Froude, who feared that South Africa was blighted by internal divisions and destined to become a second Ireland.[51]

COLONIAL HISTORIES

The discovery of South Africa by Froude and Trollope was paralleled by contemporaneous initiatives on the part of locally based intellectuals who likewise sought to make sense of the country, its history, and its peoples. In the field of history,

[46] S. J. R. Martin, 'British Images of the Zulu, c.1820–1879', D.Phil. thesis, Cambridge University (1982), 253. See also Hamilton, *Terrific Majesty*, 112: 'The Anglo-Zulu War altered fundamentally Europe's appreciation and understanding of the Zulu people and the Zulu kings. With the war, "the Zulu" achieved a distinct and recognizable identity in the eyes of the British.'

[47] S. Marks, 'Class, Culture and Consciousness in South Africa, 1880–1910', in *The Cambridge History of South Africa*, vol. 2 (forthcoming).

[48] *Trollope, South Africa*, 454–5.

[49] Ibid. 39.

[50] Ibid. 67, 74–5, 460–1.

[51] Froude, *Oceana*, 228, 59–60, 70, 82.

important examples include G. M. Theal's *Compendium of South African History and Geography* (1873, 1876, 1877), John Noble's *South Africa. Past and Present* (1877), Wilmot and Chase's *History of the Cape Colony* (1869), Wilmot's *History of the Zulu War* (1880), F. R. Statham's *Blacks, Boers and British. A Three-Cornered Problem* (London, 1881), as well as the first history of the country written in Afrikaans, the Revd S. J. du Toit's *Die Geskiedenis van ons Land in die Taal van ons Volk* (1877), which recounted a nationalist narrative of Afrikaner heroism and English duplicity.

This cluster of books, all published within a decade of each other, represents a definitive departure from the familiar genre of almanac-like compendia, aimed principally at prospective immigrants, in which brief historical digests were randomly mixed with statistical information, geographical description, and economic and social data.[52] Considered together as a body of literature, it is possible to identify the emergence of a distinctive and competing use of historical narrative and interpretation to define, survey, and explain the South African predicament. Noble and Theal, as well as Wilmot and Chase, all attempt to narrate a continuous story starting with the period of first European settlement. Du Toit's contribution marked an important step in the construction of a providential view of Afrikaner history written from an explicitly nationalist point of view. Beginning with Bartolomeu Dias and concluding with a vision of a 'United States of Africa', it invested events like the Great Trek and the Slagtersnek rebellion with the force of foundation myths.[53] The interpretations of these writers may have differed in important respects, but in all cases the history of discovery and the discovery of history were used to inscribe Europeans as a permanent fixture of the African continent.

Theal's *Compendium* is perhaps the best example of this transitional stage in South African historiography, signalling the advent of a national historical writing tradition as well as the start of disciplinary professionalization. It begins with almanac-style descriptions of the country's physical features, harbours, climate, population, electoral divisions, and so on. But chapters chronicling South African historical development, from Bartolomeu Dias's discovery of the Cape and its settlement by Van Riebeeck to the 1835 frontier war and the trek, form its core. A distinctive feature of this work is the way in which Theal interlaces the colonial historical record with ethnographic material on the Hottentots and Bushmen, as well as a chapter devoted primarily to Xhosa history. The *Compendium* was first published at the Lovedale educational mission and was intended in part to appeal to an educated African audience. Theal's concern to combine ethnography and elements of African history within the compass of colonial history does not in and of itself mitigate his pro-settler viewpoint; if anything, his ability to master, by means of narrative, the history of conquered as well as conquerors suggests the

[52] A good example of the historical compendium/almanac approach is S. W. Silver & Co.'s *Handbook to South Africa*.

[53] F. A. van Jaarsveld, *The Afrikaner's Interpretation of South African History* (Cape Town, 1964), 39.

contrary. However, by contrast with his later writings, which articulate a powerful defence of white supremacy, the *Compendium* evinces a notable sensitivity to, and engagement with, African societies. Moreover, the idiom of social Darwinist racism that so marks Theal's later works is much less developed here: the *Compendium*, for instance, makes positive comments about Bushman paintings and the expressive capacity of the Xhosa language.[54]

Theal's pre-eminence as South Africa's foremost and most influential colonial historian has tended to occlude the work of contemporaneous writers whose books were widely read at the time, and were more directly concerned than Theal was with current political issues. An important example was John Noble's *History*, which told the story of European colonization from the time of Portuguese and Dutch settlement through the prism of colonial and constitutional advance at the Cape. Noble's professional association with the *Cape Monthly*, and his connection to the Cape parliament (he served as its clerk), considerably influenced his approach. Thus, he adopted a tone of studied neutrality towards blacks and Boers, focusing his account instead on achievements such as freedom of the press, the success of local anti-convict agitation, and the acquisition of responsible government. Noble's key protagonists include Pringle and Fairbairn, George Grey, William Porter, Saul Solomon, and J. C. Molteno.

By contrast, Wilmot and Chase's *History of the Cape Colony* presented a decidedly pro-settler chronicle of complaint as seen from the viewpoint of eastern Cape separatism.[55] With their close affinities to the 1820 settlers, Wilmot and Chase thought of their book as a 'pioneer' exercise—a neat pun signalling the book's claim to be the first consecutive narrative account of the Cape from its 'discovery' to the present (1868), as well as its articulation of the providential settler viewpoint in the tradition of Robert Godlonton and Donald Moodie.[56] African agency is represented by Wilmot and Chase rather more fully than in Noble's account, but only for the purpose of documenting, in considerable detail, a story of continuous depredations and unprovoked aggression on the part of savages, abetted by the wilful deceptions of missionaries and philanthropists, and compounded by neglect and ignorance on the part of most Cape Town-based colonial governors. Key events in the liberal constitutional canon, such as the anti-convict agitation and responsible government, are thus undercut by criticisms of politically motivated boycotts in the case of the former, and low franchise qualifications in the case of the latter.

[54] G. M. Theal, *Compendium of South African History and Geography* (Lovedale, 1873).The *Compendium* was a much expanded version of Theal's 1871 pamphlet, *South Africa As It Is.* The fullest discussion of Theal's contribution to South African historiography is in C. C. Saunders, *The Making of the South African Past: Major Historians on Race and Class* (Cape Town, 1988).

[55] An abridged version was published as a school textbook by A. Wilmot, *History of the Cape Colony for Use in Schools* (Cape Town, 1871).

[56] R. Ross, *Status and Respectability in the Cape Colony, 1750–1870* (Cambridge, 1999), 63; A. Lester, *Imperial Networks: Creating Identities in Nineteenth-century South Africa and Britain* (London, 2001), 57, 66. See also Ross, *Beyond the Pale*, ch. 10.

Wilmot's contribution to the *History* is more scholarly and moderate in tone than the sections written by Chase, which makes the hardening of Wilmot's views in his later works all the more conspicuous. These books focus more fully on South Africa as a whole, their pro-imperial, loyalist, and anti-black rhetoric becoming ever more strident. As Wilmot proclaimed in his *History of the Zulu War* (1880): 'It must be borne in mind that South African Kafir wars constitute one tragedy in various acts. The Zulu campaigns are merely the last links in a chain. . . . The tide of savagery has been periodically rolled back, and it was either necessary that this should be done, or that white men should abandon South Africa.'[57]

By the close of the century Wilmot, now a prolific historian and writer of school textbooks, as well as a parliamentary member of Rhodes's Progressive Party, had moved beyond his concern with the politics of eastern Cape separatism and the negative influence of meddling humanitarians to assert the virtues of imperialist expansionism. Thus, whereas in 1869 he and Chase considered the history of South Africa to be little more than the story of the Cape colony writ large, his emphasis now shifted to a view of South Africa seen in terms of a vastly expanded imperial domain stretching over most of the subcontinent. In *The Story of the Expansion of South Africa* (1895), and even more so in its update, *The History of South Africa* (1901), Kruger-supporting Boers join blacks as the enemies of progress, while Wilmot himself looks forward triumphantly to 'a successful South African federated Dominion under the British Crown'.[58]

Standing in diametrical opposition to the British- or Cape-centred history typified by Wilmot and Chase is S. J. du Toit's *Geskiedenis van ons Land in die Taal van ons Volk* (1877), which rails against English injustice and seeks, in the words of du Toit's newspaper, *Die Patriot*, 'to awaken a sense of nationhood'.[59] (In light of the discussion above, it is of interest that one of the chapters in the *Geskiedenis* is entitled 'What Froude said of the Boers', and consists entirely of quotes in support of the Boer republics).[60] As F. A. van Jaarsveld has observed, the 1870s and 1880s were also the period in which a distinctively Afrikaner version of historical nationhood took root. The annexation of the Transvaal and the 1881 war of independence invested terms like the 'Afrikaans people' and 'South Africa' with fresh meaning and purpose. The efforts of Cape Afrikaners like S. J. du Toit to evoke a pan-Afrikaner identity were met by historian counterparts in the Boer republics such as C. P. Bezuidenhout and F. Lion Cachet, who moved

[57] A. Wilmot, *History of the Zulu War* (London, 1880), p. v. See also Wilmot's *The History of South Africa* (London, 1901) and his *The Story of the Expansion of South Africa* (London, 1895). These were designed as school textbooks and evince an increasingly jingoistic viewpoint.

[58] Wilmot, *History of South Africa*, 233; see also his *Story of the Expansion of South Africa*. Despite his growing antipathy towards Afrikaners, Wilmot remains sympathetic to the freedom-loving Voortrekkers and their descendants—who had to contend both with interfering British humanitarians and hostile Africans.

[59] Quoted in van Jaarsveld, *The Afrikaner's Interpretation*, 40.

[60] *Die Geskiedenis van ons Land in the Taal van ons Volk, uitegeé deur Die Genootskap van Regte Afrikaners*, (facsimile reproduction, Cape Town, 1975), ch. 17.

beyond the narrative boundaries of the Free State and Transvaal to evoke a shared future destiny built out of a common national memory.[61]

The crude settler triumphalism evoked by Wilmot was also challenged by the explicitly anti-imperial and pro-colonial viewpoint of the radical British journalist F. R. Statham, who edited the *Natal Witness* in 1877. Statham's horror at the annexation of the Transvaal, news of which greeted him immediately on his arrival in Cape Town, was powerfully reinforced by his disapproval of Frere's belligerence in Natal, and he duly joined Bishop Colenso in supporting the Zulu cause. In *Blacks, Boers and British*, an eloquent and passionate polemic, Statham castigated the 'Jingo' imperialism personified by Frere (and subsequently Cecil Rhodes, whom he attacked in the satirical *roman-à-clef Mr Magnus*) for stirring up conflict and inducing instability. This form of imperial arrogance, Statham explained, was typically voiced by Englishmen who casually denounced 'the Boers' in the smoking-rooms of Cape-bound mail steamers, and was based on profound ignorance of the country.[62] Its chief political exponents were Carnarvon at the Colonial Office and High Commissioner Bartle Frere, who bore joint responsibility for attempting to foist the unwanted policy of confederation on South Africa and for fomenting the disastrous conflict with the Zulu.

Statham's sympathies lay instead with those he termed 'Dutch' or 'Africanders'—a vague formulation embracing Afrikaner Bondists like J. H. Hofmeyr, colonial Englishmen like Solomon, Merriman, Molteno, and Bishop Colenso, as well as moderate republicans such as F. W. Reitz, who was soon to become president of the Orange Free State.[63] In Statham's view the country could only solve its 'three-cornered problem' if colonists were afforded the space to resolve their own difficulties without external interference.[64] Left to themselves, colonists would most likely find a solution to their country's problems and deal justly with Africans.

The title of Statham's book, which defined the South African problem as a triangular conflict, is noteworthy, not least because it singled out 'blacks' for first mention (inviting comparison with W. M. Macmillan's similarly titled work written half a century later.[65]) In fact, Statham's treatment of black society was notably less original and incisive than his reflections on the Boer and British dimensions. Caught up in the politics of imperial intervention, inter-colonial politics, and rising Afrikaner sentiment, Statham's acute journalistic eye was at

[61] *Die Geskiedenis van ons Land in the Taal van ons Volk, uitegeé deur Die Genootskap van Regte Afrikaners*, 36, 40–2. K. Smith, *The Changing Past* (Cape Town, 1988), 60–1.

[62] F. R. Statham, *South Africa As It Is* (London, 1897), 4.

[63] A more restrictive use of the term 'Afrikaner' was favoured by S. J. du Toit in *Die Geskiedenis van ons Land in die Taal van ons Volk*. See L. Witz, *Apartheid's Festival: Contesting South Africa's National Pasts* (Bloomington, 2003), 57–9; T. R. H. Davenport, *The Afrikaner Bond* (Cape Town, 1966), 32–3.

[64] F. R. Statham, *Blacks, Boers and British: A Three-Cornered Problem* (London, 1881), 259.

[65] Cf. W. M. Macmillan's *Bantu, Boer and Briton: The Making of the South African Native Problem* (London, 1929).

its most perceptive in pinpointing the convergence of three antagonistic forces by the late-1870s: white versus black; British versus Boers; imperialism versus colonial nationalism.

Underlying these competing historical understandings lies the tension between imperialism and colonialism on the one hand and, on the other, rivalry between anglophone liberal 'colonists' concentrated in the western Cape and 'loyalist' Britons in the eastern Cape (who preferred the label 'settlers' to 'colonists', not least to distinguish themselves from the 'Dutch').[66] The differences between them were expressed as competing political traditions which, by the close of the nineteenth century, coalesced within the rudiments of a party system. These animosities also mapped onto geographic and historical differences between east and west.

As Andrew Thompson has strikingly shown in his study of the regional or provincial dimensions of British identity, the eastern Cape 'prided itself on being the "cradle" or "heartland" of South African loyalism'. Thompson's identification of the 1870 Grahamstown Jubilee as a key moment in the creation of a settler loyalist tradition fits well with the idea that strong regional awareness was in part a reaction to broader developments at the national level, as well as being conditioned by specific local conditions. In the sphere of ideology and political sentiment, veneration of founding fathers and nomination of key historical events became key markers of rival political identities.[67] The liberal 'western' tradition traced its lineage back to the anti-slavery and humanitarian movement, celebrated the struggle for self-government and constitutionalism, and emphasized Anglo-Dutch cooperation. By contrast, the settler loyalist tradition of the eastern Cape disdained humanitarians and philanthropist meddling, foregrounded British descent, took strength from its role in bringing civilization to the frontier in spite of continual African aggression and adverse economic factors, and was deeply suspicious of the ethnic and economic interests of Dutch-speaking colonists.[68]

These regional tensions were inflamed by economic competition between mercantile and agrarian interests in the west and the rising merino sheep industry in the Cape midlands. They were forcefully expressed in political struggles that reached a height of intensity in the 1860s when, for a brief period, parliament sat alternately in Cape Town and Grahamstown, and calls for separatism were loudly voiced in the east. Tensions between east and west continued through the 1870s, and were exacerbated by proponents and adversaries of responsible government and confederation. On issues like attitudes to Afrikaners and Africans there was a strong convergence of opinion between the views of settlers and advocates of firm imperial intervention. In historical accounts and discussion, attitudes to figures

[66] My thanks to Alan Lester for clarification on this point.

[67] A. Thompson, 'The Languages of Loyalism in Southern Africa, *c.*1870–1939', *English Historical Review*, 118: 477 (2003), 623.

[68] Lester, *Imperial Networks*, 67. As Thompson points out in 'Languages of Loyalism', 628–9, distinct variants of loyalism emerged in other parts of southern Africa, Natal, and Rhodesia for example. Note also Thompson's distinction between 'liberal' and 'ultra' loyalists.

like John Philip and Lord Glenelg on the one hand and D'Urban, Harry Smith, and Bartle Frere on the other served as touchstones of political affiliation.

PARLIAMENT AND CONSTITUTIONALISM

The attainment of responsible government in 1872, in spite of scepticism and resistance on the part of 'easterners' who were determined to escape the 'tyranny of Cape Town',[69] underpinned the status of the Cape legislature. The enlarged scope of parliamentary jurisdiction, as well as the expansion of the Cape to incorporate African areas of the eastern Cape, had the effect of broadening the horizons of political debate beyond the parochial questions that predominated in the era of representative government.[70] It was frequently boasted that the late nineteenth-century Cape parliament set high standards of public life and that the quality of its debates were the equal of any other colonial legislature.[71]

In the last two decades of the nineteenth century the Cape parliament was held up as a paramount example of the Colony's maturation. This was given material expression in 1875 when the foundation-stone of a new Houses of Parliament, designed by Charles Freeman in the colonial neoclassicist idiom, was laid amidst 'much fanfare' in a ceremony that was presided over by notables drawn from the civil, religious, and military establishment.[72] Sited at the eastern side of the Botanical Gardens, adjacent to the South African Library, the Anglican cathedral, and the governor's residence, the new parliament was thus symbolically coextensive with the complex of civic, intellectual, religious, and legal institutions that defined Cape colonial life.

Problems with the construction of the new parliament—it took until 1885 to complete—were paralleled by political challenges to the legislature and its newly enhanced powers. Shortly after his arrival as colonial governor in 1877, Sir Bartle Frere insisted, against the express wishes of the Cape's first prime minister, J. C. Molteno, on deploying imperial reinforcements (in addition to colonial volunteers) during the Cape–Xhosa frontier wars against Kreli and Sandile. Frere's intervention provoked a clash of authority between prime minister and imperial representative which resulted in Molteno's dismissal from office in 1878. Frere now invited J. G. Sprigg, the King William's Town farmer, supporter of confederation, and proponent of a 'vigorous' frontier policy, to replace Molteno with a ministry composed of eastern Cape men. This outcome further strained existing tensions between 'westerners' and 'easterners' in the colony. Merriman, as the minister responsible for colonial volunteer forces, was a particular target for

[69] Lewsen, *John X. Merriman*, 25.
[70] Drus, 'The Political Career of Saul Solomon', 1–2.
[71] R. Kilpin, *The Romance of the Cape Parliament* (London, 1930), pp. vii, 100; S. W. Silver & Co.'s *Handbook to South Africa*, 4th edn. (1891), 55.
[72] N. Worden *et al.*, *Cape Town: The Making of a City* (Cape Town, 1998), 217; J. Rose Innes, *Autobiography*, ed. B. A. Tindall, (London and Cape Town, 1949), 23.

Frere's ire; coming soon after his confrontations with Froude and Carnarvon, the experience confirmed his growing disgust with aggressive imperial interference.

Of more general concern in the British world was the constitutional crisis provoked by the wholly unprecedented dismissal of a duly elected colonial administration by a representative of the Crown.[73] The clash between imperial and colonial interests was not easily resolved, and it continued to dog Frere throughout his time in South Africa. On the eve of the outbreak of the Anglo-Zulu War, Frere complained bitterly of lack of public support for his campaign, singling out for criticism (along with 'gun-runners' and 'pseudo-philanthropists') those 'who, from habit, mistrust all we do, because it is done by government or by what they call *"imperial" & not by Colonial people*'.[74]

The constitutional implications of the 'colonial spirit'—which were to recur in the 'suspensionist' crisis of 1902–3 (see Chapter 4)—was also evident in church–state relations, as deep-seated doctrinal and personal rifts in the Anglican communion came to be played out in the Byzantine field of ecclesiastical politics. Here, the arraignment for heresy of Bishop Colenso of Natal by Bishop Gray of Cape Town took on gladiatorial proportions and became a serious test of political and legal strength within the Anglican communion. The bitter trial, which eventually resulted in Colenso's 1866 excommunication, centred on one of the foremost philosophical and intellectual controversies of the age: the literal truth of the scriptures.[75] But this was not the principal ground on which the matter was contested. A material factor in the proceedings concerned the legal authority of the Church of the Province of South Africa, which at once claimed autonomy from the Church of England and the Crown, but simultaneously sought to defend its position as the dominant Anglican body in South Africa. The legal wrangle had direct relevance to the acquisition of representative government at the Cape, since this act limited the authority of the queen to exercise prerogative rights in the Colony and thereby constrained Gray's ability to discipline Colenso.[76]

The murky situation appeared to be resolved in 1870 when the inaugural synod of the newly constituted Church of the Province *of* South Africa, presided over by Gray in his newly enhanced role as bishop 'metropolitan', claimed constitutional separation from the Church of England *in* South Africa and rejected the legal jurisdiction of the Privy Council.[77] The arrangement was, however, deeply flawed, and tensions continued to fester. These blew up into crisis in 1879 in an unseemly battle over the right of Bishop Merriman[78] to preach in the Grahamstown

[73] Lewsen, *John X. Merriman*, 71–9; Goodfellow, *Great Britain and South African Confederation*, 153–4; Walker, *History of South Africa*, 375.

[74] Cited in Cope, *Ploughshare of War*, 226. My emphasis.

[75] On Colenso see J. Guy, *The Heretic* (Johannesburg, 1983).

[76] P. Hinchliffe, *The Anglican Church in South Africa* (London, 1963), 91; *Anglican Division in South Africa* (Pietermaritzburg, 1982), 4. Gray had been appointed as the first bishop of Cape Town in 1847 by Queen Victoria. The letters patent constituted all British possessions in Southern Africa as his diocese.

[77] Hinchliffe, *The Anglican Church*, 113–15, 119.

[78] Bishop Nathaniel Merriman was, of course, the father of John X. Merriman.

cathedral and to exercise control over its dean who, having been appointed by the imperial government, claimed a direct link to the Crown and the established Anglican Church in England. The case eventually went to Chief Justice de Villiers, whose landmark 1880 judgment confirmed that the Church of the Province was separated 'root and branch' from the Church of England.[79]

At one level these tangled disputes concerned complex theological and doctrinal matters as well as an unseemly wrangle over ownership and control of church property; at another, they reflected and foreshadowed broader questions about church–state relations. These included the relationship of the 'unestablished' Anglican Church in South Africa to the colonial government and the Church of England, the relative powers of ecclesiastical and secular law, the jurisdiction of the metropolitan bishop of Cape Town over his subordinate clergy in the vast and ill-defined 'province of South Africa', and the constitutional standing of the Cape colony relative to the imperial metropole. Some of these issues were reflected in the long-running parliamentary campaign led by Saul Solomon to separate church and state and to constitute all churches as voluntary religious bodies.[80]

The aftermath of the Anglo-Zulu War and the resumption of Transvaal political independence saw a temporary recess in imperial intervention in the political affairs of the Cape which lasted until the Jameson Raid and the arrival of Milner in 1897. Hofmeyr's moderating influence on Cape Bondsmen, and his growing political stature, helped to inaugurate a new phase of cooperation in Cape politics. This period was indeed the heyday of the Cape parliament. In an affectionately deprecating sketch, F. R. Statham likened the Cape assembly to 'a very small dog, aping the airs and manners' of its much larger British equivalent.[81] However, Statham did not consider Cape politics altogether risible. He recognized that the 'small dog with the three inch stride has yet considerable power over human life and happiness'. No politician understood this better, Statham maintained, than Saul Solomon, 'the proprietor of the *Cape Argus*, a political party in himself, a maker of Ministries, the unflinching advocate of native rights, the fearless exposer of native wrongs, the most influential man, without exception, in the whole South African continent'.[82]

Solomon may have been exceptional, but he was not unique. He inspired a mainly younger generation of parliamentarians who vigorously espoused the colonialist viewpoint, either from a liberal or a Bondist perspective—sometimes both. They included John X. Merriman, James Rose Innes, J. H. de Villiers,

[79] Hinchliffe, *The Anglican Church*, 122–5; E. A. Walker, *Lord De Villiers and his Times: South Africa 1842–1914* (London, 1925), 177–9.

[80] Solomon's Voluntary Bill was eventually passed in 1875.

[81] Statham, *Blacks, Boers and British*, 31.

[82] Ibid. 45, 46–7. Statham's claim about Solomon's influence may have been overstated but it was echoed by other contemporary commentators—e.g. Cole, *Reminiscences*, 93–4. The Solomon era ended with his retirement from active politics in 1883. Statham wrote these lines just before Rhodes's entry into parliament (1883).

J. H. Hofmeyr, J. H. Meiring Beck, W. P. Schreiner, and J. W. Sauer. The large number of biographies, autobiographies, and collections of papers devoted to these Cape Victorian eminences bears ample testimony to the spirit of moral purpose, justice, civic-mindedness, and intellectual seriousness that characterized their political lives. With all the inconsistencies and blemishes in their political records, and allowing always for their pompous self-importance and paternalism, they collectively exemplify the peculiar political tradition of Cape liberalism in its most confident, impressive, and developed form.

The stature of such individuals, who sought to rise above the parochial concerns and vested interests of lesser parliamentary colleagues, was facilitated by the absence of a developed party system.[83] In this fluid environment political allegiances were considerably determined by personal following, charismatic leadership, and displays of oratorical brilliance. The development of the Afrikaner Bond in the 1880s, under the parliamentary leadership of Jan Hofmeyr, introduced the rudiments of a party-political system that coalesced along more clearly defined lines after the Jameson Raid. Hofmeyr, like Solomon, was in control of a newspaper, and he too preferred to influence and play the role of political kingmaker than to assume ministerial responsibility himself.[84] But, whereas Solomon's authority derived from his personal standing and unrivalled status as parliament's liberal conscience, Hofmeyr's power was augmented by the fact that he could call on the support of more than a third of the votes in the House. Bondists were nevertheless careful not to press home their advantage, lest this disturb the delicate ethnic balance. In 1882, for example, Hofmeyr succeeded in having the use of Dutch officially recognized in parliamentary debate. H. J. Luttig was the first to make a speech in Dutch. But, having done so, Luttig immediately announced his intention of not repeating the exercise, and Dutch was only seldom used thereafter.[85] It was judged sufficient for the point to have been made.

This careful mixture of pragmatism and principle allowed an improbable but highly effective political alliance to flourish between Hofmeyr and Cecil Rhodes in the first half of the 1890s. Hofmeyr did not wish to risk dividing politics along 'racial' lines by taking office as head of an Afrikaner party, while Rhodes found that he could advance his own ambitions by cooperating closely with Afrikaner rural interests and aligning himself with the 'true patriotism' of the Bond. Rhodes's abandonment of free trade in favour of agricultural protectionism for wine and grain farmers, together with his newfound support of the abolition of excise duty on Cape brandy, was crucially important in cementing his alliance

[83] J. L. McCracken, *The Cape Parliament 1854–1910* (Oxford, 1967), 105–6; Yvonne Pauline Sank, 'The Origin and Development of the Cape Progressive Party (1884–1898)', MA thesis, University of Cape Town (1955), 1–2.

[84] Davenport, *The Afrikaner Bond*, 71. Hofmeyr edited *De Zuid-Afrikaan*.

[85] McCracken, *The Cape Parliament*, 28. According to figures presented by McCraken (Table 3, p. 53), the 'English language group' in parliament constituted between 50% and 60% of parliamentary representation in the period 1870–1908.

with Hofmeyr.[86] So long as the paramountcy of Cape interests was recognized in colonial affairs—and Rhodes repeatedly insisted that the Cape would remain 'the dominant state in South Africa' and that he would pursue 'northern development as a Cape colonist'[87]—the way was clear for most colonial Afrikaners to endorse continued allegiance to the British flag. As Giliomee says: 'It was Rhodes' audacious achievement to fuse his imperialism with Hofmeyr's colonial patriotism without, initially at least, giving any hint of being on a collision course with Kruger's republicanism.'[88]

Rhodes's assiduous cultivation of Afrikaner support was facilitated by his firm approach to 'native policy', his pandering to protectionist agrarian interests, and his pursuit of northward expansion and closer union which, he promised, would enhance the Cape's regional position in a future 'United States of South Africa', extending a thousand miles north of the Orange River.[89] Material interests were not the only binding factor. Rhodes inspired genuine affection on the part of Cape Afrikaners, many of whom were drawn by his power and charisma, seduced by his disarming lack of formality, and flattered by his appreciation and support of Cape Dutch furniture and architecture.[90] As Tamarkin has convincingly shown, Afrikaner Bondsmen did not seek ethnic exclusivity. They were wholly committed to a vision 'in which Cape Afrikaner ethnicity, white Cape colonialism, pan-Afrikaner solidarity, and British imperialism lived side by side, not without tension, but on the whole quite happily'.[91]

The Rhodes–Bond alliance was violently disturbed by the 1895 Jameson Raid and the revelation of Rhodes's complicity in the plot to gain control of the Transvaal. Liberal-minded stalwarts of Anglo-Afrikaner cooperation like Jan Hofmeyr, J. H. de Villiers, James Rose Innes, J. W. Sauer, and W. P. Schreiner were astonished, even shattered, by news of the plot.[92] Merriman, once a close friend of Rhodes, was the first English-speaking statesman to denounce him and the activities of his Chartered Company.[93] At a stroke the Rhodes–Bond political alliance was broken and the middle ground between pan-Afrikaner republicanism and jingo imperialism perilously eroded. As the crisis deepened in the aftermath of the Raid, politicians were impelled to make painful choices and to redefine their sense of nationality.

[86] Thompson, 'Languages of Loyalism', 634–5; M. Tamarkin, *Cecil Rhodes and the Cape Afrikaners* (London, 1996), 96–7, 99.

[87] F. Verschoyle ('Vindex'), *Cecil Rhodes: His Political Life and Speeches 1881–1900* (London, 1900), 62, 244, 272.

[88] H. Giliomee, *The Afrikaners: Biography of a People* (London, 2003), 234.

[89] Verschoyle, *Cecil Rhodes*, pp. 52, 242–4, xxiv.

[90] Tamarkin, *Cecil Rhodes*, 212–18; W. K. Hancock, *Smuts: The Sanguine Years* (Cambridge, 1962), 59.

[91] M. Tamarkin, 'The Cape Afrikaners and the British Empire from the Jameson Raid to the South African War', in D. Lowry (ed.), *The South African War Reappraised* (Manchester, 2000), 136; Giliomee, *The Afrikaners*, 227.

[92] See e.g. E. Walker, *Lord De Villiers and his Times*, 264; Davenport, *The Afrikaner Bond*, 163; H. Wright (ed.), *Sir James Rose Innes Selected Correspondence (1884–1902)* (Cape Town, 1972), 156; E. A. Walker, *W. P. Schreiner: A South African* (London, 1937).

[93] Lewsen, *John X. Merriman*, 178.

For Cape Afrikaners the choice may have been slightly easier because the way was now clear to ally themselves, out of a sense of righteous indignation, with a pan-Afrikaner and frankly republican political programme. The young Jan Smuts, who had been an ardent admirer of Rhodes prior to the Raid, pronounced in 1896 that 'blood was thicker than water'. Sensing that political leadership in South Africa was no longer 'a prerogative of the Cape', he moved to Johannesburg where he soon became a leading figure in the Transvaal government.[94] Nevertheless, many Cape Afrikaners continued to hold on to the moderate belief that 'the Bond knows no nationality', and that cooperation between the two white 'races' was still possible.[95] Loosed from the bonds of the Bond, a wave of pro-imperialist sentiment was unleashed. Supporters of the disgraced Rhodes formed the Loyal Colonial (later South African) League in 1896, a cardinal principle of which was 'loyal maintenance of the dependence of the Colony on the British Crown'. It was aptly nicknamed the 'jingobond'.[96]

English-speaking moderates had less room to manoeuvre. Innes, now uneasily positioned between the Bond and Sprigg's Progressives as leader of the newly formed South African Political Association, wrote that it made him 'sick at heart to see the moral standards which Englishmen are supposed to cherish sneered at and trampled underfoot in the name of English Imperialism'.[97] W. P. Schreiner, who was well disposed towards the Bond and was chosen by it as prime minister of the Cape in 1898, stood for election on the platform of 'true Imperialism and true Colonialism'. A year earlier he had informed the House of Commons inquiry into the Raid: 'I am South African first, but I think I am English after that.'[98]

ROMAN-DUTCH LAW

Closely allied with parliament, and mirroring the Cape's constitutional and political history in several respects, was the law, whose evolution was notably marked by the dual inheritance of Dutch and English influences. It was in its own way perhaps the finest expression of locally elaborated intellectual culture at the Cape. The development of a mixed system of English and Roman-Dutch law was a consequence of the fact that anglicization of the legal system was only achieved incrementally, in spite of concerted efforts by the incoming British administration after 1806 to rectify perceived deficiencies in colonial Dutch law. More often than not, the process of legal reform occurred through absorption and incorporation rather than by direct imposition. The elevation in 1812 of the Cape-born and

94 Cited in Hancock, *Smuts: The Sanguine Years*, 59, 62. The phrase 'prerogative of the Cape' is Hancock's.

95 Tamarkin, 'The Cape Afrikaners', 133.

96 Sank, 'Origin and Development of the Progressive Party', 120, 145, 153.

97 Innes to P. A. Molteno, 1897, in Wright (ed.), *Sir James Rose Innes Selected Correspondence*, 193.

98 *Second Report from the Select Committee on British South Africa* (HMSO, 1897), questions 4134, 4175; Walker, *W. P. Schreiner*, 93, 112.

Dutch-educated notable, Johannes Andries Truter, to the post of first chief justice of the Cape colony was an important signal of such compromise. Truter's subsequent transmogrification as Sir John, 'the first South African to be awarded a British knighthood (1820)',[99] indicates the extent to which even the most determinedly English of colonial authorities felt constrained to incorporate representatives of the Dutch elite. It also reflects the amenability of the old order to its incorporation by the incoming British administration.

Ambitious Afrikaners were especially drawn to careers in the law, and Truter was only the first of several Dutch members of the nineteenth-century South African judiciary to anglicize his name in the process of being honoured by the Crown. A number of distinguished legal and political statesmen (typically educated at the South African College before going on to further study in Holland, England, or Scotland) carried the ethnically inclusive and relatively enlightened customs of the Cape into the Boer republics and, ultimately, South Africa. In this way the colonial compact linking the Anglo-Dutch aristocracy in the politics of the Bond became cemented into a national tradition.[100]

This point is exemplified by the first two chief justices of the Union of South Africa, John Henry de Villiers and James Rose Innes, both of whom brought their experience of the late nineteenth-century Cape to bear on the politics and legal institutions of Union. As the country's pre-eminent jurists, their success in moulding Roman-Dutch law with elements of English law to create a subtle and robust body of modern South African law is widely noted.[101] So, too, is the manner in which they worked to legitimize the institutions of the new South African state and consolidate the reputation of its legal system nationally and internationally. De Villiers, who was christened Jan Hendrik in 1842, but became known as 'John Henry', stands out in particular as a prime representative of the Bondist–South Africanist tradition. Born into an agrarian western Cape Afrikaner family of Huguenot origin, he was moulded by teachers like Roderick Noble at the South African College (where he was an exact contemporary of J. H. Hofmeyr) and influenced politically by the likes of Merriman and Sauer, as well as Porter and

[99] A. Sachs, *Justice in South Africa* (London, 1973), 35.

[100] Amongst the most eminent representatives of this legal-political tradition one should include William Porter, drafter of the constitution of the Cape, barrister and attorney-general of the Cape (1839–66); J. H. Brand, South African College law professor and MP for Clanwilliam, who became president of the Orange Free State (1864–88); F. W. Reitz, SAC-educated lawyer, chief justice of the Orange Free State (1875–88) and president (1888–95); J. G. Kotzé, SA College-educated advocate, chief justice of the Transvaal (1881–98); J. H. de Villiers, drafter of the Responsible Government Act of 1872, chief justice of the Cape (1873), member of the Privy Council (1897), and first chief justice of South Africa (1910); Melius de Villiers (younger brother of J. H.), who succeeded Reitz as chief justice of the Orange Free State and became professor of Roman-Dutch Law at Leiden University; and James Rose Innes, chief justice of the Transvaal (1902) and of South Africa (1914–27).

[101] S. D. Girvin, 'The Architects of the Mixed Legal System', in R. Zimmerman and D. Visser (eds.), *Southern Cross: Civil Law and Common Law in South Africa* (Cape Town, 1996), 120–1; Sachs, *Justice in South Africa*, 44–5. Comparisons with the role of the post-apartheid constitution and constitutional court are inviting. (Innes is commonly known as the 'master builder' of South African law.)

Solomon. The twin interests that dominated his professional and political life were the closely linked concerns of South African law and political federation.[102]

A formidable, if recondite, body of scholarship has arisen to explain the doctrinal origins and subsequent development of the mixed system that is modern Roman-Dutch law. Two of the most distinguished contributors to the field, Hahlo and Kahn, characterize modern South African law as a layer of English rules and concepts superimposed through 'a process of imperceptible accretion' upon the law of Grotius and Voet; the English influence, they note, 'waxed and waned with the power and prestige of Great Britain', reaching its height in the period 1860–1910. In their more recent treatment, Zimmerman and Visser use a linguistic analogy, characterizing South African law as 'a *patois* with a mixed vocabulary operating within a strongly adapted grammar . . .'.[103]

The historian of law Martin Chanock takes a more radical view, challenging the legal formalism of traditional scholarship and its preoccupation with European pedigree and narrative. Chanock criticizes the practice of close textual analysis of original authorities (such as Justinian, Voet, or Grotius) that characterizes this formalist approach for ignoring the 'multi-vocal' nature of South African law and legal culture and for vaunting ancient and medieval European historical links and traditions. By denying the existence of African customary law, other than in strictly limited and segregated domains, the effect has been to uphold the authority and prestige of white supremacy. Chanock's signal achievement is to enlarge our understanding of law and justice by locating South African common law in terms of the specific institutional requirements of the state's exclusionary race and class system and the dynamics of white supremacy.[104] Yet, in drawing the fault-lines along these major rifts, he is inclined to glide over significant aspects of intra-white contestation over the legal system. The evolution of a hybrid Roman-Dutch legal system was one of the Cape's most distinctive contributions to colonial, and then national, state formation. In ideological terms it echoed the transition from colonial nationalist to South Africanist thinking, and can therefore be read as a charged, albeit erudite, conversation about the desirable ethnic and cultural composition of white society and authority.[105]

102 Walker, *Lord de Villiers and his Times*, pp. v, 3, 8, 15, 215.

103 H. R. Hahlo and E. Kahn, *The South African Legal System and its Backround* (Cape Town, 1968), 578; Zimmerman and Visser, *Southern Cross*, 11. Robert Ross misleadingly characterizes South African law, which was superseded after 1809 by the Napoleonic code in the Netherlands, as a living fossil or coelacanth: *Status and Respectability in the Cape Colony*, 52.

104 M. Chanock, *The Making of South African Legal Culture 1902–1936: Fear, Favour and Prejudice* (Cambridge, 2001). See also Sachs, *Justice in South Africa*. Eduard Fagan, to whom I am grateful for having read this section on the law, argues to the contrary that embedded traditions of 'fairness' in Roman-Dutch law have, in post-apartheid South Africa, often proved more adaptable to the dictates of the Bill of Rights (particularly in respect of gender) than African customary law.

105 Attempts to 'purify' Roman-Dutch law in the post-1910 period were often, but not always, associated with Afrikaner nationalist scholarship. A desire to establish clear, logical, legal principles was also a factor—but this tendency often coincided with a concern for legal form over justice, as well as the desire to maintain cultural separateness. See E. Fagan, 'Roman-Dutch Law in its South African Historical Context', in Zimmerman and Visser (eds.), *Southern Cross*, 63–4.

The centrality of the law to the formation of the Cape colonial state is self-evident, though insufficiently appreciated by historians. By way of illustration one might note the salience of magisterial districts as the basic administrative units of the Colony; magistrates, in turn, embodied local nodes of power, upheld authority, and acted as the eyes and ears of the government. The circuit court system, by which judges and barristers travelled long distances by ox-wagon or cart for months at a time in order to dispense justice to the country's most remote corners, was a visible representation of the presence and prestige of the central state. The circuit court's arrival in town was frequently an occasion for excited demonstrations of welcome on the part of a public unused to ceremonial display. Legal proceedings were closely followed by the local press. The costs of funding the law's progress through the country may have amounted to a serious drain on the colonial treasury, but the system was of incalculable value as a visible symbol of orderly and responsible government.[106]

Circuits also contributed to the law's self-image as a heroic, pioneering institution, bringing civilized values to the frontier. As the century wore on it became an essential part of the legal profession's mythology and lore. Particularly evocative were the vignettes and reminiscences of Alfred Whaley Cole, barrister and judge, founding editor of the *Cape Monthly*, and authority on Roman-Dutch Law. These offer a vivid, albeit idealized, view of life on circuit, with their cast of colourful local characters, accounts of mountain passes negotiated and swollen rivers forded, and descriptions of hospitality received on the road and in towns.[107]

The Cape's evolving relationship with the metropole can be tracked with surprising accuracy by reference to the judiciary's changing composition. From the 1830s senior judges were routinely drawn from the English, Irish, and Scottish bars. At the time of representative government, members of the anglicized Dutch elite like Ebden, Cloete, and Watermeyer joined the bench. Until 1868 official efforts were made to keep a balance between British-born and colonial-born judges. But with the advent of responsible government, the colonial state took over the function of making judicial appointments.[108] In 1873, at the age of just 31, J. H. de Villiers became chief justice of the Cape. Much was made of the fact that he was 'Colonial born', as well as his youth.[109] Indeed, in some minds these qualities were associated. By the end of the century de Villiers's prestige was such that he could take his place on the newly expanded Privy Council—a forerunner of later constitutional arrangements marking the transition from Empire to Commonwealth. In the astute, albeit hagiographic, assessment of de Villiers's

[106] Sachs, *Justice in South Africa*, 3, 42–3. Hattersley, *An Illustrated Social History of South Africa* (Cape Town, 1969), 168, points out that the costs of the circuit system to the Treasury were considerable: in 1845 a judge required 121 saddle horses and 83 mounted guides in the course of a circuit, as well as the attention of 50 veld cornets and juries wherever he sat.

[107] A. W. Cole, *Reminiscences of My Life and of the Cape Bench and Bar* (Cape Town, 1896); *The Cape and the Kafirs, or Notes of Five Years' Residence in South Africa* (1852); 'Off on Circuit', *CMM* (July and Dec. 1870); 'Reminiscences of the Cape Bar and Cape Bench', *CLJ* 5 (1888).

[108] Sachs, *Justice in South Africa*, 38, 44.

[109] Walker, *Lord de Villiers and his Times*, 68.

biographer Eric Walker: 'Long before the end of his life he had become to all South Africans what his Queen had become to all the world, a symbol and an institution.'[110]

With respect to the rest of the country, the Cape took on a sub-metropolitan role. Notwithstanding the fact that the trekkers left the Cape in part because of their rejection of 'colour blind' labour legislation (such as Ordinance 50 of 1828 and the abolition of forced labour), there was no desire to avoid the long reach of Cape law altogether. Roman-Dutch law was adopted by the constitutions of the trekker republics. The Orange Free State and Transvaal high courts, established in the 1870s, were heavily reliant on trained jurists from the Cape bar. Judgments in the Boer republics were considerably influenced by the Cape Supreme and Appeal Courts, as were their traditions and style of operating. After the South African War the Transvaal Law Department was remodelled on that of the Cape, whose principles subsequently became the foundation of law in the Union of South Africa. Natal, with its Crown colony status and its tendency to identify more closely with the metropole than with other states in the region, was something of an exception and was markedly less influenced by the Cape bar. (Interestingly, until well into the twentieth century the Natal bench was regarded as the weakest in South Africa, and its judiciary seen as more authoritarian and less independent of administrative control than elsewhere.)[111]

The pre-eminence of the Cape in legal matters was reflected and maintained by the launch of the *Cape Law Journal* in 1884, edited by W. H. Somerset Bell.[112] This publication, unparalleled elsewhere in the country,[113] marked an important step in the process of professionalizing the law and standardizing judgments and procedures throughout South Africa. Its principal objects were to report the proceedings of the law societies in Grahamstown and Cape Town and to provide a digest of important cases 'decided in South African courts'. Designed in part as a 'literary venture', the *Journal*'s remit included the broader public as well as the legal community: it duly carried book reviews and articles on more general topics, ranging from comparative jurisprudence to criminality and anthropology. Features on leading South African legal figures were a staple, as were memoirs and photographs.

The *Law Journal* operated from the assumption that the Cape was the leading law-giving authority in the country, and sought to consolidate the Colony's position as a parent of legal principle as well as a nursery for the country's lawyers and judges. This guardianship role was vividly expressed in a forthright challenge

[110] Ibid. 81.

[111] E. Fagan, 'Roman-Dutch Law', in Zimmerman and Visser (eds.), *Southern Cross*, 55; Sachs, *Justice in South Africa*, 49, 70, 71, 84–6; see also 'The Brotherhood of the Bar', *CLJ* 7 (1890), 145.

[112] The *Cape Law Journal* was first edited by Bell on behalf of the Eastern Law Districts Society and printed and published from Grahamstown by Josiah Slater. Its initial print-run was 500 copies. Demand was high, and in 1887 a second edition was printed.

[113] But note the creation in the same year of the *South African Medical Journal* under the editorship of W. Darley Hartley.

issued by the solicitor-general, A. F. S. Maasdorp, in 1886 which took issue with the Transvaal Supreme Court for posing as a court of final appeal in South Africa, thereby ignoring judgments of the Eastern District's Court and implicitly questioning the unassailable authority of the Cape's chief justice. J. G. Kotzé, chief justice of the Transvaal Republic, responded indignantly that the Transvaal Supreme Court was free to differ on an interpretation of Roman-Dutch law and that Maasdorp was himself guilty of demeaning the Transvaal's judicial status.[114]

Just over ten years later the *Cape Law Journal* sprung to the defence of Kotzé when President Kruger clashed with his chief justice over the judiciary's assumed right to test decisions of the legislature in light of the Transvaal *Grondwet* (constitution) and to reject *Volksraad* laws if necessary. Kotzé appealed directly to the inhabitants of the republic, and his case was energetically taken up by Milnerite champions of the Transvaal uitlanders as yet another example of Kruger's obduracy and authoritarianism. Justice de Villiers managed to broker a temporary compromise. However, in an atmosphere of febrile suspicion and constitutional crisis, Kruger's power prevailed and Kotzé was dismissed from office in 1898.[115]

From a legal point of view Kotzé's position was not so clear-cut as his supporters purported. Nor should Kruger's inflexibility be taken at face value.[116] Nevertheless, the highly publicized affair occasioned considerable comment and moralizing in professional circles, to the effect that the standing and independent reputation of the law in the Transvaal had been seriously compromised. By contrast, the Cape judiciary basked in the reflected glow cast by Chief Justice de Villiers's elevation to the Privy Council in 1897. His reputation was further enhanced by the fact that de Villiers had been entrusted by both sides to mediate in the stand-off between Kruger and Kotzé.

The Cape's senior legal status was also evident in periodic discussions about the desirability of creating an all-embracing South African Court of Appeal. Moves to create such a court were closely tied to the desire for broader political union. An Appeal Court for the Cape had been constituted in 1879, when confederation was still under active discussion; it was assumed that this institution could easily be expanded to embrace judges from the rest of South Africa. The Appeal Court enjoyed a measure of success in its brief existence, but the expense of maintaining it was considerable, the precedence of Cape Town judges was sometimes resented by other courts, and, as the likelihood of confederation receded, the *raison d'être* of the court diminished. In 1886 it was abolished.[117]

Talk of its revival and extension surfaced periodically, as in 1891, when James Rose Innes proposed an all-embracing South African Court of Appeal which

[114] *CLJ* 3 (2) (1886), 'The Supreme Court of the Transvaal as a Court of Appeal from the Courts of this Colony', by A. F. S. Maasdorp, 102, and response by J. G. K [Kotzé].

[115] For details see e.g. Walker, *Lord de Villiers and his Times*, ch. 17, and J. S. Marais, *The Fall of Kruger's Republic* (Oxford, 1961), ch. 6.

[116] C. Van Onselen, 'The Modernization of the Zuid Afrikaansche Republiek: F. E. T. Krause, J. C. Smuts, and the Struggle for the Johannesburg Public Prosecutor's Office, 1898–1899', *Law and History Review*, 21: 3 (2003).

[117] Walker, *Lord De Villiers and his Times*, 106–7.

would function in place of the Privy Council in London.[118] For those in favour of confederation and a diminution of direct imperial intervention in South Africa, a South African Court was an attractive proposition. However, the *Cape Law Journal* observed that, since other states in South Africa were jealous of their independence, it was more practicable as a first step to reinstate the Cape's Appellate Court, whose decisions had been received with great respect by courts elsewhere in South Africa.[119] Again, in 1894, the *Journal* opined that an Appeal Court 'should emanate from one of the older, and not one of the younger, South African communities'.[120] Anticipating parallel debates about political federation, which were to coalesce a decade later, the *Journal* took the view that moves towards closer union in the legal field should be advanced by building on the Cape's achievements, rather than by establishing a federal Appeal Court *de novo*.

The *Cape Law Journal* viewed its own expansion in much the same way. In 1901, with war still unconcluded, it retitled itself the *South African Law Journal*, announced its intention of moving its editorial offices to Johannesburg, and allied itself with the cause of confederation or unification. The key to this nation-building project lay in championing the body of modified Roman-Dutch law (sometimes known as 'Cape law'[121]) as the most appropriate legal system in a future unified white nation. Not only had this law developed out of the Cape experience, its leading practitioners and theorists were distinguished Cape men such as Innes, Solomon, and Maasdorp who, as Sachs puts it, were distinguished by being 'neither anti-African nor anti-Boer nor anti-English'.[122]

The *Journal*'s editorial support of a modernized version of the Cape legal system was cast in the judicious tones of pragmatic consensus-building. It was not averse to carrying articles arguing the case for the inherent superiority or desirability of either Roman-Dutch or British law.[123] Although it gave voice to such debates, the *Journal* seems to have sided with what Hahlo and Kahn call 'moderate South African "defenders of the [Roman-Dutch] faith" '.[124]

This tradition of quietly assured pragmatism was articulated in a 1901 article by 'Vindex' which firmly opposed concerted efforts to abolish Roman-Dutch law in the Boer republics as part of Milner's package of anglicization policies—an outcome which, it was feared, might easily have resulted in the loss of Roman-Dutch law in the Cape itself.[125] Vindex insisted that the form of Roman-Dutch

118 'A Court of Appeal for South Africa', *CLJ* 8 (1891), 233. 119 Ibid. 234–5.

120 'Notes on a South African Court of Appeal' *CLJ* 11 (1894), 75–6.

121 'Sources of Cape Law No. IV', by Victor Sampson, *CLJ* 4 (1887).

122 A. Sachs, 'The Future of Roman-Dutch Law in a Non-Racial Democratic South Africa', Institute of Commonwealth Studies, London, seminar paper (1989), 7. Sachs points out that, under the guise of purifying Roman-Dutch law, a later generation of judges combined authoritarian views supportive of the apartheid state with obscure and impractical medieval scholarship.

123 In the very first issue of the *Cape Law Journal*, for example, there had been an attack on 'the promiscuous introduction of foreign legal [i.e. English] doctrines' into the law of the Cape. Hahlo and Kahn, *The South African Legal System*, 589.

124 Hahlo and Kahn, *The South African Legal System*, 589, n.23.

125 Walker, *Lord de Villiers and his Times*, 407.

law prevailing in 'the whole of English South Africa south of the Zambesi' had stood the test of time, proved its adaptability, and 'taken deep root in the minds of the people'. It would continue to 'thrive and blossom with renewed vigour . . .'.[126] The *Journal* also carried a series of scholarly articles from 1901 written by the pioneer South African College law professor C. H. van Zyl on the 'sources of South African law', which lovingly elaborated the Roman-Dutch inheritance.[127] In the nation-building environment of the reconstruction years, the Roman-Dutch legal system became a key institutional pillar of the spirit of inter-ethnic reconciliation.

SOUTH AFRICANISM

We have seen in this chapter how ideas about South Africa took shape in the last quarter of the nineteenth century. Renewed imperial interest and activity provided one important frame of reference; responses by colonists complemented or challenged these ideas. Through the dialogue that developed between these intellectual and political brokers, conceptions of white South African identity were clarified and contested. The rough template of South Africanness set out by Froude and Trollope satisfied no one constituency and was mostly impressionistic, but it did sketch out the broad terrain with remarkable acuity.

The responses of Cape anglophone intellectuals who sought to explain South Africa by historicizing the past fell broadly into two categories: on the one hand, a civic-national tradition emphasizing economic and social progress, constitutional reform, political autonomy, and the virtues of ethnic and racial tolerance; on the other, an anti-liberal settler tradition which stressed the achievements of civilization on the frontier, espoused firm native policies, and proclaimed undying imperial loyalism. To a considerable degree these traditions coincided with and were animated by regional political tensions between the Anglo-Dutch elite of the western Cape and separatist-inclined British settlers in the eastern Cape (though of course there were supporters of liberal constitutionalism in the east and hardline ultra-imperialists in the west). Alternative conceptions of South African history and identity were also actively being articulated by the 1880s on the part of the nascent Afrikaner and African nationalist movements: the formation of organizations like the Genootskap van Regte Afrikaners (1875), the Afrikaner Bond (1880), and the Imbumba Yama Nyama (1882) were harbingers of future developments.

[126] 'The Suggested Repeal of Roman-Dutch Law in South Africa', by 'Vindex', *CLJ* 18 (1901), 160–1. Walker, *Lord De Villiers and his Times*, 406, attests to the 'risk that the anglicisation of the schools might find its counterpart in the law courts'. Zimmerman and Visser, *Southern Cross*, 22, are surely mistaken when they label Vindex an 'extremist' who promoted the repeal of Roman-Dutch law. ('Vindex' is presumably not Rhodes's biographer John Verschoyle, who used the same pseudonym: see n. 87.)

[127] The first instalment was in *CLJ* 18 (1901).

Because of the salience of opposing nationalisms in the twentieth century the lineaments of these nationalist narratives are well understood. Yet, analyses of rival pan-South African nationalisms and the class-determinants of their formation have served to occlude what, from the perspective of the late nineteenth century, were visibly more powerful forces. These were perceptively identified by High Commissioner Hercules Robinson, on the occasion of his departure from South Africa in 1889, as the competing influences of 'Colonialism, Republicanism, and Imperialism'. Robinson, it is worth noting, was a convert to the colonialist point of view, having earlier been a supporter of vigorous imperial intervention in Bechuanaland. In this speech Robinson provoked controversy by announcing that imperialism was 'a diminishing quantity, there being now no permanent place in South Africa for Imperial control on a large scale'.[128]

Current usage has so effaced the difference between colonialism and imperialism that they are almost seen as synonymous in the contemporary climate of globalization. Yet, from a late nineteenth-century perspective the colonial–imperial antinomy was all too apparent. Colonialism could well exist within a wider sense of imperial belonging, and it shared many common features with imperialism—most obviously a shared agreement that white political ascendancy should not be threatened. But those who considered themselves colonists took pride in their independence and achievements, and were resentful of unwarranted external intervention. Jingo imperialists were scornful of pretensions to independence where these might challenge metropolitan interests, and were increasingly intolerant of local nationalisms. A more subtle version enunciated the view that imperialism could be most effectively advanced through the agency of colonial sentiment.[129]

Colonial nationalism at the Cape took political, intellectual, and institutional forms. Its unique social basis was constructed upon a tacit set of agreements that drew together important components of the Anglo-Dutch elite. Anglophone culture and values suffused this *entente*, but Britishness could not be seen to predominate to the exclusion of Dutch and Afrikaner influences. The Cape's pride in its hard-won constitutional autonomy was a vital element of its colonial nationalism. In the period surveyed by this chapter, political passions were aroused in respect of Colonial Office-sponsored confederation, responsible government, and the dissolution of the Molteno government. At the conclusion of the South African War, in 1902, Milner's controversial suspension of the Cape parliament was to revive memories of the struggle for colonial independence which reached right back to the struggle for the freedom of the press under Somerset's governorship.

128 Verschoyle, *Cecil Rhodes*, p. xxxi. See also J. Benyon, *Proconsul and Paramountcy in South Africa* (Pietermaritzburg, 1980), 212 ff; D. Schreuder and J. Butler (eds.), *Sir Graham Bower's Secret History of the Jameson Raid and the South African Crisis, 1895–1902* (Cape Town, 2002), 'Editor's Introduction' and ch. 2.

129 This was the view of the apologist for Rhodes, F. Verschoyle, in *Cecil Rhodes*, pp. xxxii and *passim*. It was later taken up by Milner's Kindergarten. See my article 'Colonial Nationalism, the Milner Kindergarten and the Rise of "South Africanism", 1902–10', *History Workshop Journal*, 43 (1997).

Cape colonial nationalism drew on several elements, some of them internally contradictory. It combined a commitment to free trade with aspects of economic protectionism (in the case of the Rhodes–Bond alliance); it was resistant to taking on the financial burdens of eastern expansion if this meant fighting costly wars on the eastern Cape frontier, but amenable to Rhodes's plans for northward expansion until these transmuted into Jameson's attempted coup in the Transvaal; it contained a strong strand of colour-blind liberalism, so long as this principle did not compromise overall white ascendancy. Marxist critics of the Cape liberal tradition have made much of its inadequacies and internal contradictions, while highlighting the instrumental mercantile and political interests involved in the incorporation of a rising and aspirant black middle class. By contrast, defenders of the tradition have tended to be protective about the sterling qualities and heroic actions of its principal protagonists. Both sides of the argument have overlooked the local character and tone of Cape liberalism: if this analysis is correct, it follows that more emphasis should be laid on the *Capeness* of the tradition than the limits of its liberality.

The significance of the colonial–imperial diarchy lies also in the reverberations this had for conceptions of South Africa as a whole. Once plans for confederation had been scotched by the Cape legislature, the revival of imperialist momentum in the subcontinent sought to link up with other loyalist-inclined elements in the eastern Cape and in Natal. The conquest of the Zulu and Xhosa kingdoms in the 1880s and the campaigns to bring the Boer republics under British control were prime instances of the new imperialism at work. Radical colonial critics of imperialism resisted the intervention of men like Frere and Rhodes because they mistrusted their motives, associated them with capitalist economic interests, and considered that their actions were calculated to disturb the country's delicate ethnic and racial balance. F. R. Statham predicted in 1881 that some successor of Bartle Frere might put 'into practical force the maxim that English and Dutch are natural enemies in South Africa, just as Sir Bartle Frere himself put the maxim into force that black and white are natural enemies'.[130] He might have had Milner in mind.

Colonists and imperialists both looked forward to South African union in one form or another. Cape colonists preferred to conceive of South Africa as an extension of the Cape colony and its civic institutions (whether official or voluntary); as the oldest, largest, and most mature state in the region, this was a natural assumption to make.[131] By contrast, imperialists were prepared to bypass or overrule the Cape establishment in order to achieve the larger goal of British

[130] Statham, *Blacks, Boers and British*, 264.

[131] As Trollope assessed the situation in *South Africa*, 460–1: 'In any coming together of Colonies under one flag, one Colony must have an ascendancy. Population will give this, and wealth,—and the position of the chosen capital. It clearly was so in Canada. In South Africa I cannot conceive any capital to be possible other than in Cape Town.'

hegemony over the entire subcontinent.[132] One important reason why they felt entitled to do so was because the costs of defence were primarly borne by the British metropole.

The debate between colonialist and imperialists over confederation (which meant very different things in different parts of the country) can in part be understood as a difference of opinion as to whether moves towards unity should be achieved via annexation of outlying territory from the Cape or, alternatively, encompass the Cape from north and east.[133] Tensions between the western and eastern Cape were exacerbated by these alternative scenarios. The regional rivalries that were animated by these alternatives should also be seen as an inevitable consequence of broader national imaginings, since the act of integration implied a rebalancing of power with the attendant threat of provincial marginalization. Such fears helped to stoke hopes of secession in the mid-nineteenth-century eastern Cape, and later in Natal.

It was out of the confrontation between imperialism and colonial nationalism that competing conceptions of South Africa and South African identity emerged. In the 1871 second reading of the Responsible Government Bill, J. T. Eustace, a leading conservative opponent of the measure, argued that self-government would only serve to encourage struggles between the English- and Dutch-speaking population, 'with the natives throwing their power into one side or the other of the scale (hear, hear, and dissent)'. Indeed, government by the majority implied that the majority of people sitting in the House 'ought to be coloured gentlemen (cheers and laughter)'. Arguing the same point a year later, Eustace insisted that he was 'as good an Africander or South African as any honourable member in this House'. He was reproved by James Buchanan, who chose instead to speak 'as a South African colonist'. Buchanan questioned Eustace's claim to be a good South African, noting that he constantly used expressions like 'we Englishmen' and 'we English'. And, in the first use of the term that I have found, Buchanan suggested that Eustace was 'still in a state of transition to South Africanism'.[134]

Afrikaners were grappling with similar issues of multiple identity and conflicting loyalties. Witness the speech at the disputed diamond fields by the newly elected president of the Transvaal, the Revd. Thomas Burgers. An enlightened and idealistic modernizer, who had been embroiled a decade earlier in a heresy trial over his 'rationalist' doubts about the literal truth of the Bible, Burgers was a moderate republican and an opponent of Kruger. In 1872 he looked forward to a future South Africa that was prosperous and free, 'a great country, when all nationalities shall disappear, and the words English and Dutch become obsolete, and like

[132] It should be noted that Rhodes as well as Milner preferred to work through Cape politics until this option proved too constraining. The point has already been made for Rhodes. For Milner's brief preparedness to work with Cape politicians, see A. N. Porter, *The Origins of the South African War* (Manchester, 1980), 138.

[133] Walker, *Lord de Villiers and his Times*, 126, 130.

[134] *Cape Argus*, 6 July 1871; 25 May 1872. See also, P. A. Molteno, *A Federal South Africa* (London, 1896), 40–1.

Americans or Canadians, we shall be more proud to call ourselves Africanders than either English or Dutch'.[135] In similar vein, the 1879 programme of principles for the Afrikaner Bond, articulated by the leading cultural entrepreneur S. J. du Toit, stated its political ambitions in a broadly inclusive manner such that the term Afrikaner incorporated all those who were genuinely committed to the interests of a 'United South Africa'.[136]

Progressive forms of Afrikanerdom, often shaped by cosmopolitan experience and receptive to English as well as European and American influence, were coalescing amongst sections of the elite in the Boer republics. Cape Afrikaners were prominent among them. How extensive and deep-rooted their networks were requires more detailed research and periodization.[137] Here, Charles van Onselen's observation, in his study of 'new' and 'old' approaches to law and law enforcement in pre- and post-bellum Transvaal, is suggestive: 'changing notions of "Afrikanerness" in both the relatively anglicized Cape Colony and in the former Boer republics were beginning to blend on the margins with a sense of "British Englishness" that was also mutating.'[138]

It was only in the 1880s and 1890s that the divide between imperialists on the one hand, and colonialists and republicans on the other, widened to breaking point. The hardening attitude of imperialism in these years found expression in tones of derision towards Boers that were strikingly different from the more sympathetic or ambivalent views of imperialists like Froude or Trollope in the 1870s. With respect to Africans, the racialization of difference was palpably more marked than a generation before. Typical of this strident imperialism was William Greswell's *Our South African Empire*, a two-volume history written in uncompromisingly triumphalist tones. One of Greswell's central concerns was to restore the sullied reputation of Bartle Frere, to whom the book was dedicated. Greswell contrasted the supreme virtues of Anglo-Saxons with the unimaginative stolidity of the 'Afrikander' race and their 'jejune, coarse, and contemptible patois'. He was also at pains to correct Froude's mistaken and ill-considered 'exultation of the Boer', suggesting that in 'his partisan addresses to the Dutch themselves' Froude had 'recklessly introduced, perhaps unwittingly, nationalist ideas, and therefore trod upon dangerous ground'.[139] The use made of Froude in S. J. du Toit's *Geskiedenis van ons Land* suggests that Greswell had a point.

The racialization of South African politics was so extensive in the final years of the nineteenth century that even its critics could scarcely avoid the language of social Darwinism and widespread expectation of imminent 'race conflict', personified nowhere more graphically than by the stand-off between Milner and Kruger.

[135] *Standard and Mail*, 24 Sept. 1872 (Editorial).

[136] Davenport, *The Afrikaner Bond*, 35–6.

[137] See e.g. S. Swart, ' "An Irritating Pebble in Kruger's Shoe"—Eugène Marais and *Land en Volk* in the ZAR, 1891–1896', *Historia*, 48: 2 (2003).

[138] Van Onselen, 'Modernization of the Zuid Afrikaansche Republiek', para. 105.

[139] W. Greswell, *Our South African Empire*, vol. 1 (London, 1885), 51, 235, 240–1.

The feminist novelist and activist Olive Schreiner found it difficult to escape this discursive trap. Tortured by the march of events and desperate to avert war, she strove to imagine a future South African nation in which all could play a part. Yet she could do so only by making a virtue of racial necessity and expounding the merits of unity in diversity. As she wrote in *Thoughts on South Africa*:

> . . . there is a subtle but a very real bond, which unites all South Africans, and differentiates us from all other peoples in this world. *This bond is our mixture of races itself.* It is this which divides South Africans from all other peoples in the world, and makes us one . . . The only form of organization which can be healthily or naturally assumed by us is one which takes cognizance of this universal condition.'[140]

As we shall see in the following chapters, Schreiner's plea for racial toleration was taken up in the era following the South African War by advocates of Boer and British reconciliation, who seized on their common 'Teutonic' origins; blacks, by contrast, became ever more subject to the logic of racial segregation.

[140] Olive Schreiner, *Thoughts on South Africa* (London, 1923), 61 (emphasis in the original).

4

Science and South Africanism

The previous chapter showed how, in the tension between imperialism and colonial nationalism, incipient ideas about a broader white South Africanism began to take form. It pointed as well to underlying continuities linking the traditions and institutions that underwrote Cape colonial nationalism to the emerging South African state of the early twentieth century. In this chapter the discussion focuses more closely on the emergence of South Africanist sentiment associated with the achievement of political union in 1910. The ideological and practical role of science in promoting reason and rationality in the post-war reconstructionist era is highlighted. Further evidence is adduced to support the idea of a continuing influence of the Cape experience upon the future South African state, especially in respect of the revival of the political process and the renewal of Bondist traditions of compromise and conciliation. The chapter also shows how the centralizing dynamic of unification and statism unleashed significant forces of regionalism and provincialism. These opposing forces are clearly to be seen in the history of scientific, technical, and knowledge-based institutions. They are evident as well in aesthetic and intellectual evocations of nature and locale.

The devastating war of 1899–1902 created the conditions for the emergence of the modern South African state. This conflict marked the end of a century of British imperial expansion in the subcontinent and inaugurated a new phase of racial rule in South Africa that lasted almost a hundred years. Scholarly agreement about the origins and motivations of the war is unlikely ever to be achieved, though there is rather more consensus about its consequences. Most significantly, the conflagration cleared the terrain for the consolidation of white national rule within a unified dominion state. South Africanism was to become its governing ideology.

A major stress in recent work refutes earlier assumptions that this was a 'white man's' war, in recognition of the fact that black as well as white societies, women as well as men, were fully participant. Hence its currently preferred appellation: the South African War. This use of the adjective 'South African' is indicative of a further dimension of the conflict, namely, that the war was more than a grand showdown between the forces of British imperialism and Boer republicanism. The involvement of so many groups in the conflict, as well as its all-encompassing intensity, meant that it took on many of the aspects of a civil war. A war that was at once fought *over* possession of the country's riches, *by* what were to become South Africans, *in* what was to become South Africa, has surely to be understood as a war *for* South Africa, not only in the immediate sense of acquisition and control, but also in the forward-looking sense of making a new nation-state—in effect, a 'white man's country'. No one was more aware of this than Lord Milner himself, when he divulged his peacetime vision for the country only a month into the war and at a time of significant military reverses. 'The *ultimate* end', he stated in a confidential letter, is 'a self-governing white Community, supported by *well-treated* and *justly-governed* black labour' under 'one flag, the Union Jack'.[1]

Given the bitter nature of the war, Milner's vision of a British South African colony, powerfully forged into the imperial chain, was unlikely to prevail. In the immediate post-war period Milner's insensitive imposition of anglicization policies in the former Boer republics, together with his harsh treatment of indentured Chinese labourers in the gold-mining industry, aroused resentment and opposition.[2] Most accounts of Milner's aggressive interventionism focus on his work in the Transvaal and Orange River colonies. Yet, it should not be forgotten that he was also heavily criticized for seeking to reduce the Cape to Crown colony status and for the imposition of martial law in 1900 (which led to the resignation of W. P. Schreiner as prime minister). Milner was especially reviled for attempting to suspend the Cape's constitution in order to circumvent any domestic interference in his reconstruction plans.

Memories of Bartle Frere's high-handed dismissal of the Molteno government in 1878 were fresh, and parliamentarians like John X. Merriman were still on hand to castigate a new imperial 'raid upon the liberties of this country'. It was, indeed, from within the Cape that the political process in South Africa was revived—as a direct reaction against Milner's wartime efforts to suspend the Cape's constitution. The central character in this tussle was the premier, Gordon Sprigg, whom Frere had installed after removing Molteno in 1878. Despite or because of his reputation as a political opportunist, Sprigg now refused to accede to the demands for suspension from among the ultra-loyalists within his own Progressive Party (who had meanwhile succeeded in enlisting the declining but still influential Rhodes to bolster their case). A group of senior Cape parliamentarians managed to head off suspension. In a significant sign of Commonwealth mutuality, they were supported by other dominion leaders meeting together in 1902, as well as by several members of the British cabinet. Milner was rebuked. As a result of the confusion and rancour within loyalist circles, Sprigg lost his leadership of the Progressives to Rhodes's old ally, Jameson, and his parliamentary seat to Smartt in the 1904 election.[3]

[1] C. Headlam (ed.), *The Milner Papers: South Africa 1899–1905,* vol.2 (London, 1933), 35, Milner to FitzPatrick, 28 Nov. 1899 [very confidential].

[2] In 1906 the British House of Commons censured Milner in a 'disguised' manner—following an amendment proposed by Churchill—for the flogging of Chinese labourers. See ibid., 560.

[3] T. R. H. Davenport, *The Afrikaner Bond* (Cape Town, 1996), 236–43; P. Lewsen (ed.), *Selections From the Correspondence of John X. Merriman 1899–1905* (Cape Town, 1966), 466; E. D. Thielscher, 'The Suspension Movement in the Cape Colony, and its Effects (1901–04)', BA Hons thesis, University of Cape Town (1962). See also A. Thompson, 'The Languages of Loyalism in Southern Africa *c.* 1870–1939', *English Historical Review*, 118: 477 (2003), 633–4.

The tone of the 1904 electoral campaign was set by an unrepentant Jameson, who employed the blunt slogan: 'Down with the Dutch.' But the resurgence of imperialist sentiment that this seemed to herald proved illusory: in 1908 the South African Party (which took over the mantle of the Afrikaner Bond and eventually became the governing party of South Africa under Botha and Smuts) roundly defeated Jameson's Progressives, and it was Merriman who was called upon to form a ministry.[4] Signals of the changing political environment are not hard to find. Philip Buckner, for example, has observed, in respect of the 1901 South African Royal Tour, that loyalist sympathies, especially in the Cape, were mixed with elements of colonial nationalist self-assertion: 'the colonists were only prepared to give their loyalty to a monarch who reigned but did not rule.'[5] Rhodes's death in 1902, and his grand funeral procession by rail from the Cape to Rhodesia, occasioned impressive outpourings of loyalist grief. Yet this emotional response was perhaps as much an elegy to the demise of 'new' imperialism as it was a tribute to the passing of one of its greatest proponents.

Davenport has argued that it was as a consequence of the failure of the Cape's suspensionist movement that the new mood of conciliation in the South African colonies began to develop. This is what James Molteno was getting at when he inverted common terminology to claim that those who fought against suspension were the 'true loyalists of South Africa'.[6] The term 'loyalism' was here being claimed by constitutionalists who refused to accept that they should choose between their South African patriotism and their affinities to the Empire. Constitutional loyalism of this sort had strong popular support, and it was by no means only the preserve of established leaders of Cape liberalism like Merriman.[7]

Many senior political figures were responsive to the Cape's reassertion of hard-won colonial freedoms. The colonial secretary, Joseph Chamberlain, for one, had been strongly opposed to Milner's plans to suspend the Cape constitution. In the course of an official visit to the country in 1902–3, Chamberlain developed a close rapport with Cape Afrikaner leaders such as Frederic de Waal and J. H. Hofmeyr.[8] Even Milner began to moderate his tone when he conceded, in 1904, that although he was 'an Imperialist out and out', he did not believe this entailed 'the domination of Great Britain over the other parts of the Empire'. And in his 1905 farewell speech Milner ventured the novel idea that 'the true Imperialist is also the best South African'.[9]

[4] J. T. Molteno, *Further South African Recollections* (London, 1926), 89. One reason why the Progressives won in 1904 was that over 10,000 Afrikaner voters were disenfranchised for supporting Boer commandos. In 1910 the South African Party (merged with the Orangia Unie of the Free State and Het Volk of the Transvaal) became the first Union government under the premiership of Louis Botha.

[5] P. Buckner, 'The Royal Tour of 1901 and the Construction of an Imperial Identity in South Africa', paper presented to 'Rethinking the South African War', Unisa, Pretoria, 1988, p. 20.

[6] Molteno, *Further South African Recollections*, 19. James Molteno was the son of the Cape premier John C. Molteno. A parliamentarian himself, James became the first speaker in the Union of South Africa.

[7] My thanks to Andrew Thompson for pointing this out.

[8] Davenport, *The Afrikaner Bond*, 236–7, 242–3.

[9] Headlam, *Milner Papers*, ii. 503, 547.

In Edwardian Britain a considerable reassessment of imperial priorities was also under way. Andrew Thompson has charted a conspicuous decline in jingo fervour within loyalist, pro-imperial pressure groups in the wake of the South African War.[10] Fresh thinking was evident in Conservative and Unionist circles, prompted in part by an assessment of the costs of the South African War and the need, more generally, for the Empire to bear more of the expense of imperial defence. This was clearly manifested in the development of a 'new conception' of empire, whose core constituent members—Britain and the white dominions—would be bound together in an organic union of equal states, sustained by principles of economic reciprocity, and nurtured by a shared sense of British racial pride.[11] The victory of the Liberals in Britain in 1905, followed by Campbell-Bannerman's meeting with Smuts in early 1906, proved decisive in opening the way for self-government in the Transvaal and Orange River colonies, and it served to encourage a revival of political activity in these regions.[12] By 1908 Afrikaner-supporting parties were in power in both colonies, while the Cape was governed by the South African Party. With the exception of Natal—which had supported Milner's suspensionist attempts at the 1902 Colonial Conference of dominion premiers—a 'remarkable anti-imperialist reaction' was making itself felt in the country at large.[13]

Faced by these challenges, loyalist opinion-formers were forced to adapt their understanding of imperialism. Milner's influential young apostles, the so-called 'Kindergarten', were keenly aware of the new political climate in which their reconstruction project had to be pursued. They were especially taken with the ideas promoted by their fellow Oxford graduate, Richard Jebb, whose influential *Studies in Colonial Nationalism* described the Empire as a field of 'expanding loyalties'. Jebb described how the same forces of nationalism that had animated Europe and the United States in the nineteenth century were now beginning to emerge in Canada, Australia, and New Zealand. His underlying message was that the Empire could only be sustained by working in association with local nationalisms; to resist such forces was futile. When Jebb toured South Africa in 1906 for the *Morning Post* he called for Boers and Britons to unite on the basis of mutual interests and a common South African patriotism, and he defined imperialism as the commitment to 'closer union of the Empire, in any form'. This formulation, which entailed sharing the financial burdens and responsibilities of empire amongst the settler nations,

[10] A. Thompson, 'Pressure Groups, Propaganda and the State: The Manipulation of Public Opinion in Britain During the South African War', paper presented to 'Rethinking the South African War' Unisa, Pretoria, 1988; id., 'The Language of Imperialism and the Meanings of Empire: Imperial Discourse in British Politics, 1895–1914', *Journal of British Studies*, 36 (1997).

[11] E. H. H. Green, *The Crisis of Conservatism: The Politics, Economics and Ideology of the British Conservative Party, 1880–1914* (London, 1995), 198–201.

[12] Hancock, *Smuts: The Fields of Force 1919–1950* (Cambridge,1968), 180, 215.

[13] L.Thompson, *The Unification of South Africa* (Oxford, 1960), 29; E. H. Brookes and C. de B. Webb, *A History of Natal*, 2nd edn. (Pietermaritzburg, 1987), 212.

involved a mixture of high-minded idealism, political pragmatism, and shrewd economic calculation.[14]

SOUTH AFRICANISM

In this fluid political and ideological environment the search for a consensus ideology, capable of spurring national reconstruction and achieving 'closer union', found expression in the slogan of 'South Africanism'. The term, as we have seen, was floated as early as the early-1870s, but it was only in the first decade of the new century that it began to acquire common currency and tangible meaning. 'South Africanism', in its broadest sense, connoted an inclusive form of patriotism which English- and Afrikaans-speakers could both subscribe to, one that was sufficiently capacious to reconcile local nationalisms with continued membership of the British Empire. Though adumbrated by many English-speakers, the legitimacy of South Africanism was crucially dependent on its endorsement by Afrikaner politicians and statesmen.

One of these statesmen was J. H. Meiring Beck. A 'South Africa patriot of the very finest type',[15] Beck had been educated at the South African College (he graduated in 1874 together with W. P. Schreiner, and James Rose Innes) and then at Edinburgh, where he shone as a medical student. Back in South Africa he entered parliament in 1898 as a Bondsman, and later became a Union cabinet minister and senator. He was knighted in 1911. Rodney Davenport refers to Beck as 'an ardent yet moderate opponent of British imperialism'. Despite evincing pronounced republican sympathies during the South African War, he always considered himself a loyal British subject.[16]

In 1905 Beck delivered a lecture entitled 'South Africanism'—perhaps the first time in which the word appears so prominently. Here he deployed a tacit form of Lamarckian thinking in order to make the point that environmental and evolutionary forces were acting to create a vigorous 'South African type'. 'The problem for those guiding the destinies of the British Commonwealth', Beck argued, was thus 'not how to banish for ever the phantom of a South African Nation, but how, by fostering the national spirit, to drive it into grooves sympathetic to common interests rather than into grooves out of harmony with those interests.' Beck's emphasis was on the creation of a hybrid South Africanism born out its fundamental European racial stocks, but he allowed that Africans and coloureds were inevitably part of the evolving South Africanist mix. He concluded his lecture

[14] R. Jebb, *Studies in Colonial Nationalism* (London, 1905); J. D. B. Miller, *Richard Jebb and the Problem of Empire* (London, 1956); D. Schreuder, 'Colonial Nationalism and "Tribal Nationalism": Making the White South African State, 1899–1910', in J. Eddy and D. Schreuder, *The Rise of Colonial Nationalism* (Sydney, 1988); S. Dubow, 'Colonial Nationalism, the Milner Kindergarten and the Rise of "South Africanism", 1902–10', *History Workshop Journal*, 43, (1997).

[15] E. H. Burrows, *A History of Medicine in South Africa* (Cape Town, 1958), 355.

[16] T. R. H. Davenport's entry on Beck in *DSAB*.

thus: 'Let us recognise our responsibilities to our Native population. Let us never forget that they are Africanders; that they are the oldest Africanders in the land; and that we have great duties towards them. So shall we be able to make good our own title to our national birthright.'[17]

Grounded as this message was in the Cape's traditions of political toleration, Beck's definition of 'South Africanism' was sufficiently broad to encompass moderate republicans as well as post-Milnerite imperialists. But the suggestion that South Africanism might embrace people of different colour (Beck was a follower of W. P. Schreiner, who was then campaigning unsuccessfully to preserve the Cape's non-racial franchise) was sacrificed to the greater cause of white solidarity and the growing clamour for racial segregation. It is worth noting the striking similarity between Beck's South Africanist viewpoint and that of his Bondist compatriot and fellow Union statesman F. S. Malan. As well as being a notable advocate of bipartisan South Africanism, Malan proved effective in projecting the relatively enlightened Cape traditions of 'native administration' into the Union until the early 1920s; he was one of the very few senior liberal politicians to oppose on principle the abandonment of the Cape African franchise in 1936.[18]

'South Africanism' was first evoked in order to bring closer polarized positions and break down political mistrust, but it was still too nebulous a concept to provide any firm guide to action. The spirit of reconciliation and practice of nation-building could only take root on the basis of mutual interests, shared institutions, and broadly based functioning networks of association. However, these were conspicuously lacking. In the years after 1905 there was an effusive outpouring of political, cultural, and educational initiatives designed to give content to South Africanism and to root it in the public domain. The growth of local 'closer union' societies (around sixty of which existed by 1909) was one significant means of spreading the gospel. Yet, in a country lacking either a national newspaper or a unitary parliament, it was difficult to stimulate public opinion and debate on a countrywide basis.

One response to this lack was the launch, in 1908, of a monthly journal, fittingly titled *The State*. Its founding editor, Philip Kerr, was a member of the Kindergarten, and the publication received discreet financial backing from the Rhodes Trust as well as two well-known imperialist mining magnates, Abe Bailey and Lionel Phillips.[19] The launch issue took pains to stress its non-sectarian

[17] W. C. Scully, *Sir J. H. Meiring Beck: A Memoir* (Cape Town, n.d.), app. II, pp. 94, 96, 106. Cf. the version of this speech quoted on pp. 62–4.

[18] Davenport, *The Afrikaner Bond*, 281–2; B. Bozzoli, *The Political Nature of a Ruling Class: Capital and Ideology in South Africa 1890–1933* (London, 1981), 122; P. Kallaway, 'F. S. Malan, the Cape Liberal Tradition, and South African Politics, 1908–1924', *Journal of African History*, 15: 1 (1974); S. Dubow, *Racial Segregation and the Origins of Apartheid in South Africa 1919–36* (London, 1989), 171.

[19] In contributing funds to *The State*, Bailey reportedly said:'I am a South African. I mean to be in on this movement. The time has come to realize the dreams of Cecil Rhodes. You young men are doing the writing, but you will want funds to run these Closer Union Societies. I can't

approach and carried messages of support from prominent politicians—among them John X. Merriman, F. S. Malan, Louis Botha, George Farrar, Abraham Fischer, and Jan Smuts. It professed itself to be 'thoroughly South African in its outlook', and proclaimed that the country could only become great and united when there was 'a common sentiment and a common patriotism animating the whole and every part with vigorous life'.[20] The columns of *The State* debated a wide range of contemporary political issues, including the educational system, the role of white labour, segregation and the future of the African franchise, the 'Asiatic question', imperial naval defence, and the economy.

As the official organ of the Closer Union societies, *The State* served as a medium for the dissemination of news and discussion of constitutional issues bearing on the progress of political unification. But its emphasis, as a 'National Magazine', was on developing a sense of cultural identification with South Africa.[21] Peter Merrington has strikingly observed that *The State* lay at the centre of a developing 'aesthetics of Union'; it provided the focus for an explosion of cultural activity on the part of 'a group of architects, artists, writers, historians, archivists and photographers, all of whom were dedicated to the idea of a united South Africa within the context of the British Empire'.[22]

In the period after Union, the spirit of inclusive nationalism originally sponsored by *The State*, was carried forth by the *South African Friend*, a literary journal founded in 1913 by the South African College lecturer and Quaker idealist Arnold Wynne. It was briefly relaunched in 1915 as the *South African Quarterly*, under the editorship of the ubiquitous liberal social moderator J. D. Rheinallt Jones, and revived after a four-year pause in 1919. The *Quarterly* strove to establish a common internationalist-minded South African identity that was consciously non-sectarian and capable of transcending imperial and nationalist extremes. In the post-war era it served as a voice for Johannesburg's developing anglophone liberal intelligentsia. As well as covering literary topics and advancing original thought, the *Quarterly* advocated enlightened social attitudes on issues such as 'poor whiteism' and the 'native question'.[23]

The creation of the South African Historical Society in 1913 was another example of the nation-building impulse at this time. Its objectives included the

write books, but I can write cheques.' Obituary to Bailey in *Round Table* (Sept. 1940), cited in W. Nimocks, *Milner's Young Men: The 'Kindergarten' in Edwardian Imperial Affairs* (Durham, NC, 1968), 112.

20 'By Way of Introduction', *The State* (Jan. 1909), 1–2.

21 *Proceedings at the Annual Meeting of the Association of Closer Union Societies*', 'Introduction'.

22 Peter Merrington, 'Pageantry and Primitivism: Dorothea Fairbridge and the "Aesthetics of Union"', *Journal of Southern African Studies*, 21: 4 (1995), 644.

23 See e.g. *South African Friend*, 3: 1 (1913) 'editorial'; *South African Quarterly*, 2: 1 (1919), 'Editor's Notes'. Though originally intended for limited distribution within Quaker circles, the *Quarterly* soon expanded beyond this base. Arnold Wynne, who was born in England and educated at Leeds University, arrived in Cape Town in 1911 and became a lecturer in English at the South African College. He died in Flanders in 1917. In 1925–6 around 800 copies of the *South African Quarterly* were printed.

preservation of historical buildings and archives and the interviewing of 'old people', black as well as white. George Cory and Eric Walker were two of its leading protagonists, and G. M. Theal was invited to become its first president, with John X. Merriman as his deputy. The Society met only once on a national basis in 1914 and was defunct by the end of the First World War. Committed to the spirit of Anglo-Boer reconciliation, it never recovered from the setback it received when J. F. W. Grosskopf, who had been granted a salaried fellowship (funded by Otto Beit) in order to conduct interviews with aged people, was arrested and fined in 1915 for having joined the Afrikaner rebellion in the Transvaal.[24]

SCIENTIFIC ORGANIZATIONS

Scientific and technical agencies—ranging from professional associations, museums, and botanical gardens to transport and communication systems—were part of the intellectual and political substructure of Union. Their salience has largely been ignored by historians, perhaps because they were not obviously 'ideological'. Yet, agencies representing progress—ranging from 'pure' scientific institutions such as observatories, to applied technology in the form of railways and agriculture—had the capacity to evoke powerful aesthetic and affective responses. The ideology of progressivism, which underwrote many technical and scientific developments, was a powerful force with the capacity to transcend sectional interests. Apparent neutrality and relative political invisibility meant that technical matters could more easily be presented as the acceptable face of the imperial connection.

In the closing years of the nineteenth century a number of professional bodies were created to advance specialist knowledge as well as the interests of their members. Although some of the professional bodies created at this time had pretensions to national representivity (often indicated by the title 'South African'), as often as not they were regionally located and locally specific in their activities. Organizations created to advance specialist knowledge and the collective interests of their members included the Institute of Mechanical Engineers and the Association of Engineers and Architects (1892), the Chemical and Metallurgical Society of South Africa (1894), the Geological Society of South Africa and the Cape of Good Hope Geological Survey (1895), the South African Society of Electrical Engineers (1897), the South African Republican Medical Society (1895), and the Transvaal Medical Society (1889).[25]

[24] C. C. Saunders and B. Le Cordeur, 'The South African Historical Society and its Antecedents', *SAHJ* 18 (1986).

[25] S. M. Naudé and A. C. Brown, 'The Growth of Scientific Institutions', in A. C. Brown (ed.), *A History of Scientific Endeavour in South Africa* (Cape Town, 1977), 70. Burrows dates the Transvaal Medical Society to 1889, as do P. W. Laidler and M. Gelfand, *South Africa: Its Medical History (1652–1898)* (Cape Town, 1971), 488. There was also the South African Republican Medical Society in Pretoria, probably formed in 1895.

The last-mentioned organization merits further comment, because medicine is strongly illustrative of the difficulties that many scientific and professional bodies encountered in developing a national presence. In respect of the politics of knowledge, it is worth noting that doctors were often amateur scientific investigators. Many medical practitioners were based in country towns and, along with lawyers and teachers, were amongst the most educated members of the community. Publications like the *South African Medical Journal* (1884) and the *South African Medical Record* (1903) were principally created to promote professional interests, but they also functioned as a much-needed outlet for the discussion and dissemination of medical and associated scientific research.

Although a South African Medical Society had been established as early as 1827, and was therefore coterminous with some of the first attempts to establish scientific societies at the Cape, it was only the first organization so described and was beset by conflicts relating to the licensing, control, and distinction between different classes of medical practitioners. According to Burrows, one of the most important consequences of parliamentary government in the mid-nineteenth century was increased interest in and concern for the medical institutions of the Cape. The legislature included a number of prominent doctors, and a series of important legislative reforms soon followed in the areas of public health and sanitation.[26]

In the late nineteenth century, when the numbers of doctors seeking opportunities in a rapidly expanding economy grew rapidly, these conflicts became even more acute. No sooner was a South African Medical Association established in Cape Town in 1883 than its pretensions to national representivity were undercut by a breakaway branch in Grahamstown (the Eastern Medical Association of South Africa) and the parent organization reverted to its Cape Town core constituency. Criticism was made of its continued use of the title 'South African'. In 1888, when a branch of the British Medical Association (BMA) was established in Kimberley, there were no fewer than four other independent medical associations in the Cape. By 1896 there were a further five competing medical societies in the country.[27]

Affiliation to the BMA was seen by some as the best way of creating unity, but for others, like the champion of South Africanism and secretary of the South African Medical Association J. H. Meiring Beck, fragmentation only underlined the need to establish a genuinely South Africa-wide association. In Beck's opinion it was unreasonable to expect that doctors in the Boer republics should look to the BMA: unity would have to come from within, through forming a genuinely South African Association.[28]

Consolidation was practically achieved by appealing to common professional interests rather than institutional loyalties. The convening of the first South African Medical Congress at Kimberley in 1893 was an important development,

[26] Burrows, *A History of Medicine*, 301, 332.

[27] Ibid. 350–1, 352, 361. Burrows reports that one of the cofounders of the South African Medical Association, C. L. Herman, 'was a staunch, even fanatical Capetonian'.

[28] Ibid. 354–5.

as were decisions to publish the *South African Medical Journal* on a monthly basis and to convene an annual congress. After much politicking, a South African Medical Association was eventually formed in 1897. These processes of professionalization and integration were consonant with a growing consciousness of the need to address 'National Disease Problems'. The formation of a specialized colonial government department, along with moves to establish research institutes, was regarded as essential to tackling problems of public health and veterinary disease on a broad basis.[29] In the case of medicine, however, the cause of unity was arrested by the war. It was only well into the twentieth century that consolidation of the still fragmented profession took place. The creation of a government Department of Health in 1919 and the establishment of the Medical Association of South Africa in 1926 represented important landmarks.[30]

The cluster of museums which developed throughout the country in the second half of the nineteenth century were also important nodes of local scientific activity.[31] In Natal, museums were established at Durban in 1877 and Pietermaritzburg in 1904. The year 1877 also saw the creation of a National Museum of the Orange Free State, while the South African Republic resolved to create a Staats Museum in 1892.

Rather like the example of the South African Museum in Cape Town, these institutions were objects of considerable civic and national pride. In the case of the Boer republics they were repositories for republican national aspiration. Competition between the South African and the regional museums could be keen. One indication of such rivalry—as well as a measure of developing professional standards—can be seen in the production of scientific house journals, beginning with the *Annals of the South African Museum* in 1898. Within a decade other aspiring institutions commenced publication of their own titles: the Albany Museum (1903), Natal Government Museum (1906), and the Transvaal Museum (1908).

The origins of the Free State's National Museum were prompted by the country's participation in an 1876 exhibition in Philadelphia, organized to mark the centenary of the United States' independence. President Brand saw this event as an ideal way of promoting his new republic in the outside world. Moves were soon afoot to house the exhibits sent to America in a new museum in Bloemfontein (at least those which had not already been sold). In response to a circular issued by the museum's committee chair F. W. Reitz (a future president

[29] Ibid. 335–7, 356–7; 359–60. See also Laidler and Gelfand, *South Africa: Its Medical History*, ch. 14.

[30] Burrows, *A History of Medicine*, chs. 16, 17; Laidler and Gelfand, *South Africa. Its Medical History*, 505.

[31] In light of the discussion immediately above, it is worth noting that doctors were instrumental in establishing the first museums in the country: Andrew Smith in the case of the South African Museum, and William Atherstone and a handful of fellow medical men in the case of the Albany Museum (whose immediate progenitor was the Grahamstown Medical-Chirurgical Society, formed in 1855).

of the republic), the proposal for a museum was warmly supported by *De Express*. As this newspaper astutely observed, such a museum would help to bestow upon the new state 'a past'.[32]

In the case of the Transvaal Staats Museum, the initiative for its creation came from state secretary W. J. Leyds, a Dutch-born cultural entrepreneur and a close associate of Kruger. (Dutch immigrants or 'Hollanders' proved to be some of the most avid Afrikaner nationalists.) Along with its stated commitment to promote education and science, an important objective of the museum was 'fostering the love of the country' and 'stimulating national pride'. Considerable emphasis was therefore laid on the collection of historical relics, which included items illustrative of the Voortrekker heritage and of the recent, failed, Jameson Raid.[33] The Staats Museum proved a highly popular attraction, drawing 14,000 visitors in 1895. After conquering Kruger's capital city the British military government kept its original director, the Dutch-born medical doctor and zoologist J. W. B. Gunning, but changed the composition of its management board. Alterations to the institution's name (it reopened after just a few days' closure in 1900 as the Pretoria Museum, and changed again in 1903 to become the Transvaal Museum) mirrored the Transvaal republic's loss of independence and its rebirth as a province of the future Union.[34]

THE SOUTH AFRICAN ASSOCIATION FOR THE ADVANCEMENT OF SCIENCE

By the outbreak of the South African War, then, a rudimentary scientific infrastructure—ranging from museums and medical societies to botanical gardens and a professional scientific association—was in existence throughout the country. Yet, apart from the elite South African Philosophical Society, there was no single overarching body capable of coordinating scientific activity. The creation of the South African Association for the Advancement of Science (generally known as the S2A3) in 1903 was intended to answer this perceived need. Its emergence, following a proposal by Theodore Reunert, in 1901, to organize an annual congress of engineers, would seem to be an unlikely beginning; but the term

[32] D. A. van der Bank, 'Nasionale Museum. Deel 1. Stigting en vroeë Geskiedenis, 1877–1921', *Navorsinge van die Nasionale Museum Bloemfontein*, 13: 2 (1977), 47–8.

[33] J. W. B. Gunning, 'A Short History of the Transvaal Museum', *Annals of the Transvaal Museum*, 1: 1 (1908), 4; E. Grobler and R. Dixon, 'The State Museum of the South African Republic 1892–1900', in N. J. Dippenaar, *Staatsmuseum 100* (Pretoria, 1992), 2, notes that the idea for a museum in the ZAR originated with President T. F. Burgers in the 1870s, who established a short-lived natural history museum in Potchefstroom.

[34] Grobler and Dixon, 'The State Museum'; also R. Bigalke, The *National Zoological Gardens of South Africa* (Pretoria, 1954). The Transvaal Museum's early bias towards cultural artefacts soon shifted towards natural history. In 1899 a zoological garden was created, and from 1900 to 1913, when the two institutions were separated, they were jointly directed by Gunning.

'engineer' was interpreted sufficiently broadly so as to encompass those who were involved in the 'pursuit of Science for its own sake', as well as those who were engaged in the direct 'utilization of Science'.[35] Further meetings in Cape Town, where many engineers and scientists had chosen to congregate during the war, resulted in a decision to form a scientific society modelled on the British Association for the Advancement of Science.

The outstanding figure in the formation of the S2A3 was a dynamic Aberdonian clockmaker-turned-astronomer, David Gill (now a somewhat forgotten figure), who became its founding president. Gill's scientific and social credentials placed him in an unrivalled position to lead the S2A3. He arrived at the Cape as astronomer royal in 1879, whereupon he took on the task of rebuilding and modernizing the Observatory after years of neglect. Gill immediately assumed the presidency of the elite South African Philosophical Society, a position he held again in 1901–2. Under his direction the Observatory rapidly recouped its scientific reputation: Gill's innovative use of celestial photography to record the 'Great Comet' of 1882, and his role in organizing the observation of Venus' transit across the sun, brought his work to international attention and prompted his election to a fellowship of the Royal Society in 1883. Gill's involvement in Cape Town's prominent civic and intellectual institutions (he was a trustee of the Library and of the Museum) further augmented his local standing. Under his custodianship the Observatory again became a natural port of call for visiting dignitaries, as it had been during the time of Thomas Maclear; in Brian Warner's estimation, the Observatory was now 'the principal intellectual meeting place in South Africa'.[36]

In the unpromising context of post-war devastation and regional fragmentation, the newly formed S2A3 was an ambitious venture. Amongst its declared constitutional aims were the objectives 'to promote the intercourse of Societies and individuals interested in Science in different part of South Africa' and 'to obtain a more general attention to the objects of pure and applied Science'.[37] From the outset, therefore, the S2A3 sought to include in its constituency amateurs and enthusiasts as well as professional scientists. Its catholic approach and relatively modest subscription fee (£1 per annum) resulted in rapid expansion: 268 'foundation members' in 1902; well over 700 members and associates at the end of the first meeting in 1903; and 1,322 in 1906—a figure never surpassed in the twentieth century.[38] Care was taken to ensure that S2A3 could not challenge the premier

[35] Address by Sir David Gill to the S2A3, in *Report of the South African Association for the Advancement of Science. First Meeting Cape Town 1903* (Cape Town, 1903), 17.

[36] B. Warner, *Astronomers at the Royal Observatory Cape of Good Hope* (Cape Town, 1979), 85, 88–9, 105. See also D. Gill, *A History and Description of the Royal Observatory, Cape of Good Hope* (HMSO, 1913).

[37] *Report of the South African Association for the Advancement of Science. First Meeting Cape Town 1903* (Cape Town, 1903), 9.

[38] 'Report of the Council for the Period Ended May 2nd 1903', in *Report of the South African Association for the Advancement of Science 1903*, 517; B. J. F. Schonland, 'The South African Association for the Advancement of Science, its Past and Future', in *SAJS* 49: 3–4 (1952), 61. Membership was drawn from the whole of South and southern Africa.

status of the far more exclusive South African Philosophical Society, whose superior standing was confirmed when it received the Royal Charter in 1908 to become the Royal Society of South Africa.

Leaders of the S2A3 were conscious of the Association's role in sewing together the sinews of science and, by analogy, in forging a common South African identity and assuaging the bitterness of the war. In 1905 Reunert, now the president of the S2A3, recalled that its creation 'at a time of great and universal despondency, was an act of faith which has been amply justified by events'.[39] Two years later, at a time when closer political union between the four South African colonies was being actively pursued, the Grahamstown-based *African Monthly* observed that the S2A3 exemplified the desire to break down barriers of 'petty localism' and provided 'evidence of true progress towards nationality'. The existence of organizations such as these, it explained, was essential preconditions to the creation of a future parliament in a united South Africa that was British in character and imperial in tendency.[40] (There are clear parallels here with the creation of the Australasian Association for the Advancement of Science (est. 1888) which was viewed by contemporaries as a concrete step towards Australian federation.[41])

Although it is not clear whether the *African Monthly* envisioned the S2A3 as a metaphorical South African parliament, had it done so it would only have been echoing a recurrent image of its model, the British Association for the Advancement of Science, as a 'parliament of science'. This view reflected the British Association's desire to serve a range of scientific constituencies and interests; it underlined the fact that, in acting as an advocate for science, the British Association inevitably participated in the realm of politics, often in support of social reform.[42] The S2A3 may not have seen itself in quite the same terms at this stage, though J. H. Hofmeyr referred to it as the country's 'Parliament of Science' in 1939.[43] The S2A3's contribution to the politics of symbolic unity, particularly for anglophone South Africans, was implicit in its very structure, which consisted of devolved local committees with an overarching council composed of members drawn from the major centres of the country. The annual meeting of the peripatetic body alternated between the major cities of southern Africa. By the advent of Union, the S2A3 had met in all four colonies. It was thus one of only a few nationwide organizations (including, perhaps, the churches and the medical profession) which could claim extensive local, national, and international links.

[39] Presidential address by T. Reunert, in *Report of the SAAAS. 3rd and 4th Meeting 1905–6* (Cape Town, 1906), p. i.

[40] *African Monthly*, 2: 8 (1907), 227.

[41] R. Macleod, 'On Visiting the "Moving Metropolis": Reflections on the Architecture of Imperial Science', in N. Reingold and M. Rothenberg (eds.), *Scientific Colonialism: A Cross-Structural Comparison* (Washington, DC, 1987), 239.

[42] R. Macleod, 'Retrospect', in R. Macleod and P. Collins (eds.), *The Parliament of Science: The British Association for the Advancement of Science 1831–1981* (Northwood, 1981), pp. v, 17.

[43] But see J. H. Hofmeyr, 'Science and South Africa's Essential Poverty', *The Forum*, 22 July 1939: 'The South African Association for the Advancement of Science is the Union's Parliament of Science. Following in the footsteps of the British Association, it aims at uniting workers in all branches of science.'

Fig 5. The Royal Observatory, Cape of Good Hope. Founded in 1820 by the British Admiralty, the Observatory was the first official scientific organization in South Africa. The building was based on a design by J. Rennie and completed in 1827. (Courtesy of the South African Astronomical Observatory)

As the country's most prominent scientific citizen, Gill used his high-level social and political contacts to pursue a frankly imperial scientific agenda, albeit one that was sensitive to the need for cooperation and nation-building. Gill cultivated the patronage of leading officials and statesmen assiduously in order to advance his ambitious scientific plans. He was called upon for advice on a variety of political, administrative, and military matters, services which undoubtedly played a part in his being awarded a KCB in 1900.

Gill's most ambitious project was his inauguration of a programme of geodetic surveys. This included the triangulation of the 30th east meridian, a prestigious and costly scheme with important implications for assessing the size and shape of the earth and its distance from the sun and the stars.[44] Gill's work was by no

[44] A. C. Brown (ed.), *A History of Scientific Endeavour in South Africa* (Cape Town, 1977), 417, 442–3. Gill's project to measure, by triangulation, the 30th east meridian (the longest measurable arc of meridian) stretching from South Africa to Norway, was only completed in 1955. See also O. J. R. Howarth, *The British Association for the Advancement of Science: A Retrospect 1831–1931* (London, 1931), 130; D. Gill, 'On the Origin and Progress of Geodetic Survey in South Africa, and of the African Arc of Meridian', in *Report of the Seventy-Fifth Meeting of the British Association for the Advancement of Science. South Africa 1905* (Cape Town, 1906), 228–48.

means only a matter of blue-skies thinking. Geodetic surveying had vital strategic implications for the accurate mapping of South Africa and the African continent. It was crucial for the determination of land-title and national boundaries—a point fully appreciated by imperial politicians and military planners.

The high-level patronage that Gill secured was commensurate with the imperialist scope of his ambitions: it was Frere who authorized Gill to begin his geodetic map of the Cape, and Milner who ensured in 1902 that it should encompass the newly acquired Crown colonies. Thereafter, Gill proposed to extend the survey to Egypt, commenting that it was 'a fit precursor of [Rhodes's] great scheme for a Cape to Cairo railway'. The financial and political support that Rhodes gave Gill made possible his extension of the geodetic survey to the northern limits of Rhodesia.[45]

The prominence which the Cape Observatory gained in the public mind was augmented by its expanded role in determining and disseminating time.[46] This had been achieved at a local level through the practice of firing a noonday gun from Signal Hill, Cape Town, which resounded through the city. But the national context posed different challenges. Prior to 1903, when standard South African time was first introduced on the basis of Gill's proposal of Greenwich mean time plus 2 hours, large towns maintained two clocks: one providing local time, the other 'railway' time. The adoption of a single time zone in 1903 was thus a powerful metaphorical statement of national integration or, in Benedict Anderson's phrase, a contribution to the 'experience of simultaneity'.[47] Agreement on the question of South African time necessitated high-level negotiations, and it was secured only after representations from the governments of Natal and the Transvaal.[48] Gill's role thus serves as a further reminder of the persistent influence of Cape-based institutions on the South Africa that was in the making. But, as we shall see, the Transvaal was increasingly affirming its centrality in the nation-building enterprise.

THE VISIT OF THE BRITISH ASSOCIATION FOR THE ADVANCEMENT OF SCIENCE

In 1900 Gill was invited to attend a meeting of the Council of the British Association for the Advancement of Science which had earlier expressed interest in meeting in South Africa. The Cape's premier, Gordon Sprigg, agreed to the proposal to bring out the British Association, and backed this up with the promise of material assistance.[49] The cessation of hostilities now made the plan realizable.

[45] Gill, 'On the Origin and Progress', 230–2; Warner, *Astronomers at the Royal Observatory*, 104–5.

[46] The Observatory provided a daily time signal to ships in Table Bay from its inception.

[47] B. Anderson, *Imagined Communities* (London, 1983),132, 41.

[48] Warner, *Astronomers at the Royal Observatory*, 103; Gill, *History of the Cape Observatory*, p. cxliv.

[49] Notwithstanding the fact that Gill took a high-profile role in arguing against Sprigg for the suspension of the Cape constitution.

From 1902 Gill became actively involved in canvassing support for the proposed visit amongst civic and industrial leaders in Cape Town and Johannesburg. Senior administrative figures were duly appointed as vice-presidents of the British Association. These included the high commissioner for South Africa, Lord Selborne, and his predecessor, Lord Milner; the governors and lieutenant-governors of the four colonies of South Africa; the administrator of Southern Rhodesia; and the mayors of cities included in the Association's itinerary.[50]

Gill also proved successful in securing the support of Joseph Chamberlain, informing the colonial secretary during his visit to South Africa that the British Association would be going '6000 miles' to help the 'work of conciliation in South Africa'.[51] In his 1903 presidential address to the S2A3, Gill argued that science lay 'at the foundation of our civilization'. It could play a vital role healing the bitterness of war and cementing the imperial connection. 'Science', he concluded, 'knows no nationality, and forms a meeting-ground on which men of every race are brethren, working together for a common end—and that end is truth.'[52]

The 1905 visit of the British Association to South Africa was an enormous undertaking. Nearly 400 scientific visitors were transported by rail to the centres hosting the visit—Cape Town and Johannesburg were given equal status as the principal receiving cities—while side tours were arranged to Bulawayo and Beira (cynics spoke of the meeting as a glorified 'picnic'). Sir Donald Currie of the Union Castle Line offered reduced fares for the Atlantic crossing, on the grounds that the enterprise was 'thoroughly in accord with the spirit of Imperialism'. Further subventions were granted by the South African colonial governments. Enthusiastic civic receptions and social events were laid on to greet the dignitaries as they progressed through the country, while towns vied with each other for the privilege of acting as hosts. There was extensive press coverage of the official proceedings and regular reports of endless speeches and toasts proposed in honour of the visitors.[53]

The presence of the British Association was widely commended as a gesture of political conciliation in a country recently racked by war. A recurring theme in the many speeches at public functions was the healing role that science could play and the advantages of maintaining the imperial connection. The tone was set by

[50] Archives of British Association for the Advancement of Science (BAAS), Bodleian library, Oxford, box 201, Gill to secretary British Association, 5 Aug. 1902; *The Times*, 16 Aug. 1905; 2 Feb. 1905.

[51] BAAS box 201, Gill to J. G. Garson, Asst. Gen Sec BA, 9 Mar. 1903.

[52] Address by Sir David Gill to the SAAAS, in *Report of the South African Association for the Advancement of Science. First Meeting Cape Town 1903*, 20, 36. Cf. the words of Gill's fellow astronomer John Herschel, who was also a leading light in the early years of the BAAS (serving as president in 1845): 'Let selfish interests divide the worldy; let jealousies foment the envious; we breathe a purer Empyrean air. The common pursuit of truth is of itself a brotherhood.' Cited in Macleod, 'Retrospect', in Macleod and Collins (eds.), *The Parliament of Science*, 4.

[53] For a full account see my article 'A Commonwealth of Science: The British Association in South Africa, 1905 and 1929', in S. Dubow (ed.), *Science and Society in Southern Africa* (Manchester, 2000).

Sir Walter Hely-Hutchinson, governor of the Cape. In welcoming the British Association, he expressed the hope that 'an important step had been taken in drawing closer together the bonds of the brotherhood of science' and, in so doing, 'promoting brotherly feeling between his Majesty's subjects in South Africa and the motherland'.[54]

The notion that science had no political boundaries sent out a message of political inclusivity on the basis of continued membership of the imperial fraternity. Some speakers reached out to Afrikaners and expressed the hope that science would help to break down barriers and prejudices, because science 'knows no nationality'. Others could not resist speaking in tones of condescension, and there were those who indulged in displays of jingoistic triumphalism. Yet, such remarks were secondary to the promise that scientific reason and enlightenment would yield practical benefits for all mankind. Under the guise of scientific detachment and objectivity, a close equivalence was being drawn between the universality of science and the progressive global role of the British Empire.

The British Association was gratified by the success of the meeting in South Africa, which fully vindicated its dedication to a broad 'imperial mission'. Its commitment to imperialism had developed from the 1880s, when greater specialization and professionalization of science in Britain threatened to erode the British Association's established role as a popularizer of scientific method and culture in the provinces. A precedent for overseas visits had been set in 1884 when the Association met for the first time in Canada. This occasion reaffirmed the Association's importance as an organizing centre, helped to promote colonial scientific activity, and demonstrated the advantages of federal cooperation in science as well as politics.[55] In scientific terms the joint meeting of the British and South African Associations underlined the depth of metropolitan knowledge, but it also revealed deficiencies in South Africa's scientific base. The subordinate status of the South African Association was all too evident, and deference to the metropolitan visitors was routinely registered. Consider the view of the *Bloemfontein Post*: 'We are indeed very sensible of the honour and privilege of entertaining... the representatives of old-world learning though we regret that our own advancement makes us worthy of being little else than a *pied-à-terre* between greater centres.'[56]

Notwithstanding forelock-tugging of this sort, the effects of the British Association visit on South African science were salutary. In time the S2A3 was credited with giving important support to the creation of institutions like Kirstenbosch, the Union Observatory, and national parks and game reserves.[57] Local achievements in fields like geology, agronomy, astronomy, and geodesy

[54] *The Times*, 16 Aug. 1905.

[55] Macleod and Collins (eds.), *The Parliament of Science*; A. Gale, 'Science at the Margins: The British Association and the Foundations of Canadian Anthropology, 1884–1910', Ph.D thesis, University of Pennsylvania (1986).

[56] *Bloemfontein Post* (editorial), 2 Sept. 1905.

[57] Schonland, 'The South African Association for the Advancement of Science, 64.

Fig 6. The 1905 Joint Meeting of the British and South African Associations for the Advancement of Science: geology excursion led by A. W. Rogers (Courtesy of Historical Papers, William Cullen Library, University of the Witwatersrand)

were afforded considerable attention, and Gill was personally rewarded when he secured funding from the Association to continue his survey work. A considerable boost was given to anthropological and ethnological activity by the presence of the Cambridge expert A. C. Haddon. In the section devoted to educational science, a need for new universities geared to practical and technical needs was expressed. The visit of the British Association therefore amounted to more than an occasion for imperial promenading. The fact that the British and South African Associations were meeting jointly was indicative of the rebalancing process that was beginning to take place between metropole and colony.

RECONFIGURING RACE

One area of growing scientific interest which the 1905 joint meeting of the British and South African Associations helped to stimulate was discussion of the 'native question'. Here, Alfred Haddon's address proved to be a key intervention. The renowned Cambridge anthropologist used the occasion to survey the state of ethnological work in South Africa, and he called for state assistance to undertake a comprehensive programme of research. In several respects Haddon's address recalled Bartle Frere's suggestion, in his 1878 address to the South African Philosophical Society, of an ethnological section 'devoted to the languages of ethnology, history and characteristics . . . of the native African races'.[58] Also at the 1905 meeting, and on the urging of Sir Godfrey Lagden (the author of the

[58] B. Frere, 'The Native Races of South Africa', *Transactions of the South African Philosophical Society*, 1 (1877–8), p. xvii.

South African Native Affairs Commission report), Howard Pim outlined an overall scheme of racial segregation by way of an address to the 'economic science and statistics' section on 'Some Aspects of the Native Question'.[59] The quest for theoretical and applied anthropological knowledge was by now closely interlinked, and it was out of this developing dialogue that modern sociological conceptions of South African society—or, more to the point, societies—were to emerge.

Science played a key role in redefining the South African problem in racial terms. But it could only be effective as a white nation-building exercise if the damaging application of eugenic nostrums, so often deployed by imperialists to explain the alleged innate backwardness and degeneracy of Boers, could be undone. In the age of reconciliation, Milner's frank description of himself as a British 'race patriot' and his many public and private slights on Boer characteristics—for example, his assessment that they could continue fighting as guerrillas for some time, 'just as low types of animal organisms will long survive injuries which would kill organisms of a higher type outright'[60]—were no longer politically acceptable.

Rudyard Kipling was one of those imperialist ideologues who attempted, in deliberately opaque terms, to reassess deep-seated assumptions of Boer racial inferiority. But if Kipling was caught in an 'ideological limbo' and proved unable to confront his own disillusionment with 'adventurist imperialism',[61] other minds proved more supple in explaining British military reverses and pointing the way to a common white future charged by the youthful energy of the imperial periphery. Such thinking was evident in the language of the landmark Selborne Memorandum of 1907, which was intended as a blueprint for the closer union movement. It had been conceived with strong input from members of the Milner Kindergarten, Lionel Curtis in particular, and underwritten with the bipartisan support of the Bondist politician F. S. Malan. The opening passage of the Memorandum spoke of the need for the two principal 'races' of South Africa—British and Dutch—to overcome their historic differences. Inverting standard eugenic ideas so as to portray racial mixture as nationally advantageous (in a manner that bears interesting comparison with Olive Schreiner's diagnosis of the South African body politic and also with Meiring Beck's account of South Africanism), the Memorandum declared that both white nationalities were Teutonic in origin: '. . . the fusion between them is merely a matter of time, as it was with the Saxons and Normans, who were related to one another in a similar degree of kinship.'

The Selborne Memorandum went so far as to suggest that divisions of race and language in South Africa were rather less severe than those existing in Ireland or Canada, where the physical separation of the rival nationalities acted to check

[59] *Report of the 75th Meeting of the British Association for the Advancement of Science. South Africa 1905* (London, 1906).

[60] Milner to Haldane, 21 Jan. 1901, 'very confidential', in Headlam (ed.), *Milner Papers*, ii. 206.

[61] M. van Wyk Smith, 'Telling the Boer War: Narrative Indeterminacy in Kipling's Stories of the South African War', paper presented to 'Rethinking the South African War', Unisa, Pretoria, 1988, p. 17.

the process of fusion.[62] Even more fulsome in praise of the Boers was the creator of Sherlock Holmes, Arthur Conan Doyle, who had served as a physician during the South African War. Conan Doyle described the Boers as 'one of the most rugged, virile, unconquerable races ever seen upon earth', emphasizing too their deeply felt Protestant faith and the simple nobility of their pre-modern ways.[63] Froude would have heartily agreed.

If such reappraisals of the underlying racial affinities between Boers and Britons were intended to assist in the goal of securing the future of a united 'white man's country', this objective entailed a corresponding stress on the need for more systematic separation of whites and blacks. As Lionel Curtis confided to his fellow Kindergarten colleague Patrick Duncan, in respect of closer union: 'The fact is we have all been moving steadily from the Cape idea of mixing up white brown and black and developing the different grades of colour strictly on the lines of European civilisation, to the very opposite conception of encouraging as far as possible the black man to separate from the white and to develop a civilisation, as he is beginning to do in Basutoland, on his own lines.'[64]

The term applied to this strategy was 'segregation'. As a policy it was increasingly adumbrated by self-appointed 'experts' on the 'native question' from the early years of the twentieth century. Segregation received close attention in the semi-secret counsels of the Fortnightly Club, in the South African Native Affairs Commission (1903–5), as well as in newspapers and journals. The 'Asiatic menace' was another preoccupation. Several, though not all, of these early 'experts' were English-speaking members of the Milner Kindergarten, or their associates.[65]

The trend to pronounce on the 'native question' with greater technical precision and scientific detachment spurred interest in the newly developing field of social anthropology, the first example of modern social science in South Africa. Although as yet unprofessionalized and lacking in disciplinary specificity, anthropologically minded investigations into the condition of Africans and the future relationship between whites and blacks were rapidly gaining currency. The transition from nineteenth-century travellers' 'manners and customs' depictions of African societies, towards an altogether tougher and more prescriptive treatment of the native

[62] *A Review of the Present Mutual Relations of the British South African Colonies. Memorandum prepared by the Earl of Selborne at the request of the Government of Cape Colony, 1 January, 1907.* Repr. in A. P. Newton, *Select Documents Relating to the Unification of South Africa*, vol. 2 (1968), 54–5. Recall Schreiner's comment that the bond which unites all South Africans is '*our mixture of races itself*'. See also O. Schreiner, *Closer Union* (London, 1909): 'The white race consists mainly of two varieties, of rather mixed European descent, but both largely Teutonic, and though partly divided at the present moment by traditions and the use of two forms of speech, the Taal and the English, they are so essentially one in blood and character that within two generations they will be inextricably blended by inter-marriage and common interests, as would, indeed, long ago have been the case had it not been for external interference' (p. 42).

[63] A. Conan Doyle, *The Great Boer War* (London, 1900), 1–2.

[64] Patrick Duncan papers BC 294 C23.3.8, University of Cape Town, Manuscripts, and Archives, Curtis to Duncan, 26 Nov. 1907.

[65] For details see my *Racial Segregation and the Origins of Apartheid in South Africa.*

problem, can be seen in Dudley Kidd's trilogy: *The Essential Kafir* (1904), *Savage Childhood* (1906), and *Kafir Socialism and the Dawn of Individualism* (1908). Pitched for a popular as well as a more specialist audience, these books drew on familiar stereotypes as well as fashionable new eugenic theories and helped to prepare the ground for segregationist policies.[66]

Dudley Kidd laid considerable emphasis on the supposed irrational nature of African 'mentality', highlighting belief in magic and sorcery, in order to argue that Africans were inherently unsuited to the spirit of democratic individualism that defined western civilization. The implication of such thought was that Africans, albeit suitable *subjects* for scientific investigation, were innately different from whites and therefore incapable of living on equal terms with whites in a rational scientific world. As the discourse of segregation took shape, the association of Africans with pre-scientific and non-progressive ways of thinking was to prove as important in asserting their ineligibility for common citizenship as was spurious evidence purporting to demonstrate directly their alleged intellectual inferiority. There is surely no better example of how the universalizing vision of science could be put to use for highly particularist political purposes.

TECHNOLOGY AND PROGRESS

Just as science could be deployed to emphasize intrinsic differences between blacks and whites, so it could be used to incorporate willing Afrikaners within a prosperous imperial community. Civilization, progress, and improvement were deeply established and still-powerful Victorian tropes, which built variously on the language of social evolutionism, racial superiority, and moral obligation. The secularization of the European idea of progress during the nineteenth century is a process requiring more careful periodization and explanation than it has so far received.[67] In South Africa, the attractiveness of the civilizing mission, especially in its moralistic and humanitarian guise, was waning. For Afrikaners who had just suffered the war, it spelt rank hypocrisy, and for many English-speakers it was too closely associated with the meddling or misguided idealism of mid-nineteenth-century evangelical liberalism. An inviting alternative was to neutralize the idea of civilization by stripping it of its overt moralisms so as to reposition it more centrally in the idiom of scientific and technical proficiency. In the view of many anglophone modernizers, technical competence was proof of the transformative and progressive capacity of a reawakened British imperialism/civilization. Equally importantly, it represented the politically acceptable face of state intervention.

These technocratic and statist tendencies were amply evident in the case of agriculture. Karen Brown has shown how the ideology of progressivism took hold

[66] Kidd's usage of photographic plates—many of them of fine quality—must have added to his books' appeal.

[67] R. Nisbet's *History of the Idea of Progress* (London, 1980) is only a start.

at the Cape at the turn of the twentieth century. Associated in particular with anglophone agrarian interests, but drawing on substantial support from improving African and Afrikaner farmers as well, progressives advocated a modern, scientifically based approach to all aspects of rural development, environmental reform, and conservation. The formation of an Agriculture Department in 1887, the launch of the official *Agricultural Journal of the Cape of Good Hope*, as well as innovations in forestry control, wildlife management, and veterinary practices were amongst the progressivist movement's most important progenitors.[68]

During the reconstruction era the Cape model of scientific agronomy spread northwards, where it found enthusiastic bipartisan support amongst capitalist farmers in the old Boer republics. Indeed, agrarian reform was the one dimension of Milner's work that attracted fulsome praise from Louis Botha, the first post-Union prime minister. It was, as Jeremy Krikler has pointed out, a point of convergence capable of drawing English- and Afrikaans-speaking landowners together in a class alliance.[69] The creation in the Transvaal of a Department of Agriculture after 1902, under the direction of F. B. Smith, attracted the support of commercial farmers. It provided an important stimulus to the adoption of modern scientific methods and, crucially, opened up career paths for ambitious young Afrikaner agronomists.

One such champion was H. S. du Toit, a 'bitterender' war hero, raised in the rural Cape midlands, whose sense of displacement after the British victory led him to travel widely in Europe and the United States, until his eventual return in 1908. Du Toit's political outlook was shaped by an anti-imperial amalgam of Hertzogite nationalism at home and Rooseveltian progressivism in the United States (topped off by an inspiring meeting with Tolstoy in Russia). As a senior official in the agriculture department du Toit became an energetic champion of scientific irrigation and soil management, a message he targeted especially towards Afrikaner farmers. In doing so he was able to counter entrenched nineteenth-century stereotypes of the anti-progressive 'backveld' Boer so as to demonstrate that Afrikaners were fully committed to modernization and progress.[70]

[68] K. Brown, 'Progressivism, Agriculture and Conservation in the Cape Colony circa 1902–1908', D.Phil. thesis, Oxford University (2002). Brown points out that although the ideology of progressivism was closely associated with the Progressive Party, it enjoyed far broader support.

[69] D. Denoon, *A Grand Illusion: The Failure of Imperial Policy in the Transvaal Colony During the Period of Reconstruction 1900–1905* (London, 1973), 68–9; J. Krikler, *Revolution from Above, Rebellion from Below: The Agrarian Transvaal at the Turn of the Century* (Oxford, 1993), 66, 76 ff. Krikler's work reminds us that land had to be reappropriated from Africans who had made use of a wartime power vacuum to re-establish themselves in white farming areas of the Transvaal. The need to reassert white authority (the 'revolution from above') helped to bridge Anglo-Afrikaner animosities and fed into the pressure for rural segregation.

[70] For a full discussion of Du Toit, see W. Beinart, *The Rise of Conservation in South Africa: Settlers, Livestock, and the Environment 1770–1950* (Oxford, 2003), ch. 7. On the tradition of anti-modernizing Afrikaner agriculture see D. Nell, '"For the public benefit": Livestock Statistics and Expertise in the Late-Nineteenth Century Cape Colony, 1850–1900', in Dubow (ed.), *Science and Society in Southern Africa*; M. Tamarkin, 'Flock and Volk: Ecology, Culture, Identity and Politics Among Cape Afrikaner Stock Farmers in the Late Nineteenth Century', paper presented to African Environments conference, Oxford, 1999.

Veterinary science was another area in which close cooperation between science and state was evident. Inter-colonial initiatives designed to contain the spread of animal disease and depredations by locusts (an Inter-Colonial Agricultural Union was formed in 1905) helped to make the case for 'closer union'. Immediately following the grant of self-government to the Transvaal, and on the invitation of Botha, now prime minister of the Transvaal, Dr Arnold Theiler, a Swiss-born zoologist and bacteriologist, created a new and well-funded research institute at Onderstepoort near Pretoria. After Union, Onderstepoort came to coordinate veterinary research work in South Africa, in the process absorbing the existing laboratories in Grahamstown and Natal as satellites.[71] A young cohort of South Africans was dispatched to Europe to study, but Theiler was convinced that South African solutions had to be found for its own unique conditions. In the post-Union era Onderstepoort provided a paradigm example of the ways in which localized scientific knowledge contributed to a sense of growing national self-assurance.[72] It was to become the locus of some of the country's most important scientific achievements.

The successes of Onderstepoort in applied agricultural and veterinary research (most notably the containment of rinderpest and East Coast Fever), together with the creation of a centralized Department of Agriculture in 1911, encouraged greater emphasis on specialized research and training. This impetus was maintained by the country's growing network of agricultural colleges, including those at Elsenburg (1898) and Grootfontein (1911) in the Cape; Potchefstroom (1909) in the Transvaal; Cedara (1906) in Natal; and Glen (1919) in the Orange Free State. Training provided by these colleges was supplemented by the creation of faculties of agriculture at what were to become the universities of Stellenbosch and Pretoria.[73] In ideological terms the demonstrable benefits of applied research work in agriculture and veterinary science made a significant contribution to the modernizing narrative of man's improving technical mastery over nature and, by extension, white South Africans' special scientific knowledge of, and control over, a hostile African environment.

Railway policy was another sphere in which ideological and political concerns came to be sublimated within a discourse of technical rationality. Ever since the 1880s disputes relating to the completion of railway lines and haggling over railway rates and customs agreements had been amongst the most divisive issues in inter-colonial politics. Yet, railway politics also offered a powerful incentive to overcome obstacles to unity: it was the desire to resolve deadlock on the issues of customs and railways that helped to persuade senior statesmen like John X. Merriman that, politically, 'we must either unite or break'.[74]

[71] B. C. Jansen, 'The Growth of Veterinary Research in South Africa', in Brown (ed.) *A History of Scientific Endeavour*, 166–70; F. Stark (ed.), *Pretoria: 100 Years* (Pretoria, 1955), 161–5.

[72] K. Brown, 'Tropical Medicine and Animal Diseases: Onderstepoort and the Development of Veterinary Science in South Africa 1908–1950', *JSAS* 31: 3 (2005).

[73] D. M. Joubert, 'Agricultural Research in South Africa: An Historical Overview', in Brown (ed.), *A History of Scientific Endeavour*, 266–8.

[74] Thompson, *Unification of South Africa*, 94.

Just as in the United States or Australia, railways were a powerful icon of the modernizing South African national state. During the South African War troop trains had been prized targets for Boer commandos, but in peacetime the railways came to serve as a visible demonstration of the material benefits of progress (not least in the service of agriculture). As we have seen, the logistics of railway travel were afforded great importance during the visit of the British Association: for instance, care was taken to ensure that seating on the trains used to transport the scientific visitors permitted local hosts to assemble on the platform in the order in which their guests would alight. The prestige that attached to railway engineering was such, that to mark the conclusion of its visit Professor George Darwin, mathematician son of Charles and president of the British Association, was honoured by an invitation to open the Victoria Falls railway bridge constructed by Rhodes's British South Africa Company for passenger traffic. Stoked by enthusiasm, Darwin delivered the predictable panegyric to imperial achievement and the wonders of steam.[75]

In the reconstructionist era, members of Milner's 'Kindergarten' devoted considerable time to the study of railway and transportation systems in a future Union of South Africa. Mindful of the virtues of centralized planning, and also of the need to foster political cooperation in a quiet, behind-the-scenes manner, they contributed essential documentary evidence on these topics in a supplement to the 1907 Selborne Memorandum and also in *The Government of South Africa* (1908).[76] Philip Kerr, who worked on the railways submission for the Selborne Memorandum, sat on the influential Inter-colonial Council, and subsequently served interests of closer union as a literary and political journalist, was also keenly attuned to the aesthetic and ideological significance of railways.

It was from the perspective of a train traversing the Karoo in 1905 that this key member of the Kindergarten first became aware of the singular beauty of the dry South African landscape and of the potential of this new, young country. Kerr's initial impression of the Karoo 'as "a most ghastly place, looking like 'the morning after Creation . . . just Nature all red and naked' " changed in the cool of the evening. "Everything was still and the line wound away like a snake in the distance, unfenced and unembanked, a light streak in a barren land. The sunset was a novelty to my Northern eyes . . . ." '[77] Afrikaners who had endured the devastation of the war may have viewed the landscape rather more ruefully. In his classic poem 'Dis Al' (That's All), Jan F. E. Cilliers evoked the pitiful spectacle of a bird circling in solitary flight over the veld, a grave in the grass, and a falling tear.[78]

For Kerr, by contrast, the sinews of steel track, along with railway cuttings, gradients, and curves, gave form to a sometimes featureless physical terrain and helped to invest the 'empty' interior with meaning and purpose. Jeremy Foster has

75 Dubow, 'A Commonwealth of Science', 67–8; 72.

76 Dubow, 'The Milner Kindergarten'.

77 J. R. M. Butler, *Lord Lothian (Philip Kerr) 1882–1940* (London, 1960), 12, 22. See also pp. 19–20.

78 My thanks to Jeremy Krikler for reminding me of this poem.

demonstrated the significance of railways in reshaping ideas and images of national space and identity in the first decades of the century. His work highlights the singular role played by railways in promoting tourism and refiguring the country within recognizably pastoral conventions of European landscape aesthetics. Trains, proceeding at a sedate but steady pace, 'made it possible to grasp, for the first time, South Africa as a coherent and characterful whole'. They also facilitated the characterization of different landscape types and regions. The railways' highly active publicity department, and the use which was made of photography and journalism in its highly popular *Magazine*, helped to promote an array of visual and cultural motifs which served to tie nature to nationhood. Considerable care was taken to emphasize the railways' role as an agency for the common good: established within the terms of the Act of Union as a state-owned enterprise, the South African Railways effectively operated semi-autonomously as a 'government within a government', in a manner that was avowedly 'above party politics'.[79]

If pride in technical achievements could enhance aesthetic appreciation of the landscape, scientific knowledge contributed to the esteem in which the natural environment was held. This association is apparent in the growing awareness of South Africa's floral heritage and, especially, in the world of botany which entered a new phase of professionalization in different parts of the country during the 1880s and 1890s. Locally based researchers such as Harry Bolus, Peter MacOwan, and Rudolph Marloth at the Cape, John Medley Wood in Natal, and Selmar Schonland in the eastern Cape, now worked to broaden the scope of botanical science, in particular by building up and enhancing plant collections and herbaria.[80]

The decision to replace the old Company Garden in the centre of Cape Town with a botanical garden designed to encourage recreation and research, as well as the cultivation of exotic plants from all over the world, proved unsuccessful.[81] When the noted Indian botanist J. S. Gamble visited the Cape in 1890, he commented unfavourably on the unkempt appearance of the Botanical Gardens, and expressed disappointment at the absence of indigenous South African vegetation: '. . . the Gardens has not even a single silver-tree to show a stranger, and the heaths, and indeed all flowering plants, were conspicuous by their absence.'[82]

[79] J. Foster, '"Land of Contrasts" or "Home we have always known"?: The SAR&H and the Imaginary Geography of White South African Nationhood, 1909–1930', *JSAS* 29: 3 (2003), 665 and *passim*; also id., 'The Poetics of Liminal Places: Landscape and the Construction of White Identity in Early 20th Century South Africa', D.Phil. thesis, University of London (1998), 242–62. Like all activity claiming to be above party politics—notably the 'native question'—this was a dubious claim. The Union railway network was designed to serve the overriding interests of a newly unified country in which the needs of white farmers were given high priority. The requirements of black peasants were sacrificed, as they were in the Union compromise on the franchise question: railheads which extended towards areas of high-density African population were built to transport men to the mines, not to support agricultural production in the reserves.

[80] E. Percy Phillips, 'A Brief Historical Sketch of the Development of Botanical Science in South Africa and the Contribution of South Africa to Botany', *SAJS* 27 (1930).

[81] R. H. Compton, *Kirstenbosch: Garden for a Nation* (Cape Town, 1965), 34.

[82] Cited in Pearson, 'A National Botanic Garden', 40.

In subsequent years, amid increasing concern that indigenous flora were under threat, there were intermittent calls for the creation of a botanical garden dedicated to the cultivation of indigenous flora. Such demands were most eloquently expressed by Harry Bolus (1834–1911) who, along with Marloth, was probably the outstanding figure amongst the country's late nineteenth-century resident botanists. Bolus did a great deal to establish botany as a professional science in South Africa: he endowed the first South African chair of botany at the South African College in 1903, and became a leading proponent of the need to preserve and appreciate the ancient and unique floral kingdom of the Cape, which was feared to be 'doomed to extinction'.[83]

Writing in *The State* in 1909, Bolus couched his preservationist appeal in terms of the awakening of 'a higher patriotism in South Africa'. He feared that Table Mountain was being denuded of heaths, orchids, and anemones by flower-sellers and casual pickers. It was ironical that while money was being spent on attractions such as museums, exhibitions, public halls, and musical entertainment, 'we sit quietly and allow one of our finest natural attractions to be ruthlessly destroyed'. Citing the examples of the Yellowstone Park in the United States and the National Trust in England, Bolus duly called for the creation of a special reserve in order to protect the native flora of South Africa.[84]

This plea was reinforced by Harold Pearson, who arrived from Kew Gardens in 1903 to take up the new Bolus Chair of Botany at the South African College. Along with Neville Pillans, Pearson was the prime figure in the move to establish the Kirstenbosch Botanical Gardens. He couched his appeal for a state botanical garden in the familiar language of utilitarian objectives (commercial and economic possibilities, medical and agricultural research), while stressing as well the idea that the pursuit of 'science for its own sake' was a mark of civilized society. Tying this ideal to the quest for national identity, Pearson portrayed the natural world as uniquely expressive of the nation, arguing that there was 'nothing more truly "South African" than its indigenous flora'.[85] Thus:

> The South African Botanic Garden cannot be merely an economic undertaking; it must also be an expression of the intellectual and artistic aspirations of the New Nation whose duty it is to foster the study of the country which it occupies, to encourage a proper appreciation of the rare and beautiful with which Nature has so lavishly endowed it.[86]

The creation of Kirstenbosch as a National Botanic Garden (encouraged by the S2A3 and aided by a small grant from the Botha government) gave expression to

[83] Bolus cited in L. van Sittert, 'From "Mere Weeds" and "Bosjes" to a Cape Floral Kingdom: The Reimagining of Indigenous Flora at the Cape, c. 1890–1939, *Kronos*, 28 (2002), 108.

[84] H. Bolus, 'The Native Flora of South Africa and its Preservation', *The State*, 2: 7 (1909), 105–6. See also 'The Proposed National Botanic Garden', *Report of the Eighth Annual Report of the South African Association for the Advancement of Science. Cape Town 1910* (Cape Town, 1911), 421–3, for a posthumous letter by Bolus to Harold Pearson on this subject.

[85] H. H. W. Pearson, 'A State Botanical Garden', *The State*, 5: 5 (1911).

[86] H. H. W. Pearson, 'A National Botanic Garden', *Report of the Eighth Annual Report... of the South African Association*, 54.

this sentiment and reflected the brief blaze of optimism that characterized the immediate post-Union period. From the outset, the cultivation, display, and study of the indigenous flora of South Africa were the *raison d'être* of Kirstenbosch. In this sense it was especially distinctive because, unlike botanical gardens elsewhere in the Empire which were typically devoted to the collection and study of foreign flora, Kirstenbosch was additionally intended 'to be a Garden for the study of the flora of South Africa itself—the country in which it was situated'.[87]

The establishment of Kirstenbosch can thus be understood as the culmination of a process whereby the Cape flora of South Africa, which had previously attracted the enthusiastic attentions of botanists and collectors as 'marvels' or 'curiosities', now became truly indigenized and came to be seen as valuable in their own right.[88] As a garden dedicated to the cultivation and display of the unique and the local, rather than of exotic botanical specimens, Kirstenbosch proudly proclaimed South African nationhood. Yet it also expressed a more restricted and specific sense of Cape-ness. In addition to being a showcase for Cape flora, its creation testified to the involvement of leading Cape dignitaries and institutions, and it was located on the verdant side of Table Mountain, bordering the most salubrious English-speaking suburbs of the city.[89] The adoption in 1895 of an indigenous orchid, the *Disa grandiflora*, as the emblem of the elite anglophile Mountain Club (est. 1891) was a further illustration of an emerging confluence which was to bring localized Cape loyalism into contact with a broader stream of South Africanism.[90]

LANDSCAPE AND BELONGING: NORTH AND SOUTH

Perhaps no one exemplified the Cape tradition of anglophone South Africanism more clearly than the writer and historian Dorothea Fairbridge, a leading proponent of the need to fuse imperial loyalism with local patriotism. Fairbridge, who was a

[87] Compton, *Kirstenbosch*, 33, 50. [88] Ibid. 31.

[89] Although Kirstenbosch was founded as a National Botanic Garden, this status was only uneasily maintained. Cape dignitaries provided a great deal of impetus for its creation and it was more closely linked for research purposes to the University of Cape Town than to a government department (like that of agriculture). At a meeting on 8 March 1912 attended by many Cape notables it was resolved to establish 'a National Botanic Garden within the Cape Peninsula'. In 1923 a new herbarium was established under the aegis of the department of agriculture at Pretoria as an embryo 'Kew'. This initiative was regarded as a snub to Kirstenbosch. Local Cape opinion was indignant at what it saw as regional competition and an unwelcome instance of state centralization. See Compton, *Kirstenbosch*, 43, 80–1; also D. P. and E. McCracken, *The Way to Kirstenbosch* (Cape Town, 1988), chs. 11, 12. Van Sittert, 'From "Mere weeds"', 114–15, makes an even stronger case for Cape botany's insertion into the 'soil of urban, English Cape Town' and its close links to the imperial botanical establishment.

[90] L. Van Sittert, 'The Bourgeois Eye Aloft: Table Mountain in the Anglo Urban Middle Class Imagination, c. 1891–1950', *Kronos*, 29 (2003). My thanks to Lance van Sittert for an early sight of this paper and for providing me with the date of the Disa's adoption as an emblem by the Mountain Club.

close friend and admirer of Milner, placed herself at the centre of an interlocking group of Edwardian artists and cultural enthusiasts resident at the Cape. Amongst her many activities, she was a founder member of the Guild of Loyal Women (which assumed responsibility for looking after war graves), a devoted founder of the National Society for the Preservation of Objects of Historical Interest and Natural Beauty (1905), and a member of the Botanical Society of South Africa (est. 1913), which was instrumental in the management of Kirstenbosch. She was also a founding trustee of the Michaelis collection of art. As the daughter of Charles Aken Fairbridge (1824–93), the noted lawyer, parliamentarian, bibliophile, and trustee of the South African Museum and South African Library, Dorothea was deeply imbued with the refined values of Cape colonial nationalism.[91]

Dorothea Fairbridge was particularly drawn to the idea of the south-western Cape as an old, civilized society relocated in a new and dynamic country. Her evocation of the southern tip of Africa as a cultivated offshoot of Europe recalled the cultural politics of the *Cape Monthly Magazine*, and allowed her to elaborate a vision of Anglo-Dutch cooperation rooted in a deep sense of colonial history. This was precisely the form of South Africanism that gentrified Afrikaners like J. H. Meiring Beck, in his capacity as president of the National Society for the Preservation of Objects of Historical Interest, could readily encompass. As well as writing the South African College anthem and 'The National Song for South Africans' (performed at the Union Pageant), Beck composed a patriotic ode to 'Table Mountain'.[92]

Dorothea Fairbridge's selective historical appropriations from the Dutch colonial period were deployed to link the colonial aspirations of Van Riebeeck and Willem Adriaan van der Stel to those of Rhodes and Milner. Her fascination with the 'Cape Malays' and the 'Bushmen' allowed her to represent the Cape as an exotic locale that, through myth, climate, romantic association, and landscape was more closely linked to northern than sub-tropical Africa. The Cape's 'Mediterranean' affinities and the 'Cape to Cairo' ideal, both championed by Rhodes, gave form to these conceptions.[93]

In the Transvaal a complementary, if grittier, sense of English South African identity took shape, with variants in the pastoral regions of the Cape midlands where moves to rehabilitate the countryside and reappraise the malign reputations of avian and faunal predators were important dimensions of Cape

[91] P. Merrington, 'Carrying the Torch: Dorothea Fairbridge and the Cape Loyalist Imagination', paper presented to' Once there was a War: Conference on the South African War in Fiction', Unisa, Pretoria, 2002.

[92] Scully, *Sir J. H. Meiring Beck*, 65.

[93] A typical statement of Fairbridge's views is contained in her mock travelogue *Along Cape Roads* (Cape Town, 1928), in which she seeks to introduce the distinctive indigenous qualities and scenic beauties of the Cape to an English audience. Fairbridge's underlying message of the expansive potential implicit in a hybrid English identity is delivered by the fictional 'Bryan', a South African-born Englishman, who avoids having to choose one country over the other by saying: 'It isn't a question of choosing between the two but of enlarging one's inheritance. It is all a matter of realising that the whole is larger than the half' (p. 75).

progressivism.[94] Rooted in the idea of an African interior, and drawing on the romantic qualities of a conquered, if not entirely tamed, frontier, this rural awareness was pre-eminently focused on what the young writer Sarah Gertrude Millin referred to as the 'cult of the veld, a real or fancied delight in the vast desolateness of it all'.[95] Millin's observation is borne out by the increasing frequency with which the word was used after the South African War in the titles of English novels, travelogues, stories, and volumes of verse. Fascination with the veld as a signifier of authentic locality and symbol of youthful nationhood (comparable in some respects to the Australian bush or the Canadian prairies) suggests that the word came to connote the idea of 'open country' in a figurative as well as a literal sense.[96]

The iconic anglophone expression of the emerging landscape-based South Africanist ideology was Percy FitzPatrick's classic novel *Jock of the Bushveld* (1907), the publication of which seems to have coincided with the author's metamorphosis 'from the hammer of the Boers to the passionate advocate of moderation'.[97] Set in the heyday of the transport-riding and hunting era of the eastern Transvaal, populated by a cast of rugged folk characters, and spiced with idiomatic Afrikaans expressions, the eponymous hero of the book is an otherwise nondescript bull-terrier (the runt of the litter) whose pluck and determination neatly captured the ethos of a closing frontier. The meaning of *Jock* was aptly captured in a contemporary review which welcomed it as a rare and significant contribution to South African national literature: 'It breathes the very soul of the country; it is the spontaneous outpouring of a heart full of sympathy with those whom the "wanderlust" drives forth into the Silent Places, the haunts of untamed beasts, the fascinating regions of the mysterious veld.'[98]

Just as it had previously inspired nineteenth-century progressives in the Cape, so scientific agriculture had considerable appeal for anglophone South Africanists like FitzPatrick. Spurred by Rhodes's interest in the possibilities of scientific

[94] For the eastern Cape dimension, see Beinart, *The Rise of Conservation*, and Brown, 'Progressivism, Agriculture and Conservation',

[95] S. G. Liebson, 'The South Africa of Fiction', *The State* (Feb. 1912), 135.

[96] Further research on this topic is needed. A selective sample of such literature would include H. A. Bryden, *From Veld Camp Fires: Stories of Southern Africa* (London, 1900); J. H. M. Abbott, *Plain and Veld: Being Studies, Stories and Sketches of My People, in Peace and War* (London, 1903); H. R. Haggard, 'Lost On the Veld', *Windsor Magazine* [1902–3]; W. C. Scully, *By Veldt and Kopje* (London, 1907); F. Dawson, *Fact and Fancy from the Veld* (Cape Town, 1909); K. Fairbridge, *Veld Verse and Other Lines* (London, 1909); L. Lyster, *Ballads of the Veld-Land* (London, 1913); M. Byron, *A Voice from the Veld* (London, 1913); C. Mansfield, *Gloria: A Girl of the South African Veld* (London, 1915). Having firmly established itself as an anglophone trope for South Africa during the war of 1899–1902, it became almost a clichéd term by the 1930s.

[97] Editorial introduction by Deborah Lavin to *FitzPatrick's South African Memories: Scraps of History* (Johannesburg, 1979), 15.

[98] *African Monthly*, 3: 15 (1908), 332. On the elaboration of these tropes in the case of anglophone projections of South African identity in the low-veld, see D. Bunn, 'An Unnatural State: Tourism, Water and Wildlife Photography in the Early Kruger National Park', in W. Beinart and J. McGregor (eds.), *Social History and African Environments* (Oxford and Ohio, 2003).

irrigation methods, and ever mindful of Milner's hope that South Africa could be anglicized by settling English-speaking farmers on the land, FitzPatrick sought not only to extol the mysteries of the veld but also to transform it into a productive resource. In 1903 he pronounced: 'I believe in the British Empire . . . I believe in this my native land.' Now he set out to prove the point—literally—by devoting the last part of his life to a grandly ambitious British land-settlement scheme in the eastern Cape.

This involved large-scale irrigation of the dry lands surrounding the Sundays River for the purposes of citrus cultivation, which led FitzPatrick to study and acquire new technology in Australia and the United States. The financially disastrous scheme marked the culmination of a varied career that included FitzPatrick's youthful experience as a trader and transport rider in the goldfields of the eastern Transvaal, close involvement in the Jameson conspiracy, a spell as president of the Chamber of Mines (1902–3), followed by his participation as a deputy-leader of the Transvaal Progressives and a prominent role in the Union convention as a standard-bearer of 'equal rights' for English-speakers. FitzPatrick subsequently campaigned with passion to establish the Delville Wood memorial in honour of South Africa's participation in the Great War.[99]

That this romantic and charismatic doyen of South African Englishness invested so much energy and money in agricultural experimentation and modern irrigation techniques towards the end of his life exemplified his belief that British loyalism could be shown to be compatible with selfless South African patriotism. Although short on contrition for his involvement in imperialist escapades, FitzPatrick evidently considered that a demonstrable concern with the rehabilitation of the country would help to further the more acceptable forms of Rhodesian and Progressivist ideals. Fittingly, FitzPatrick's irrigation plans were only realized with the help of former wartime enemies: by the early 1920s the troubled Sundays River scheme (having stretched the considerable resources of the Rhodes Trust in England) finally gained the support of Smuts and Deneys Reitz, the minister of lands, who recognized FitzPatrick's project as a 'cornerstone of development' and conveniently arranged for a government bail-out.[100]

James Stevenson-Hamilton, who was effectively the founder of that exemplar of white South African national identity, the Kruger National Park, was also a recipient of Deneys Reitz's ministerial patronage. Albeit less politically minded than FitzPatrick, and more reticent about revealing his private thoughts, Stevenson-Hamilton wrestled with similar questions to those exercising the creator of *Jock*. It was while sleeping out in the open in 1899 that he experienced a moment of epiphany that led him to record his appreciation of the 'immensity', 'loneliness', and 'grandeur' of the veld. At the end of the South African War,

[99] J. P. R. Wallis, *Fitz: The Story of Sir Percy FitzPatrick* (London, 1955); Bill Nasson, 'Delville Wood and South African Great War Commemoration', *English Historical Review,* 119: 480 (2004).

[100] Deborah Lavin, 'The Price Of Fidelity: Smartt, FitzPatrick And Agricultural Settlement', paper delivered to Rhodes Conference, Oxford, 2002.

during which he was able to hone his bushcraft skills, Stevenson-Hamilton was given a military posting as game warden of the barely developed Sabi Game Reserve. Stevenson-Hamilton was a staunch imperialist, whose attitudes were strongly conditioned by his backround as a Lanarkshire aristocrat and his experience as a military and hunting man. Yet, like so many Scots before him, he quickly developed a passion for his adopted country. By 1910 he was thoroughly committed to the preservation of South African fauna and began to think of himself at least in part as South African. His enormously successful book *Animal Life in Africa* (1912) formalized his evolving ideas and brought them to public notice; it is regarded by his biographer, Jane Carruthers, as marking a 'milestone in the development of protectionist philosophy'.[101]

John Buchan was another lowland Scot (as well as a key Kindergarten associate) who sought in his writings and memoirs to soften the image of Milnerism by evoking a landscape-based British colonial identity.[102] Buchan arrived in South Africa in 1901 at Milner's personal invitation, and soon found himself in charge of the Land Settlement Board, a position that provided him with a unique opportunity to explore the Transvaal countryside. The depth of Buchan's engagement in the South African natural environment cannot be explained by his official duties, for he genuinely thrilled to the possibilities of the new colonial frontier and its invigorating potential for the Empire as a whole. His well-developed Scottish identity also played a role. Like fellow lowland borderers, going all the way back to Thomas Pringle in the 1820s, Buchan's immersion in the works of Wordsworth and Walter Scott, together with his awareness that Scotland was itself a colonized country, provided ready comparisons between rural South Africa and his native country. It also facilitated Buchan's exploration of the nature of multiple, overlapping national identities and loyalties.[103]

The landscape of the Transvaal provided Buchan with a repertoire of metaphors which caused him to reflect on the confrontation between what he saw as white civilization and black barbarism and, within this oppositional dynamic, to project the idea of a unified white South African nation. The mysterious and romantic veld in all its different manifestations (bush, high and low) contained attributes which he saw as key to a nation-building project: vigour, sport, patriotism, and indigenous belonging. It was on the veld that men of British stock could join together with the hardy and patriarchal Calvinist Boers who so reminded Buchan

[101] J. Carruthers, *Wildlife and Warfare: The Life of James Stevenson-Hamilton* (Pietermaritzburg, 2001), 50, 94, 103, 115–16. *Animal Life in South Africa* carried a foreword by Theodore Roosevelt.

[102] In his autobiography, *Memory Hold-the-Door* (London, 1940), 109, Buchan wrote: 'The hope of breaking down the racial barrier between town and country was always very near to Milner's heart. He wanted to see the Dutch share in the urban industries, and men of British stock farming beside the Boers on the veld.'

[103] See e.g. D. J. Shaw, 'The Writings of Thomas Pringle', Ph.D thesis, Cambridge University (1996); R. Grove, 'Scotland in South Africa: John Croumbie Brown and the Roots of Settler Environmentalism', in T. Griffiths and L. Robin (eds.), *Ecology and Empire: Environmental History of Settler Societies* (Edinburgh, 1997).

of the Presbyterian lowland Scots amongst whom he had grown up. Indulging in the fashionable anti-industrial rural nostalgia which he shared with many of his peers, Buchan contrasted life in the mining centres ('cosmopolitan, money-making, living at a strained pitch') with 'this silent country' that offered so much in the way of 'a leisured country life' and which would serve as the ideal repository for the growth of a patriotism rooted in the soil:

> For those who are truly South Africans at heart, and do not hurry to Europe to spend their wealth, there is a future, we may believe of another kind than they contemplate. All great institutions are rooted and grounded in the soil. There is an art, a literature, a school of thought implicit here for the understanding heart,—no tarnished European importation, but the natural, spontaneous, fruit of the land.[104]

Buchan's reference to genuine South African institutions 'rooted' in the soil must surely be seen as a response to the allegation, frequently levelled, that the randlords or 'goldbugs' (brutally satirized as 'Hoggenheimer' by the cartoonist Boonzaier[105]) extracted huge wealth from the mines but failed to give anything back to the country. Both in Britain and in South Africa there were frequent sneering comments directed to those *nouveau riche* magnates who decamped to Europe with their new fortunes when war broke out, and then tried to enter into the ranks of the English gentry by purchasing country estates and accumulating art collections.[106] In the era of reconciliation, a handful of public-spirited magnates felt impelled to reinvest some of their wealth back in the country in which it was made, in an effort to preserve the imperial connection in the new South Africa. Abe Bailey, for instance, announced in 1908 that 'The true El Dorado of South Africa is not in the Rand or Johannesburg, but in the soil of the country', and proceeded to buy up farms in the Colesberg district of the Cape.[107]

The randlord couple Lionel and Florence Phillips provide an outstanding example of this trend. Sir Lionel personified the image of the imperialist mining plutocrat, owing to his involvement in the Jameson plot, his prominent position in the Corner House mining group, and his role as head of the Chamber of Mines. In the post-Milner environment Phillips worked hard to offset this image by demonstrating his, and the mining industry's, commitment to the best interests of South Africa. Phillips developed a special interest in rural affairs, reckoning this to be an exemplary way of bringing English- and Afrikaans-speakers together and,

[104] J. Buchan, *The African Colony: Studies in the Reconstruction* (Edinburgh and London, 1903), 86. My understanding of these themes has been shaped by readings of Bill Schwarz, 'The Romance of the Veld', unpublished paper kindly sent to me by the author; J. Foster, 'The Poetics of Liminal Places', ch. 4; P. B. Rich, 'Milnerism and a Ripping Yarn: Transvaal Land Settlement and John Buchan's Novel *Prester John*, 1901–1910', in B. Bozzoli (ed.), *Town and Country in the Transvaal* (Johannesburg, 1983)'; P. Henshaw, 'John Buchan from the "Borders" to the "Berg": Nature, Empire and White South African Identity, 1901–1910', *African Studies*, 62: 1 (2003).

[105] M. Shain, *The Roots of Antisemitism in South Africa* (Johannesburg, 1994), ch. 3.

[106] Michael Stevenson, 'Old Masters and Aspirations: The Randlords, Art and South Africa', doctoral dissertation, University of Cape Town (1997), 12 ff.

[107] K. Brown, 'Progressivism, Agriculture and Conservation' 30.

equally importantly, a means of signalling his social achievements. The British aristocratic ideal of a residence in the city and a seat in the country provided Phillips with a ready model: in 1897–8 he purchased a substantial property in Grosvenor Square, London, and an estate in Hampshire, whose 'feudal traditions' he sought to reinstate. On his return to South Africa in 1906, Phillips reproduced this arrangement by having a grand Baker-designed mansion (which he called Villa Arcadia) built for him in Johannesburg's prime suburb of Parktown; he also bought a block of farms in the Woodbush region of the northern Transvaal, about which Buchan had written so lyrically, and thereafter devoted much time to reviving the Witwatersrand Agricultural Society.[108]

In 1908 Phillips became the first president of the South African Union, an organization dedicated to the showcasing of local artefacts and agricultural products. Its philosophy bore the strong imprint of the Arts and Crafts Movement that was so enthusiastically taken up by his overbearing wife, Florence, 'an ardent South African'[109] who also served on the South African Union's executive and headed its 'ladies' committee'. The Phillipses were both active in promoting Kirstenbosch as the country's national botanical gardens, and they commissioned Marloth's important six-volume *Flora of South Africa* (1913–32).[110] Their energetic interest in rural traditions, and their decision to retire to the historic Cape Dutch homestead Vergelegen, represented a further statement of their South African affinities and social ambitions. At great expense they embarked upon the restoration of Vergelegen, which was used to entertain fashionable representatives of Cape society and culture, Florence's mentor Dorothea Fairbridge amongst them.[111]

Other grandees who invested in old Cape Dutch homesteads included politicians and statesmen like John X. Merriman, J. H. de Villiers, J. H. Beck, and J. W. Sauer. Cape Dutch architecture was a key trope of Cape colonial identity and a visible reminder of the country's mixed European heritage. Its aesthetic appeal for the suburban as well as the landed elite was underlined by Herbert Baker, a protégé of Cecil Rhodes and close associate of the Kindergarten, whose domestic and official buildings did so much to define the architecture of 'new' imperialism in South Africa and India during the Edwardian era.[112] In his 1909 manifesto on the 'Architectural Needs of South Africa', Baker sought to create a distinctive and monumental architectural style which could be imperial and South African at one and the same time.

[108] M. Fraser and A. Jeeves, *All that Glittered: Selected Correspondence of Lionel Phillips 1890–1924* (Cape Town, 1977), 5, 209, n. 45; T. Gutsche, *No Ordinary Woman: The Life and Times of Florence Phillips* (Cape Town, 1966), 198–203. Also Lionel Phillips, *Some Reminiscences*, ed. M. Fraser (Johannesburg, 1986), ch. 7.

[109] Phillips, *Some Reminiscences*, 141.

[110] M. Arnold, *Art Meets Science: Flowers as Images*, Standard Bank National Arts (1992), 11.

[111] P. Merrington, 'Pageantry and Primitivism: Dorothea Fairbridge and the "Aesthetics of Union" ', *JSAS* 21 (1995) Stevenson, 'Old Masters and Aspirations', 168–9; Fairbridge, *Along Cape Roads*, 81 ff.

[112] T. R. Metcalf, *An Imperial Vision: Indian Architecture and Britain's Raj* (London and Boston, 1989), 181 ff.

Fig 7. Entrance to Johannesburg Art Gallery, designed by Edwin Lutyens, and opened in 1915 (Courtesy Neil McCarthy)

The virtues of Cape Dutch architecture lay in the fact that its 'grand manner' of building echoed and reiterated the defining nature of the southern African landscape itself: in Kipling's words, 'Great spaces washed with sun'. Baker was particularly impressed by the buildings of the early Dutch and Huguenot settlers who had '. . . handed down to us very noble examples of how to build South African homesteads'.[113] He was strongly influenced by Ruskinian aesthetics and recognized in the often disparaged work of 'rustic boers' a simplicity of design that 'corresponded exactly with the ideals of domestic building laid down by Morris and the crafts movement'.[114] Rhodes, as ever, was a forerunner in this regard. Not only was he Baker's first patron in South Africa, Rhodes also devoted

[113] H. Baker, 'The Architectural Needs of South Africa', *The State* (May 1909), 522. I am grateful to Donal Lowry for pointing out the Kipling reference.

[114] Metcalf, *An Imperial Vision*, 182. According to Olive Schreiner, *Closer Union*, 23: 'John Ruskin has said the only type of absolutely new and beautiful architecture evolved in the last two centuries has been the old type of Dutch farmhouse at the Cape.'

considerable resources to growing fruit and wine in the western Cape, and he engaged Baker to restore the Groote Schuur estate in Cape Town before turning it over to the nation.

Florence Phillips, who was also greatly enthused by the Arts and Crafts Movement, spearheaded the creation of the Lutyens-designed Johannesburg Art Gallery (1910–15) built to house a collection of modern British and European fine art selected by Hugh Lane.[115] As Michael Stevenson and Jillian Carman have shown, Lady Phillips's bold intervention in South African public culture can best be understood as a multiple effort to serve the cause of nation-building, to secure cultural respectability for Johannesburg and its English-speaking financial and civic elite, and to expedite the social and moral 'upliftment' of the politically volatile white working class.[116] It formed part of a wider strategy to reshape the jingoistic brand of anti-Krugerite 'British South Africanism' that had developed amongst the Transvaal 'uitlanders' in the lead-up to war, as well as the powerful brand of imperial 'white labourism' that took root after 1902, into a style of patriotism that was both more responsible and responsive to reconciliation.[117]

The desire to consolidate British influence and civic-minded values was broadly evident in the selection of artwork from overseas (works by local artists were conspicuous by their absence). Specially commissioned portraits of portentous randlords and imperial personages further underlined the sense of British predominance. The donation of a diminutive bronze of *Kruger in Europe* by the Afrikaner sculptor Anton van Wouw did little to counter this impression. Therefore, in a bid to rebalance the message of inclusive South Africanism, the portrait section was augmented by sculptures of leading Boer generals—de Wet, de la Rey, and Louis Botha. These were paid for by the ubiquitous entrepreneur Sammy Marks, a republican sympathizer and friend of Kruger.[118]

Having successfully solicited contributions from more or less willing mining magnates for the Johannesburg Gallery, the unstoppable Florence Phillips was instrumental, too, in persuading the randlord Max Michaelis to present a collection of Dutch and Flemish old masters in 1913 (also assembled by Lane) as the

[115] The Anglo-Irish connoisseur and critic Sir Hugh Lane had been responsible for assembling an impressive art collection for the Dublin Municipal Gallery which opened in 1908. He had a considerable influence on Florence Phillips (see Carman, next note), and was sensitized to rising colonial nationalist sentiment.

[116] This discussion, and the paragraphs which follow, are based on two excellent recent doctoral theses: Michael Stevenson, 'Old Masters and Aspirations', chs. 3 and 4, and Jillian Carman, ' "Modern Art for South Africa": The Founding of the Johannesburg Art Gallery', University of the Witwatersrand (2002).

[117] J. Hyslop, 'The Imperial Working Class Makes Itself "White": White Labourism in Britain, Australia and South Africa Before the First World War', *Journal of Historical Sociology*, 12: 4 (1999); also D. Denoon, *A Grand Illusion*, 5–6; A. A. Mawby, 'The Political Behaviour of the British Population of the Transvaal, 1902–1907', Ph.D thesis, Witwatersrand University (1969).

[118] R. Mendelsohn, *Sammy Marks: 'The Uncrowned King Of The Transvaal'* (Cape Town, 1991).

nucleus of the long-awaited South African National Gallery.[119] This plan did not turn out as envisaged. The collection was instead displayed in the eighteenth-century Burgher Watch House in Cape Town, which was now remodelled, with the help of Herbert Baker and his assistant J. M. Solomon, on the Mauritshuis in The Hague, and embellished with old Cape furniture. With prompting by Smuts, amongst others, the Michaelis collection was conceived of as a conscious reminder of the links that brought English- and Afrikaans-speakers together in the spirit of Union. The fact that many of the paintings were coterminous with the date of the first European settlement at the Cape was duly noted, and specious allusions were made to the religious and political affinities shared by the seventeenth-century Dutch and English nations. It was somehow fitting that modern art should have been selected for the industrial heartland of Johannesburg, while European old masters were located in the country's historic centre.

'UNITY AND DIVERSITY'?: NORTH AND SOUTH

In this discussion of the Phillipses and their interests, it is possible to see how gentrified concerns with agronomy, landscape, identity, and aesthetics segued into the cultural politics of Union. The process was by no means smooth. In respect of the Michaelis collection, for example, there was considerable debate as to whether it should be housed in the Cape or the Transvaal.[120] Such rivalry mirrored competition over where the capital of the country should be situated, and pointed to a growing divide between the western Cape and the Witwatersrand. It was a tension, fuelled by ethnic nationalism, that grew more intense as the Union's industrial heartland increasingly challenged the cultural authority of the 'mother city'. The conflict was played out in multiple, overlapping tensions between south and north, a distinction that was to become a convenient (if misleading) political shorthand for opposing political 'traditions'—liberal versus reactionary, or English versus Afrikaner, respectively. The developing north–south axis also had the effect of leaving other regions of the country feeling increasingly marginal—the eastern Cape and Natal in particular. In the case of the former, separatist hopes were long extinguished, but in Natal the possibility of secession was to be seriously, albeit unrealistically, entertained in the first half of the twentieth century.

Milner's move to base his post-war reconstructionist effort in Johannesburg, even before the conclusion of the war, was an early and clear signal that political power would henceforth be more closely aligned with the country's financial and industrial centre. The decision at Union to split the capital of South Africa between Cape Town and Pretoria signalled that only by acknowledging the

[119] Stevenson points out that Michaelis's generosity may have been motivated by the hint of a baronetcy. The South African National Gallery was established by legislation in 1895, though its origins go back to 1871 when the Fine Arts Association was founded (see ch. 2). The National Gallery only acquired a dedicated building in 1930.

[120] Stevenson, 'Old Masters', 242–3.

existence of competition between north and south could such tensions be contained. Recognition of provincial governing structures and capitals in the Union constitution, based on the boundaries of the former British colonies and Boer republics, also bore testimony to the fact that regionalism had to be accommodated within a centralizing nation-state. But regionalism was not easily surmounted. Provincial identity was often aroused and asserted all the more forcibly in reaction to pressures in favour of national unity—to the extent that many argued that a federation rather than unification would have been the preferable outcome in 1910. Such tensions were intensely felt in the political domain: they affected African as well as Afrikaner nationalist movements to a considerable degree.

The drive towards Union offered a ready platform for the consolidation of existing institutions or the creation of new ones. In the civil service, the gospel of uniformity was often heard as a justification for merging systems and practices. But political centralization was as likely to exacerbate underlying tensions as to resolve them. The country's motto, 'unity in diversity', was a recognition of this fact, and the process of establishing a truly unified nation-state was therefore slow and sometimes difficult. The creation of new scientific organizations with pretensions to national status—such as the Geological Survey of the Union (1910),[121] the Union Astronomical Observatory (1912), the Meteorological Office (1912), the South African Ornithologists' Union (1904), the Botanical Society of South Africa (1913), the National Botanic Gardens in Kirstenbosch (1913), the Transvaal-based Botanical Survey (1918), and the Associated Scientific and Technical Societies of South Africa (1920)—is revealing of the complex interrelationship between provincial and national power and identity.

In the field of medicine, the formation of the South African Medical Association (1906) and the South African Institute of Medical Research (1912) were significant developments. But, as we have already seen, the Medical Association of South Africa was only created a full fifteen years after Union. In agriculture the creation of a strong, centralized government department, supported by a regional network of research institutes, has also been noted. The railway network and tariff system, which reflected deep intra-state rivalries in the nineteenth century, continued to exert powerful fissiparous forces on the new nation-state.[122] Consolidation of road transport through the development of a national roads system in the twentieth century proved almost as divisive.

[121] Like the Museum, the Geological Survey underwent a name-change during the South African War to become known as the 'Geological Survey of the Transvaal'. After 1910 it became the Geological Survey of the Union, following a merger between the Cape and Transvaal Surveys. The Geological Survey of Natal and Zululand closed before Union. See R. D. Dixon, 'The Museum of the Geological Survey 1897–1992', 61–2.

[122] S. H. Frankel, 'Road and Rail Transport', in E. H. Brookes *et al.*, *Coming of Age: Studies in South African Citizenship and Politics* (Cape Town, 1930).

The history of museums, considered earlier in this chapter and elsewhere, provides a good example of the unequal manner in which processes of national assertion and consolidation impacted on regional institutions. By 1910 there were around ten public museums in South African towns and cities, several of which were grouped in towns of the eastern Cape. Natural-history collections tended to predominate, supplemented by holdings concerned with local history and ethnography. Union secured the premier position of the museums located in the provincial capitals—Cape Town and Pretoria, followed by Pietermaritzburg and Bloemfontein—which henceforth benefited from a central government grant and were treated as 'national' museums. However, this dispensation worked to the detriment of local institutions, like the Albany Museum in Grahamstown, which now came under the aegis of the provincial administrations: as 'second-class' institutions, they experienced reductions in funding and status.[123] In the view of one museum director, assessing the situation in 1936, country-wide cooperation was bedevilled by 'parochialism and insularity'.[124]

In the case of Kirstenbosch, whose creation as the country's national botanical gardens in 1913 was proposed in parliament by Lionel Phillips, its Cape-centric orientation 'had the disastrous consequence of dividing South African botany into north and south'. As Compton explains, Durban had some claim to the new institution on account of its established grounds and excellent herbarium, but it was eclipsed from consideration. In the Transvaal, after 1903, the Department of Agriculture developed a strong research-oriented botanical section under Joseph Burtt Davy and I. B. Pole-Evans, a protégé of Smuts. It received formal recognition in 1913. But the creation of a National Herbarium in Pretoria, which Smuts likened to a South African Kew (and which enjoyed direct links with this metropolitan centre), was much resented in Cape Town.[125]

A not dissimilar form of turf rivalry, albeit more subdued, concerned the establishment, in 1903, of the Transvaal Observatory in the Johannesburg suburb that bears its name. This represented a particular success for the S2A3 which, under Gill's influence, had urged its creation upon the Transvaal government. Beginning as a meteorological observatory with a time-keeping department (Milner considered that weather forecasts for farmers were a higher priority than

[123] R. F. H. Summers, *A History of the South African Museum, 1825–1975* (Cape Town, 1975), 92; N. Fowler, 'A History of the Albany Museum, 1855–1958', unpublished MS (1968), Albany Museum, 183–4; J. Hewitt, 'The Albany Museum, Grahamstown', *South African Museums Association Bulletin*, 4: 4 (1947); H. A. Miers and S. F. Markham, *A Report on the Museums and Art Galleries of British Africa* (Edinburgh, 1932).

[124] E. C. Chubb, quoted in C. K. Brain and M. C. Erasmus, *The Making of the Museum Professions in Southern Africa* (Pretoria, 1986), 5

[125] Compton, *The Way to Kirstenbosch*, 93, 110, 112. The aspect of regionalism is not discussed in P. Anker *Imperial Ecology: Environmental Order in the British Empire, 1895–1945* (Cambridge, Mass., 2001), ch. 2, who nonetheless correctly identifies the key role of botany in the politics of Union. Unfortunately, Anker's illuminating work is marred by a surprising number of errors.

stargazing), the scientific standing of the Transvaal Observatory was considerably enhanced when the Transvaal government provided it with a large refractor telescope. Its status was confirmed when it became known as the Union Observatory (1912) and its director, R. T. A. Innes, was promoted to the post of Union astronomer. Innes's elevation was facilitated by the consolidation of various meteorological services of the four provinces into the Union Meteorological Department at Pretoria, leaving the Transvaal Observatory free to concentrate on pure astronomical research. But this posed a challenge to the pre-eminence of the Royal Observatory in Cape Town. The ascendancy of the Union Observatory was achieved at the cost of the demise of the Natal Observatory in 1911, a story that echoed the eclipse of the Durban herbarium.[126]

Examples drawn from geology and ornithology are further illustrative of such dynamics. In the case of the former, a Geological Survey Museum was first created in 1897, following the transfer to it of the Transvaal Museum's extensive mineralogical and geological holdings. According to Levitz, the Geological Museum offered the opportunity to 'promote the Transvaal Republic over the other three provinces', and it may also have been helpful in attracting investment into the extractive industries. The Geological Museum expanded rapidly under the curatorship of Arthur Hall, whose energetic collecting policy in the 1920s showcased the Transvaal's immense mineral resources as well as displaying evidence for the theory of continental drift. The sense that geology was naturally the preserve of the north fed into tensions between the Cape and the Transvaal over custody of the country's palaeontological and fossil holdings. These came to a head in 1926 when E. C. N. van Hoepen, a champion of Afrikaner scientific nationalism, attempted to force the South African Museum in Cape Town to hand over its fossil material to the Transvaal.[127]

Ornithology was another field in which the Transvaal began to challenge the traditional museum-based expertise of the Cape. The creation of the South African Ornithological Union at Johannesburg in 1904, as well as the publication, under its aegis, of Alwin Haagner and Robert Ivy's *Sketches of South African Bird Life* in 1908, has been addressed by Jane Carruthers in terms of the story of South Africa's 'emerging nationhood'.[128] Up to this time the most comprehensive work on South African ornithology was Stark and Sclater's *Birds of South Africa* (1900–6), but this compilation was heavily biased towards collections held in the Cape and Natal. By contrast, *Sketches* was generously illustrated with photographs

[126] *Standard Encyclopedia of South Africa*, vol. 1 (Cape Town, 1970), 584; P. Moore and P. Collins, *The Astronomy of Southern Africa* (Cape Town, 1977), 29, 96–8, 135–6; R. H. Stoy, 'Astronomy in South Africa', in Brown (ed.), *A History of Scientific Endeavour*, 420–2. See also J. D. Laing (ed.), *The Royal Observatory at the Cape of Good Hope 1820–1970* (Cape Town, 1970).

[127] C. Levitz, 'The Politics, Ideology and Social Practices of South African Natural Science and Technology Museums', MA dissertation, Wits University (1995), 26–8; also Dixon, 'The Museum of the Geological Survey', 65.

[128] Jane Carruthers, '"Our Beautiful and Useful Allies": Aspects of Ornithology in Twentieth Century South Africa', *Historia*, 49: 1 (2004), 92.

and intended as 'a popular [rather] than a scientific book',[129] that avoided technical detail in favour of a broadly ecological approach. It also represented an effort to promote ornithology as an applied, field science, with important economic value for farmers.

The scope of the new Ornithological Union was avowedly pan-South African, and its founder members included representatives of the main museums in the country. Yet, the new organization was practically based in the Transvaal, as was the majority of its membership.[130] Two of the organization's leading figures, Alvin Haagner and J. W. B. Gunning, were at various stages directors of the Transvaal Museum and Zoological Gardens. Here, too, as Carruthers explains, a distinct north–south divide was emerging. The Transvaal Museum prided itself on its unrivalled egg collection and, in particular, its aviary. It made a virtue of its leadership in avian ecology and observational field-work, rather than systematics and taxonomy, which had been the preserve of natural history at the Cape ever since the days of Andrew Smith and Edgar Layard. Over the next two decades the research centre of South African ornithology—as opposed to the many local and amateur birdwatching societies which supported professional endeavours—'passed firmly from the South African Museum in Cape Town into the domain of the Transvaal Museum . . .'.[131]

The history of tertiary education, so central to the politics of knowledge production, provides some of the most striking examples of regional, provincial, and political tensions. We have already seen how the 1905 joint meeting of the South African and British Associations for the Advancement of Science exposed the limitations of South Africa's scientific infrastructure. At this time there was only one tertiary educational institution in the country, the University of the Cape of Good Hope, whose foundation in 1873 was one of the first tangible rewards for the grant of responsible government to the Cape. The Cape University was, however, really little more than an examining body, modelled on the old University of London. It was principally designed to serve the disparate teaching colleges spread about the country, and its role was increasingly vestigial. In 1905 there were around fifty teaching staff and fewer than 600 students throughout South Africa.[132] Students with ambitions and means were encouraged to study in Europe.

129 A. Haagner and R. H. Ivy, *Sketches of South African Bird-Life* (London, 1908), preface.

130 Of the 59 members of the Ornithological Union in 1905, 33 listed Transvaal addresses. 'Remarks relating to the formation of the Union', *Journal of the South African Ornithologists' Union*, 1: 1 (1905), pp. x–xiii.

131 Carruthers, '"Our Beautiful and Useful Allies"', 92, 97, 105. The interpretation and bulk of the evidence in this and the foregoing paragraphs is drawn from Carruthers.

132 These figures should be treated as indicative rather than entirely accurate; there are inconsistencies in the sources from which they have been drawn: Hofmeyr, 'Africa and Science', 4; E. G. Malherbe, Memo on 'Educational Development in the Union 1905–1930', Killie Campbell Library, KCM 56973 (305) file 427/7, p. 5; E.G. Malherbe, *Education in South Africa Vol. II: 1923–75* (Cape Town, 1977), 727.

Just as colonial nationalists had urged the need for the University of the Cape of Good Hope from the 1870s, so propagandists of closer union made much of the need for a national policy on university education in the first decade of the century. For some, like Chief Justice J. H. de Villiers, the creation of a national teaching university was 'a necessary first step under Union'. Similarly, John X. Merriman, speaking at the Cape University's graduation ceremony in 1907, argued powerfully for the necessity of a strong, research-oriented national university in order to promote the 'union of the European races in South Africa'; at the same time he cautioned against the dangers of academic localism (as in the United States).[133] These issues were extensively aired in journals like *The State* and the *African Monthly*, building on earlier discussions in the *Cape Monthly*.[134]

The requirements of science, rather more than the arts, were an important factor in pressures for university reorganization. Its needs were often articulated in terms of the country's international standing or competitiveness. This prerogative could be, and was, articulated from different political standpoints. According to Boucher, a strong current of imperialism was countered by 'a far from muted note of South Africanism', which stressed the need for colonies to create their own institutions and which was apt to resist anglicization.[135] In 1908 an inter-colonial conference was called in Cape Town in order to discuss the future of university education and, in particular, the creation of a national university. The conference proposed a federal university with all colonies represented on the governing body, but the initiative failed, largely because the commission insisted that the language of such a university should be English. Predictably, this provoked an indignantly hostile reaction from Dutch-speakers.[136]

In the post-Union period various government commissions considered what had become known as the 'University Question', though with a conspicuous lack of success. The individual aspirations of more than a half-dozen university teaching colleges in the four South African colonies ensured that unanimity as regards the construction of a single national university was almost impossible to achieve; regional, institutional, and linguistic rivalries were sharpened by competition to secure the Wernher-Beit bequest of £500,000 intended for the establishment of a single national university.[137] Finally, in 1916 several

[133] *Address Delivered Before the University of the Cape of Good Hope, February 23rd 1907*, by the Hon. John X. Merriman (Cape Town, 1907).

[134] M. Boucher, *The University of the Cape of Good Hope and the University of South Africa 1873–1946* (Pretoria, 1974), 124; R. P. B. Davis, 'University Reconstruction in the Cape Colony', *African Monthly*, 3: 18 (1908); J. Edgar, 'Union and the University Question', *The State*, 3: 4 (1910); 'The Case for a National University', *The State*, 4: 1 (1910).

[135] Boucher, *The University of the Cape of Good Hope*, 120.

[136] M. Boucher, *Spes in Arduis: A History of the University of South Africa* (Pretoria, 1973), 107.

[137] Phillips, *The University of Cape Town*, 2–4; Boucher, *Spes in Arduis*, ch. 5; E. A. Walker, *The South African College and the University of Cape Town 1829–1929* (Cape Town, 1929), 78–9.

key university acts were passed as a result of which the modern distribution of anglophone and Afrikaner white tertiary educational institutions was established.[138]

The growth in university education in the post-Union period and the failure to secure consensus with regard to the creation of a single national university reflected growing sectarian conflict between English- and Afrikaans-speakers, as well as ongoing regional competition for cultural and political primacy between the Cape and the Transvaal. A reawakening of ethnic sectarianism during the First World War exacerbated existing divisions. In the years after 1918 the University of Cape Town consciously projected itself as embodying an ethos of 'broad South Africanism', drawing on the traditions of the South African College from which it had originally evolved, but its anglophone traditions and imperial origins failed to satisfy Afrikaner nationalist aspirations.[139] The old Cape of Good Hope University (and competitor for the status of national university) likewise sought to portray itself as a non-sectarian institution. However, its commitment to English as the supreme language of instruction, and its long-standing British inheritance meant that its pretensions to be a truly South African university were unconvincing.[140]

The difficulties experienced in establishing a unitary university reflected many of the passions that had been aroused by Milner's attempts to anglicize school education, and the fierce reaction to this which helped to propel J. B. M. Hertzog's launch of the National Party between 1912 and 1914. In the inter-war years this translated into a fierce battle around the question of bilingualism, a struggle that would eventually result in the development of a parallel system of education for white South Africans. A decade after the landmark university acts, the 1928 Van der Horst Commission considered that the ideal of a single national university in South Africa was impractical, and came to the view that the historical trend was in the opposite direction. It also argued that university expansion had been too rapid, and questioned whether all those institutions conferred with university status were worthy of the name.[141] The historian Eric Walker commented at the

[138] For details see E. G. Malherbe, *Education in South Africa Vol.I: 1652–1922* (Cape Town, 1925), 420–1. Briefly: in 1918 the South African College became the University of Cape Town; Victoria College became the University of Stellenbosch; and the various other colleges—Grey, Rhodes, Huguenot, Natal, the School of Mines, and Transvaal University College—came to be incorporated under the University of South Africa. This federal institution based in Pretoria was, somewhat confusingly, the successor to the old University of Good Hope. In 1922 the School of Mines became the University of the Witwatersrand and in 1930 the Transvaal University College became the University of Pretoria. Potchefstoom University College became part of the University of South Africa in 1922. Fort Hare (the South African Native College), whose students worked externally for University of South Africa degrees, was recognized as an institution of 'higher education' in 1923, though it was not a full constituent of the federal university. See Boucher, *The University of the Cape of Good Hope*, 200.

[139] Phillips, *The University of Cape Town*, 172, 368.

[140] Boucher, *The University of the Cape of Good Hope*, 65, 89, 105, 148.

[141] UG 33-'28, *Report of the University Commission* (Pretoria, 1928), paras. 48, 105–6, and 'summary of conclusions'.

time: 'Politically and economically, South Africa had just moved centripetally; academically, in the very act of political union, it had rushed confusedly in the opposite direction.'[142]

CONCLUSION

Walker's observation about the countervailing tendencies of Union was experienced in many other knowledge-based institutions as well, and the process of creating a unitary South Africa out of different states and colonies gave regionalism new salience. The most acute fault-line developed between north and south, as the Transvaal became not only the industrial heartland of the Union but also its political centre. Parity, if not leadership, in intellectual and cultural life took longer to achieve, but the economic and social vibrancy of Johannesburg meant that this process was already well under way.[143] The north–south divide also became a handy way of characterizing different political traditions. By the 1930s it was commonplace for liberal commentators to explain Afrikaner obduracy and instinctive political conservatism by reference to the 'frontier' tradition of the Voortrekker north.[144]

In the period surveyed by this chapter, appeals to South Africanism served as the glue holding white South Africa together. It was a key legitimating ideology for an embryonic and still-fragmented state, which marvelled over its own creation but couldn't quite believe the reality. The precedent of cooperation amongst the Anglo-Dutch elite of the nineteenth-century Cape fed directly into the language of Union. Many of the institutions of governance that had developed in the Cape were replicated or adapted to suit the needs of the new South African Union. The refined, not to say gentrified, nature of Cape colonial nationalism had been founded on the commercial, agrarian, and professional interests of a well-established middle-class elite. Yet, for modern South Africanism to succeed in the twentieth century it had to incorporate not just the legacy of the Cape, but also to accommodate itself to the industrial-based politics of the Witwatersrand.

A vital task, therefore, was to convert the amalgam of English-speaking workers, artisans, and industrialists who subscribed to British South Africanism, into a broader South Africanism that was capable of reaching out to leading Afrikaners as well. Despite the South African War, there were many Afrikaners who were receptive to such initiatives. Belinda Bozzoli has traced the process of creating a shared South African patriotism in terms of a class alliance between agrarian and industrial interests, led by a rising 'national bourgeoisie' with close links to

[142] Walker, *The South African College*, 78.

[143] The establishment of Witwatersrand University (1922) and publication of journals like the *South African Quarterly* are indicative of Johannesburg's claims to cultural leadership.

[144] See e.g. E. Walker, *The Frontier Tradition in South Africa* (London, 1930).

manufacturing.[145] Historical fashions change, and this class-based account of South Africanism now seems overly instrumental. Yet Bozzoli's underlying insight remains sound. It helps to explain why mining magnates like Phillips and FitzPatrick, both exemplars of militant British South Africanism, chose in the moment of Union to demonstrate their material and aesthetic commitment to the land.

The lure of scientific progress was one of the ways used to defuse ethnic antagonisms in the reconstruction era; it was also a means of diffusing the transcendent values of a supposedly benevolent Empire devoted to universal truth and shared brotherhood. This was the spirit that helped to inspire the formation of the South African Association for the Advancement of Science. Technical achievements, particularly those related to agriculture, were eagerly welcomed by modernizing Afrikaners, some of whom began to look to the United States for inspiration. Agronomy and veterinary science formed one of several means by which the natural environment was tamed and put to the service of South African nationhood. New understandings of landscape, as well as moves to foster appreciation of fauna and flora, contributed significantly to the process of defining a shared South African identity and promoting the wider cultural politics of Union.

The almost euphoric support for South Africanism that emerged in the immediate post-Union era did not, however, go unchallenged. C. J. Langenhoven, the Afrikaner nationalist poet who went on to write the country's national anthem, was unfashionably sceptical about the 'roseate hue of enthusiasm' that flourished in 'the first flush of the dawn of our Union'. Writing in the premier intellectual organ of Union, *The State*, he issued this warning in 1910: 'It is true, no doubt, that the racial bitterness of the past has been buried; but since neither of the two races has been buried, it is equally true that racialism is not dead but sleepeth. That this must be so is abundantly taught us by the history of our own past.'[146]

This prediction was soon vindicated. Hertzog's formation of the National Party in 1914 was widely interpreted as evidence of a resurgent 'racialism'—that is, anti-English sentiment—and as a repudiation of the Union settlement. Full-scale political crisis followed South Africa's entry into the First World War. Hertzog's Nationalists were in the forefront of those who objected to a South African invasion of German South West Africa. Unreconciled Boer War generals like Beyers, de la Rey, de Wet, and Maritz either refused to countenance the occupation of South-West Africa or plotted to foment a coup against the Botha government in order to reinstate a republic. They and the thousands of supporters who joined the brief Afrikaner rebellion were defeated. Government troops led by generals Smuts and Botha took over South West Africa. Had the conflict not

145 B. Bozzoli, *The Political Nature of a Ruling Class Capital and Ideology in South Africa 1890–1933* (London, 1981).

146 C. J. Langenhoven, 'The Problem of the Dual Language in South Africa', *The State* (Mar. 1910), 403.

been contained so effectively, a full-scale Afrikaner civil war might have erupted.[147] This was not the only violent episode. Just months before, serious labour unrest involving white mineworkers exposed underlying political tensions in the urban areas, and resulted in Smuts declaring martial law. If the aftermath of the South African War showed the need to heal divisions, the advent of the First World War demonstrated that social and ethnic fissures within white South Africa were deep, and possibly unbridgeable.

One might expect that the fresh uncured glue of South Africanism would not be able to withstand such challenges. In fact, the crisis made it all the more imperative that South Africanism should prevail. Paradoxically, the decision to recruit over 20,000 black non-combatants to support the war effort in Europe had the effect of uniting whites in fear: politicians across the ethnic divide lined up to condemn a move which might encourage blacks to demand political rights and, equally worryingly, provide opportunities to fraternize across the colour bar. 'Central to white objections was the argument that the venture would, in general terms, be harmful to South African "interests".'[148] More positively, pride was a unifying sentiment. The heavy military losses incurred by English- and Afrikaans-speaking soldiers on the Somme and in Flanders 'lubricated the spring of a common, pro-British patriotism'. In particular, the heroic sacrifice of young lives at the Battle of Delville Wood near Longeuval 'personified an idealized image of gritty South African dependability whatever the odds'.[149]

Afrikaners may have been only a small minority of the Springbok Brigade contingent; and many, if not most, of their compatriots were implacably opposed to participation in an imperialist war. But, as far as anglophone pro-war patriots were concerned, the fact that South Africa was led militarily and politically by Botha and Smuts more than compensated for this deficiency. By 1919 Smuts was internationally renowned as an imperial statesman, and his political fortunes were henceforth irrevocably tied to the Commonwealth idea. As we shall see in the chapter that follows, a central theme of Smuts's Commonwealth persona was the link between science and statehood.

[147] S. B. Spies, 'South Africa and the First World War', in B. J. Liebenberg and S. B. Spies (eds.), *South Africa in the Twentieth Century* (Pretoria, 1993).

[148] A. Grundlingh, *Fighting their Own War: South African Blacks and the First World War* (Johannesburg, 1987), 45. It is widely accepted that the involvement of blacks did indeed have a pronounced effect on the rise of African nationalism in the post-war era.

[149] Bill Nasson, 'Delville Wood and South African Great War Commemoration,' *English Historical Review*, 119: 480 (2004) 59, 61.

5

A Commonwealth of Knowledge

The First World War provided a strong impetus to the government's centralizing tendencies and, in the following years, concerted efforts were made to consolidate the South African nation-state. An enabling ethos stressing national efficiency, technical expertise, and scientific training and management was much in evidence in South Africa (Britain and the United States offered useful reference points). Some illustrative examples may be provided to indicate this trend. In 1920 the Associated Scientific and Technical Societies of South Africa was formed in Johannesburg, with financial support from the Chamber of Mines, in order to further the professional interests of its member organizations. Within the state, the growing authority of the Public Service Commission reflected attempts to develop a professional and efficient civil service, echoing the view of modernizers that 'democracy must be completed by science'.[1] The central government also moved to strengthen its authority by harmonizing regional and provincial forms of administration, variations within which were now seen as unwarranted residual hangovers from the Union settlement. Pressures to achieve rationalization and 'uniformity' were especially strongly evident in the field of 'native administration', while empirically based science was frequently invoked as the most reasonable and disinterested way in which to approach questions of race.[2]

Centralization was also strongly to the fore in economic matters, where factual knowledge was seen as crucial to the efficient utilization of resources. It was during the war that personal income tax was first introduced in South Africa as a whole. An Office of Census and Statistics was established in 1917 which made possible the systematic collection of economic and population data on a national basis. The first computation of the country's national income, sector by sector, was produced by the Wits economist R. A. Lehfeldt, in 1922. Faced with recession, inflation, and monetary chaos in the aftermath of the war, a Reserve Bank, operating independently of London control, was established in 1921. A Board of Trade and Industries was formed in the same year.[3]

[1] E. H. Brookes, 'The Public Service', in E. H. Brookes *et al.*, *Coming of Age: Studies in South African Citizenship and Politics* (Cape Town, 1930), 335.

[2] T. J. Haarhoff and R. F. Currey, 'Education and Research', in ibid. 401.

[3] D. Hobart Houghton and J. Dagut, *Source Material on the South African Economy: Vol. 3, 1920–1970* (Cape Town, 1973), 101–3; R. A. Lehfeldt, *The National Resources of South Africa* (Johannesburg, 1922); S. H. Frankel, *Capital Investment in Africa* (Oxford, 1938), 138–9.

The war had provided a degree of natural economic protection, and this, in turn, stimulated a notable increase in the capacity of local manufacturing industry. Patriotism combined easily with calls for more aggressive forms of tariff protectionism. Economic nationalism was everywhere in the air. The language of 'South Africanism' was enthusiastically adopted by spokesmen for secondary industry. It resonated, too, with small- and medium-scale farmers, many of whom were Afrikaners. In addition to their mutual desire to expand the domestic market and claims to represent the interests of 'the nation', common cause was found in attacks on 'imperial' capital as represented by commerce and the mines.[4]

The war was followed by a period of political instability and intense social upheaval. Widespread labour militancy, involving black as well as white workers, raised sharp questions about the constitution of the nation. African nationalism underwent considerable radicalization at this time, in response to labour unrest on the Witwatersrand and the populist campaigns led by the Industrial and Commercial Workers' Union in urban and rural areas. White workers claimed social protection from the state: on the one hand from the greed of rapacious capitalists, on the other, from competition posed by cheap and semi-skilled black labour. These closely related demands came to a convulsive climax in the 1922 Rand Revolt, when a strike by white miners led to popular insurrection in Johannesburg that was eventually put down through the combined use of the police, army, and air power. The ensuing political crisis impelled the state to intervene directly in the labour market. It did so by means of a series of measures designed to guarantee privileged living standards for 'civilized' (that is, white) labour which involved close regulation of wage rates and trade-union rights.

Calls for economic self-sufficiency and efforts to ameliorate white unemployment encouraged the formation of new parastatals. Collective state enterprises, though fiercely resisted by some, won the backing of many more. In 1923 the Electricity Supply Commission (ESCOM) was created, followed in 1928 by the highly controversial Iron and Steel Corporation (ISCOR), in which the government maintained a controlling interest. Both these capital-intensive industrial ventures, designed to provide the country as a whole with ample supplies of electricity and steel, were guided by the ambitious Afrikaner engineer, Hendrik van der Bijl, who had been invited by Smuts to serve as the government's first scientific and technical adviser in 1919. Over the next thirty years van der Bijl became the principal exponent of a technologically led industrial strategy which new government agencies like the Advisory Board of Industries and Science, the Board of Trade and Industries, and the Research Grants Board did much to facilitate. As Bozzoli has pointed out, this ' "scientific" brand of South Africanism... depicted the process of industrialisation as a matter of "policy", not "politics", of "science" and not "ideology" '.[5]

[4] B. Bozzoli, *The Political Nature of a Ruling Class* (London, 1981), 121–3, 149–50.
[5] Ibid. 163–4, 219–21.

Political consolidation was also a striking feature of these troubled years, new allliances indicating the fractured and labile nature of national politics. An attempt to reunite the South African and National parties in 1920 ended in failure on the issue of republicanism and secession from the Empire, but soon led to other mergers.[6] Later that year the imperial-oriented Unionists were absorbed within the South African Party. In 1923 the English-dominated Labourites formed an electoral 'Pact' with the National Party under Hertzog's leadership which proved successful at the elections the following year. Between the growing corporatism of the South African Party and the worker protectionism of the Pact government there was substantially more agreement than readings of *Hansard* or editorial columns suggest. A decade later, in 1933–4, the abortive 'hereeniging' movement of 1920 came to pass when the dominant political parties amalgamated to form the ruling United Party. This landmark political arrangement, known as 'Fusion', was conditioned by intense rivalry between Smuts and Hertzog. The two South African War generals, who dominated national politics throughout the inter-war era, performed a complex political dance as they wove in and out of power—mostly as adversaries, for a period in tandem, but always as combative opponents.

One way to read these political mergers and alliances is as a series of efforts to create and re-create the political centre ground. For all their political differences and personal animosities, the tensions between Hertzog and Smuts reveal just how much they shared in common. Crucially, both were South Africanists—albeit of a different stripe. Hertzog's South Africanism, as defined by his 'two-streams' and 'South Africa first' policies, was strongly oriented towards Afrikaner nationalism. Yet he fully recognized the rights of English-speaking South Africans, and by the mid-to-late 1920s had become reconciled to South Africa's membership of the Commonwealth so long as the country's sovereign independence was recognized through its 'dominion status'.

Smuts's South Africanism, by contrast, was characterized more by patriotism than by nationalism, which is to say that it was articulated principally by reference to the country and its land rather than through romantic attachment to its people. Smuts's version of South Africanism was therefore decidedly less parochial than Hertzog's. Smuts's expansive approach to South Africanism was expressed in his philosophy of holism which, while presented as a unifying scientific theory, had a political dimension too. Holism implied an organic conception of belonging and interconnection that accepted the notion of multiple national loyalties so long as these were founded on the common patriotism of English- and Afrikaans-speakers; moreover, it presumed the continued existence of an outward-looking, modern, secular state that was constitutionally independent of, but firmly tied to, the British Commonwealth of nations.

The contentious questions that so absorbed the attentions of the white electorate, and which featured so strongly in party-political realignments between 1924 and

[6] For details on this little-known attempt, see 'Hereeniging', *South African Quarterly*, 2: 5 (1920).

1939, included matters such as the choice of national flag and anthem, the status of the Afrikaans language, the indivisibility of the Crown, constitutional definitions of political sovereignty, and the nature of the Commonwealth connection. The prominence of such questions was testimony to the fact that most white South African politics, however passionately expressed, were the factional squabbles of siblings struggling for predominance within the greater white family.

On the greatest issue of the period—racial segregation—all the parliamentary parties (ranging from Labour, Unionists, and Dominionites on the margins to the United Party dominating the centre) were substantially in agreement. It took a decade to achieve unanimity about the precise manner in which segregation was to be achieved, until in 1936–7 Africans lost their residual franchise privileges and were confirmed as third-class citizens, with few rights to live in the cities and no opportunities to purchase land outside of designated 'reserves'. Notwithstanding appearances, Hertzog and Smuts were not fundamentally opposed on the matter of segregation. The noise and rancour generated by the segregationist debates had less to do with underlying principles than with the convoluted jockeying and positioning for political advantage.

The internecine intensity of white politics and the strains of state-making were reflected in the politics of knowledge production. Questions of South Africa's status as a nation-state were powerfully to the fore during the inter-war period. In respect of the ideologically loaded problem of South Africa's relationship to the Commonwealth, patriotism and pride was expressed through an understanding of science as a universal project underwritten by a shared commitment to the principles of European 'civilization'. Both the politics of intra-white ethnic competition, and the ideology of racial exclusivity, were played out in the particular challenge of rooting such a civilization in an African context.

For the purposes of our story, it is significant that social and scientific knowledge began to acquire a secure base within government during and after the First World War. It also gained a firm foothold in the expanding university system. Whether located in tertiary education or the civil service, social and scientific knowledge was increasingly subject to professionalization and treated as an aspect of national development. Research units were first located under the aegis of government departments such as agriculture, education, and mines and industries. Subsequently they developed as semi-autonomous agencies, where they were charged not only with the task of dispensing funds to university-based researchers, but also with generating and directing knowledge. The evolution of official agencies spanning the Research Grants Board (1918), the Council for Scientific and Industrial Research (1945), and the National Council for Social Research (1946) is an important theme of this chapter. Attention to their institutional histories casts new light on the country's national concerns and its changing international profile. In ideological and institutional terms, there is clear evidence that South Africanism was becoming more and more embedded in the state.

JAN SMUTS AND THE 'GREAT DIVIDE'

Jan Smuts was the most powerful champion of South African science and its role in the creation of nationhood. His understanding of this link was bound up in his theory of holism, which located the local and the particular as an aspect of the transcendent and the universal. Smuts's 'patriotism of place' was expressed through his love of nature, his expertise in botany, and the key role he played as a patron of ecological thinking. This newly developing field of interdisciplinary study, in which South Africa excelled and had much to contribute, helped to cement the country's position within the evolving network of imperial or, more properly, Commonwealth knowledge systems.[7]

After losing political office to Hertzog in 1924, Smuts devoted considerable energy to scientific affairs, the S2A3 in particular. In 1925 he served as president of the Association, an honour that recalled the age when influential political figures were routinely invited to act as intellectual patrons. Four years later, on the occasion of the British Association's second visit to South Africa (to be described in the next section), Smuts was given the opportunity to outline his theory of holism to an international scientific audience. His paper, grandly entitled 'The Nature of Life', was afforded prominent coverage in the press. It was gratifying that scientists of the stature of J. S. Haldane, Wildon Carr, A. S. Eddington, and Lancelot Hogben were on hand to present formal responses.[8]

Smuts's 1925 address on 'South Africa in Science' considered 'the South African point of view' by sketching the contribution the country could make to the sum total of human knowledge. Smuts declared a desire to reorient science away from the 'habits of thought and the viewpoints characteristic of its birthplace in the northern hemisphere', pointing instead to the country's distinctive position in the African subcontinent and the southern hemisphere more generally. Employing a geological metaphor, Smuts posited Africa as the 'great "scientific divide" among the continents, where future prospectors of science may yet find the most precious and richest veins of knowledge'.[9]

Smuts began by describing the implications of the 'Wegener hypothesis', named after the German geophysicist whose *Origin of Continents and Oceans* (1915) pioneered the theory of continental drift. His innovative use of Wegener's ideas—which were far from commonly accepted at the time—served as a fitting leitmotif. It allowed Smuts to posit Africa as the southern hemisphere's 'mother

[7] See S. Dubow and S. Marks, 'Patriotism of Place and Race: Keith Hancock on South Africa', in D. A. Low (ed.), *Keith Hancock: The Legacies of a Historian* (Melbourne, 2001); P. Anker, *Imperial Ecology: Environmental Order in the British Empire, 1895–1945* (Cambridge, Mass., 2001).

[8] *The Times*, 26 July 1929; W. K. Hancock: *Smuts: The Fields of Force 1919–1950* (Cambridge, 1968), 190–1. Hogben, in typically pugnacious fashion, was less respectful than most. He dismissed Smuts's metaphysical approach, maintaining that holism could only be treated seriously if it could be presented in a form capable of being tested.

[9] J. C. Smuts, 'South Africa in Science', *SAJS* 22 (1925), 3–4.

continent' from which South America, Madagascar, India, and Australasia originally split away or 'calved off'. By placing southern Africa at the centre of this 'great "divide" ', it became possible to correlate scientific developments across a range of disciplines in new and creative ways, with potentially far-reaching implications for 'universal science'.[10]

In the field of geology, for example, several Cape formations appeared to be mirrored by equivalents in India and South America; the pattern of mineral deposits, such as diamonds and coal, illustrated a similar symmetry.[11] Botany and palaeobotany posed especially interesting comparative questions. Whereas most South African flora was evidently of tropical origin, the south-western Cape was characterized by a distinct temperate flora which could not—as was then widely believed—have derived from northern Europe. Instead, Smuts suggested, Cape flora might have come from ancient Gondwanaland, the continent which was now largely covered by the South Atlantic ocean but which used to encompass much of Africa, Australia, India, South America, and Madagascar. This theory could well explain botanical affinities between flora in South Africa and other countries of the southern hemisphere.[12]

Smuts proceeded to apply the Wegener hypothesis to zoology, climate, and meteorology, drawing on the evidence of past ice ages to explain rainfall patterns and desertification. He also extended his field of vision to the heavens, referring to the pioneering work of Lacaille, Henderson, and Gill. South African astronomy, Smuts reminded his audience, 'has the distinction of being responsible for the determination of both great astronomical standards of measurement—the distance of the sun and the distance of the fixed stars'. Finally, Smuts paid tribute to the new field of human palaeontology which, he said, was bringing South Africa 'right into the centre of the picture'.[13]

Palaeontology had been given strong impetus as a result of A. C. Haddon's 1905 address to the joint meeting of the British and South African Associations. The pioneering Cambridge anthropologist, who had risen to international prominence as the organizer of the 1898–9 Torres Strait Expedition, used the occasion to call for more work on the 'interesting subject of comparative physiology'. Duly inspired, the director of the South African Museum, Louis Péringuey, now focused attention on its extensive collection of indigenous skulls, skeletal material, and stone artefacts. He also inaugurated a major project to model life-size casts from surviving 'Bushmen' and 'Hottentots'. The discovery in 1913 of a large-brained and thick-boned fossil skull, which Robert Broom described as 'Boskop Man' (*Homo capensis*), opened a new era in palaeontology: most notably, it lent credence to the suspicions of those nineteenth-century investigators who had long suspected that human origins in South Africa might have a deep past.

Broom hypothesized that 'Boskop Man' (a highly racialized formulation, and one that has long since been discredited) might be a direct ancestor of the 'more or

[10] J. C. Smuts, 'South Africa in Science', *SAJS* 22 (1925), 4–5. [11] Ibid. 5–6.
[12] Ibid. 6–7. [13] Ibid. 14–15, 16.

less degenerate Bushman of recent times', and he posited connections with Neanderthal and Cro-Magnon skulls found in Europe. Further discoveries, such as that of 'Rhodesian Man' in 1921, sharpened public interest in the prehistoric hominids of southern Africa and fed into a growing fascination with human origins and race typology. This momentum was sustained after the First World War by the appointment of professional comparative anatomists at the university medical schools of Cape Town (Matthew Drennan) and Johannesburg (Raymond Dart). Having been based in the museum sector up to this point, physical anthropology now had a firm institutional foothold in the universities.[14]

This presence was powerfully entrenched after 1925 when Raymond Dart announced in the journal *Nature* that a newly discovered fossil skull, which he dubbed *Australopithecus africanus* ('the Man-Ape of South Africa'), marked a crucial transitional 'missing link' in hominid evolution. Dart suggested that *Australopithecus* might vindicate Darwin's prediction that 'Africa would prove to be the cradle of mankind'.[15] His audacious claim was greeted with scepticism and outright hostility on the part of leading international experts, who were disinclined to accept that either South Africa or its unknown tyro anthropologist merited the honour that attached to such a major discovery.

In his 1925 address Smuts went out on a metaphorical limb by giving support to Dart's theory. He speculated that 'South Africa may yet figure as the cradle of mankind, or shall I rather, say, one of the cradles?'[16] In referring to 'cradles' in the plural form, Smuts probably had in mind the strength of rival claims, but he might equally have been gesturing towards the polygenetic or multilinear theories of evolution that continued to exert influence. It should be remembered that, whereas South Africa's contribution to hominid evolution is today joyously read as confirmation that Africa's contribution to the human race is nothing less than the gift of humanity itself, in the 1920s and 1930s such discoveries were invariably interpreted to support theories of innate racial difference. Smuts certainly lent his own intellectual authority to the racialized form of comparative anatomy that Dart and his followers did so much to promote. For instance, in arguing for South Africa's importance as a field of anthropological research, he ventured the opinion that 'Our Bushmen are nothing but living fossils whose "contemporaries" disappeared from Europe many thousands of years ago'. As such, he said, they were analogous to the country's cycads, which were fossil survivors of the botanical world.[17]

The notion that the aboriginal races of the country represented the end of an evolutionary line, that Bushmen could be likened to the indigenous flora and fauna of the country, and that they should be preserved primarily as evolutionary curiosities, was commensurate with broader racial attitudes during the

[14] S. Dubow, *Scientific Racism in Modern South Africa* (Cambridge, 1995), 39.

[15] R. Dart, '*Australopithecus africanus:* The Man-Ape of South Africa', *Nature*, 7 Feb. 1925, pp.195–9.

[16] Smuts, 'South Africa in Science', 17.

[17] Ibid. 16–17.

segregationist era.[18] In this connection, it is as revealing as it is paradoxical that Smuts prefaced his 1925 presidential address by affirming the ethnically inclusive spirit of the S2A3, in which 'both official languages of the Union enjoy equal privileges' and whose purpose was to 'bring together and unite all South Africans' with an interest in South Africa's scientific culture, 'irrespective of race and language'.[19] At a time when problems of 'race' customarily referred to Anglo-Afrikaner relations (in contradistinction to the 'colour problem'), and at the very moment when segregationist legislation threatened to deny Africans the rights of citizenship, this message of differential inclusiveness made perfect sense to his white audience. In political as well as scientific terms, this too was the 'Great Divide'.

THE 1929 VISIT OF THE BAAS

In 1929, at the invitation of the South African Association for the Advancement of Science, the British Association returned to South Africa. On this occasion there were more than 500 overseas visitors, funding having been secured from the South African government, individuals, and commercial organizations based in Britain, as well as a contribution from the Rhodes Trust in Oxford. The delegation was high-powered in scientific terms with some forty-four Fellows of the Royal Society attending, including its president, the atomic physicist Sir Ernest Rutherford.[20] Ten 'representative' scientists were invited to accept the hospitality of the Union. Thus, '[w]ith punctilious regard to the Union's racial traditions', three guests each were asked from England and Holland, the remainder coming from France, Austria, Italy, and the United States.[21]

There were superficial resemblances to the 1905 meeting: as before, the official programme was split between Cape Town and Johannesburg (with optional excursions further afield), there were enthusiastic civic welcomes, extensive press coverage, and evidence of considerable public interest in the proceedings. But the tone, context, and intent of the 1929 meeting were markedly different from 1905. In the intervening quarter-of-a-century, South Africa had undergone a significant degree of political change. General Hertzog was now leader of a unitary state which had just achieved the status of full sovereignty within the British Commonwealth. In the period before the Wall Street Crash, the economy seemed

[18] See e.g. R. J. Gordon, *The Bushman Myth* (Boulder, Colo., 1992); S. Dubow, 'Human Origins, Race Typology and the Other Raymond Dart', *African Studies*, 55: 1 (1996).

[19] Smuts, 'South Africa in Science', 1. In *Imperial Ecology*, Anker argues that Smuts's holistic view of ecology and botany was commensurate with his political views: 'His ecophilosophy served him as a glorification of white supremacy, with a division of society into high and low personalities, while still defending unification of South Africa, the British Empire, and ultimately the world through the League of Nations' (p. 75).

[20] O. J. R. Howarth, *The British Association for the Advancement of Science: A Retrospect 1831–1931 (London, 1931)*, 143; *The Times*, 22 July 1929.

[21] *Cape Times*, 11 Mar. 1929.

buoyant. And the focus of white politics appeared to be turning away from tensions between English- and Afrikaans-speakers and towards the 'solution' of the 'native question'; only a few months before, Hertzog had fought—and won—a general election which revolved around the 'black peril'.

By comparison with 1905, a burgeoning sense of national pride and purpose was widely evident. The composition and orientation of the nation may have been a matter of dispute, but substantial consensus existed over two fundamental principles: that South Africa would continue to remain as an independent state within the Commonwealth, and that racial segregation between white and black was to be upheld and entrenched. Within these parameters, the symbolic power of science and technology could easily be accommodated in so far as it exemplified the dynamism of a new society with independent standing in a wider community of nations. The British Association was keenly sensitive to the delicateness of the situation. If, in 1905, it could afford to be imperious, in keeping with its imperial role in the world, in 1929 it was now much more conscious of functioning within the emerging consensus politics of the Commonwealth, and of its leading, but somewhat less elevated, position within the commonwealth of knowledge.

This spirit was evident from the outset. In his first address, the president of the British Association, Sir Thomas Holland, chose to quote from W. J. Viljoen's Afrikaans translation of Percy FitzPatrick's *Jock of the Bushveld*, adding that 'the translation is in a language capable of expressing virile sentences'.[22] This gesture did not only play to the literary icon of South African Englishness; Holland's willingness to speak in Afrikaans struck an obvious chord with his audience, as did his tribute to the recently deceased Viljoen, a man widely regarded as an exemplar of pragmatic moderation in the fraught arena of language and educational policy.[23]

So as to reinforce the message of conciliation, Sir Ernest Rutherford made the point that the British Association gathering was 'representative', noting that Sir Thomas was a Canadian while he himself was a New Zealander.[24] This formula was evidently successful for, on the Johannesburg leg of the journey, Holland again quoted in Afrikaans from *Jock*, adding that the book accurately 'reproduced the spirit of the veld'. And, at a mayoral reception attended by some 4,000 guests a couple of days later, the city hall stage was hung with the Union Jack and the new South African flag, below which the words 'Welcome/Welkom' were tastefully outlined in coloured lights.[25]

The 1929 meeting provided an ideal opportunity to reflect on South Africa's achievements since the beginning of the century. The fact that sectional meetings of the British Association could be held on the imposing new Groote Schuur campus of the University of Cape Town was indicative of progress in tertiary education. It was also a source of considerable civic pride, and the excellence of

[22] *The Times*, 23 July 1929.
[23] *Cape Times*, 20 July 1929 (editorial).
[24] *The Times*, 23 July 1929.
[25] *RDM*, 23 July 1929; 25 July 1929.

the university's facilities and architectural design were widely remarked upon.[26] The university sector had indeed undergone major transition. Although its research capacity was still undeveloped, there were now close to 500 university teachers in the country, based in four universities, and the total residential student body exceeded 7,000. Over the same period total expenditure on university education had increased from £15,000 to £660,000.[27]

Jan Hofmeyr, appointed principal of the soon-to-be Wits University in 1919 (and who subsequently entered politics as administrator of the Transvaal), was in 1929 president of the S2A3. A loyal and admiring protégé of Smuts, Hofmeyr shared his mentor's view of the importance of the Commonwealth connection, as well as the hope that European civilization could be extended northwards along the spine of central and eastern Africa. It fell naturally to Hofmeyr to welcome the British Association. His widely reported address (which was hailed by the *Rand Daily Mail* as brilliant, inspirational, and visionary) was devoted to the topic of 'Africa and Science'.[28] Hofmeyr began by recalling the remarkable intellectual progress that had taken place in South Africa since the previous visit of the British Association, and he went on to summarize developments in the university system, the growth of specialist and professional scientific societies, and the expansion in state-sponsored and applied scientific work.[29]

This was all backround to Hofmeyr's central theme: the 'South Africanisation' of science. Whereas in 1905 science in South Africa could be called 'exotic', in the sense that it depended on imported expatriates, its personnel was now, whether by adoption or birth, 'essentially South African'. Yet, although South Africans evinced great pride in their science, which was 'distinctively ours', Hofmeyr noted that there was 'nothing narrow about its South Africanism. Were it otherwise, it would have been false to the spirit of Science.' He instanced Smuts's concept of holism as proof of this expansiveness. Science, Hofmeyr maintained, had been 'consolidated', 'South Africanised', and recognized as being 'of great national value'.[30]

Turning to the future, Hofmeyr outlined an even bolder task, namely, the need to 'Africanise' science. This ambition was expressed by means of a curious brand of intellectual sub-imperialism which charged whites, as the natural representatives of European civilization in Africa, with the responsibility of disseminating enlightened values to Africa as a whole. In tones redolent of the nineteenth-century civilizing mission, he declared: 'It is by way of this Southern gateway that Science itself can most effectively be made to permeate Africa. And to you, having so come, to you, the ambassadors of Science, I present—Africa. It is Africa and

[26] See e.g. *Cape Times*, 27 July 1929 (editorial); Phillips, *The University of Cape Town*, 8.

[27] E. G. Malherbe, *Education in South Africa Vol. II: 1923–75* (Cape Town, 1977), 727; KCM 56973 (305) file 427/7 'Educational Development in the Union 1905–1930', by E. G. Malherbe.

[28] J. H. Hofmeyr, 'Africa and Science', *SAJS* 26 (1929); *RDM*, 23 July 1929 (editorial). See also *Cape Times*, 23 July 1929 (editorial); *The Times*, 5 Aug. 1929 (editorial).

[29] Hofmeyr, 'Africa and Science', 3–6. Cf. press reports of Hofmeyr's talk in e.g. *The Times* 27 Mar. 1929.

[30] Hofmeyr, 'Africa and Science', 6–7, 8.

Science, which, I would like to think, are to-day met together. Happy indeed should be the fruits of the mating.'[31] Hofmeyr proceeded to detail the ways in which Africa could expand scientific understanding. Amongst these, it offered a 'great laboratory' which could 'put to the test diverse social and political theories as to the relations between white and coloured races'.[32] The optimism of his broad African vision may have been misplaced, but it was not entirely without foundation, as the concurrent opening in Pretoria of the Pan-African Agricultural Veterinary and the International Geological congresses suggested.[33]

The sense of national pride and mission evident in Hofmeyr's conception of South African science was enthusiastically taken up in the English-language press. It was reiterated in Johannesburg by Smuts, who reminded his audience of the social and intellectual transformation that had taken place over the past quarter-century. The British Association first arrived to a South Africa 'devastated and laid bare in a great war . . .'. Since then, science had been largely responsible for re-creating a 'new South Africa'. In activities such as mining, agriculture, and manufacturing it was providing the technological means to overcome animal and plant diseases and to exploit natural resources in new and original ways. Whereas members of the general public had previously been sceptical, they were now enthusiastic converts to the cause of science. Indeed, Smuts went so far as to suggest, with double-edged irony—reflecting his personal suspicion of the mass technological age—that science had become 'the new magic . . . the great intangible power, the imponderable which in the long run carries everything before it'.[34]

Smuts also took the opportunity to underline the drift of Thomas Holland's presidential address, which stressed the importance of mineral resources as a factor in the maintenance of civilization and world peace. Holland's message was that South Africa was a valued member of the Commonwealth and that the gold-mining industry was of pre-eminent significance from both a domestic and an international perspective.[35] This idea was carefully calculated to appeal to Johannesburg's industrial and business community; it served as well to remind South Africans of their international obligations and responsibilities.

Holland's pitch may also be seen as the culmination of a wider effort to secure a broad base of support for scientific research. In a public debate on 'science and industry' held in Cape Town, he emphasized science's role as a democratic and popular, rather than an arcane and elitist, activity. He thereby sought to undercut the condescending attitudes of pure scientists towards applied work, as well as to allay suspicions expressed by 'captains of industry' that scientific research was

[31] Ibid. 9. As a bachelor who remained unusually close to his mother and who was obviously uncomfortable with sexual issues (witness his clumsy handling of the Stibbe 'affair' during his time at Wits), this was an odd metaphor for Hofmeyr to use.

[32] Ibid. 15.

[33] *RDM*, 3 Aug. 1929 (editorial); *Cape Times*, 29 July 1929. Some of the meetings of the Geological and Agricultural sections of the BAAS were held in Pretoria in conjunction with these other international conferences.

[34] *RDM*, 1 Aug. 1929. See also 2 Aug. 1929 (editorial).

[35] *RDM*, 1 Aug. 1929.

Fig 8. *Cape Times* cartoon by Wyndham Robinson, July 1929, showing Nationalist politicians F. W. Beyers and D. F. Malan devouring a piece of meat denoting racial hatred. The caption reads: 'Those members of the British Association especially interested in relics of the Stone Age will not have far to look.' (Courtesy of *Cape Times*)

undeserving of their support.[36] In this regard Holland was not only addressing a South African constituency: like previous leaders of the British Association, he was alert to the need to reposition the BAAS in the public mind. As Collins and Macleod have shown, the British Association found itself under renewed pressure at this time. It had to respond to post-war public scepticism towards the idea that scientific research was an unqualified good; it needed to be seen to be serving the public interest; and, within the profession, it had to maintain the delicate balance between a commitment to pure and applied research.[37]

Just as South Africa's geopolitical importance within the Commonwealth was highlighted in 1929, so South African scientists were welcomed as part of the international community. The production and dissemination of knowledge was presented as a reciprocal and cooperative process. Fittingly, the South African content of the 1929 conference was far more extensive, and its profile significantly higher, than had been the case in 1905. In fields ranging from zoology, botany, and geography to economics, anthropology, psychology, and education South African material was to the fore.

[36] *RDM*, 24 July 1929; 25 July 1929 (editorial) 2 Aug. 1929 (editorial); 5 Aug. 1929 (interview with Holland); among the participants in this discussion were Richard Gregory, D'Arcy Thompson, and Daniel Hall. They covered subjects such as science's contribution to industry, fisheries, and soil fertility. See *Cape Times*, 24 July 1929.

[37] R. Macleod, 'Retrospect: The British Association and Its Historians' and P. Collins, 'The British Association as Public Apologist for Science, 1919–1946', in R. Macleod and P. Collins (eds.), *The Parliament of Science: The British Association for the Advancement of Science 1831–1981* (Northwood, 1981), 5–6, 211–15.

PROBLEMS IN PREHISTORY—AND WITH POOR WHITES

Along with developments in physical anthropology and palaeontology, cultural anthropology and prehistory were accorded close attention. Several international luminaries in the field of prehistory were present at the 1929 meeting, including Leo Frobenius, Henri Breuil, Gertrude Caton Thompson, John Myres, and Henry Balfour, curator of the Pitt Rivers Museum in Oxford and president of the Anthropology Section. Balfour paid tribute to the advances which had been made in South African prehistory since his first visit in 1899, and he stressed the national and international significance of South Africa's archaeological heritage.[38]

There were two especially keenly watched papers in the Anthropology Section, both with the potential to excite controversy. The talk by the German ethnologist and exponent of *kulturkreis* theory, Leo Frobenius, was well trailed in the press: he was expected to offer startling evidence for the existence of a great and civilized race which had apparently lived for over 7,000 years in the area between Great Zimbabwe and Lake Malawi, worshipping the stars and the moon. However, Frobenius may have received advanced warning of Gertrude Caton Thompson's findings about the origins of Great Zimbabwe, the great archaeological complex in Rhodesia that had long been attributed to foreign influence. In the event his talk on the different styles and types of Bushman rock art—while suggesting Carthaginian influence in characteristic diffusionist mode—was presented in careful academic style and caused no sensation.[39]

The presentation of Caton Thompson's findings, by contrast, was no damp squib. A respected archaeologist and Egyptologist, she had been commissioned by the British Association, with funding from the Rhodes Trustees, to settle the long-running popular and scholarly debate about the origins of the Zimbabwe ruins. Building upon Randall-MacIver's report, as presented to the 1905 meeting, Caton Thompson was firmly of the opinion that all aspects of the Zimbabwe civilization were the results of African endeavour.[40] This finding, which unambiguously rejected theories ascribing the authorship of Zimbabwe to mysterious ancient visitors from the Near East, threatened to puncture a key prop of white settler mythology, namely, that Africans were incapable of creating a significant material culture on their own. Unsurprisingly, Caton Thompson's views were not accepted by all. A determinedly dissenting note was sounded by Raymond Dart, who launched an intemperate attack during her talk.[41] In following the

[38] *The Times*, 2 Aug. 1929. Balfour also visited South Africa in 1905 as part of the BAAS.

[39] *RDM*, 20 July 1929; 31 July 1929; *Cape Times*, 1 Aug. 1929

[40] *Star*, 2 Aug. 1929; *RDM*, 3 Aug. 1929.

[41] For more on Dart and the racial paradigm he promoted, see S. Dubow, *Scientific Racism in Modern South Africa*; also id., 'Human Origins, Race Typology and the other Raymond Dart', *African Studies*, 55: 1(1996).

tradition of his Australian compatriot and patron Grafton Elliot-Smith, Dart was a committed proponent of cultural diffusionism.

In the related field of South African archaeology the year 1929 saw important developments as well. John Goodwin, the first archaeologist to hold a university post in the country, collaborated with his Wits colleague, Peter van Riet Lowe, on an important work entitled *The Stone Age Cultures of South Africa* (1929). This asserted a specifically southern African agenda for prehistory. The central idea hinged on their proposal of a 'new terminology' for the southern African stone age (i.e. earlier–middle–later) which eschewed conventional European sequences such as 'Acheulean' or 'Mousterian'. Goodwin's and van Riet Lowe's use of terms like 'cultures', 'industries', and 'folk'—in preference to 'race' and 'type'—also signified a reaction against the racial determinism of physical anthropology, and of Raymond Dart in particular.[42] Echoes of such terminological controversy were also evident in the field of zoology and ornithology, where Austin Roberts of the Transvaal Museum became embroiled in an argument with the British Museum's Michael Oldfield Thomas. The spat, which had strong overtones of metropolitan–colonial friction, revolved around the vexed issue of the naming of a groove-toothed swamp rat. Scientists at the British Museum were also dismissive of Roberts's attempts to propose an entirely novel taxonomy for many African birds. In this regard, Roberts felt that his colonial status counted against him.[43]

Immediately preceding the publication of *Stone Age Cultures*, a wide-ranging summative statement on South African archaeology was produced by Goodwin's Cambridge teacher and mentor Miles Burkitt, entitled *South Africa's Past in Stone and Paint*.[44] This book has recently attracted the attention of those concerned with the politics of colonial knowledge. During the apartheid cultural boycott of the 1980s, when South African archaeology was subjected to savage international condemnation, several prominent liberal archaeologists were moved to defend their discipline. One of the points they made was that Burkitt, fifty years earlier, had viewed South African archaeology in condescending terms and that he had abused the generosity of his colonial colleagues, especially Goodwin, by appropriating their research findings and rushing his own work into print before theirs.[45]

Close inspection of Goodwin's private papers does not support this view of a metropolitan–peripheral divide. As Nathan Schlanger has shown, the greater

[42] S. Dubow, 'Opération coup de poing', *Archaeological Dialogues*, 10: 1 (2003), 26–33.

[43] J. Carruthers, ' "Our Beautiful and Useful Allies" ', 105. On Roberts and British ornithologists, I am grateful to information supplied by Nancy Jacobs.

[44] M. C. Burkitt, *South Africa's Past in Stone and Paint* (Cambridge, 1928); A. J. H. Goodwin and C. van Riet Lowe, 'The Stone Age Cultures of South Africa', *Annals of the South African Museum*, 27 (1929).

[45] N. Schlanger, 'The Burkitt Affair Revisited: Colonial Implications and Identity Politics in Early South African Prehistoric Research', *Archaeological Dialogues*, 10: 1 (2003); see also 'Making the Past for South Africa's Future: The Prehistory of Field-Marshal Smuts (1920s–1940s)', *Antiquity*, 76 (2002).

rivalry was with a disputatious, Dutch-born palaeontologist Egbert van Hoepen, who joined the Transvaal Museum under Gunning's directorship. In 1922 he became the first full-time director of the National Museum in Bloemfontein. Like several other Hollander intellectuals, van Hoepen was a committed convert to the republican, and then the nationalist, cause. A leading member of the Akademie vir Wetenskap en Kuns (he was awarded its Havenga Prize in 1945), van Hoepen insisted on publishing his prolific research output in Afrikaans. He clashed frequently with English-speaking members of the Bloemfontein Museum council.

From 1925 van Hoepen began to advance an alternative typological scheme for South African palaeontology, centred on his concept of the *'Koningse Kultuur'*. In fiery exchanges with Goodwin and van Riet Lowe (whom he accused of 'scientific piracy'), van Hoepen claimed that his ideas had been appropriated and repackaged. Leaving aside the substantive scientific issues at issue, and indeed the personal animosities and jealousies, Schlanger is correct in judging that Goodwin and van Riet Lowe perceived van Hoepen, rather than Burkitt, as their real threat. Moreover, in contradistinction to van Hoepen's more narrowly based ethnic nationalist agenda, Burkitt's appreciation of the role that prehistory could play in 'the process of forming a new South African nation within the British empire' was closely in accord with the strongly South Africanist outlook of Goodwin and van Riet Lowe.[46]

Another controversy that spurred Afrikaner nationalist resentment was the accusation of plagiarism levelled in 1927–8 by Eugène Marais against Maurice Maeterlinck, the internationally famous Belgian writer and philosopher. Marais, a well-known Afrikaner poet and amateur naturalist, who was nearing the end of his life in a sad state of drug-induced decline, had been a prominent opponent of Kruger before the South African War. Educated in the Cape and in England, Marais did not conform to the stereotypical image of puritanical rural Afrikaner conservatism. Not only was he strongly opposed to religious fundamentalism of the 'dopper' variety, he was receptive to radical ideas in philosophy and psychology. As a young man he had been a follower of De Quincey (he was reputed to have recommended opium-taking at a meeting of an Afrikaner youth movement). Marais's claim that a foreigner had stolen his theories concerning the organic life of termites prompted indignation among his influential circle of friends, and his (less than clear-cut) cause was duly taken up with enthusiasm by leading Afrikaans-medium newspapers and journals. As Sandra Swart explains, Marais's theories and accomplishments resonated powerfully with nationalist intellectuals like his principal supporter, Gustav Preller. They underscored the dual notion that Afrikaner national identity, while organically and

46 I am summarizing here the discussion prompted by Nathan Schlanger's article 'The Burkitt Affair Revisited' and responses by S. Dubow, N. Shepherd, and H. Kuklick. On van Hoepen, see also his entry in the *DSAB* and D. A. van der Bank, 'Nasionale Museum. Deel II. Die van Hoepen-Era', *Navorsinge van die Nasionale Museum Bloemfontein*, 4: 5 (1998).

aesthetically linked to the soil, was also fully capable of developing along modern, scientific lines.[47]

The ethnic nationalist sensitivities that so aggrieved van Hoepen and Marais were publicly aroused at the 1929 scientific meeting, following E. G. Malherbe's discussion of 'Education and the Poor White'. 'Ernie' Malherbe, the scion of a well-established multilingual Cape Afrikaner family, was an ambitious young Columbia University-trained educational psychologist, now based at the University of Cape Town. Drawing on the preliminary results of the landmark Carnegie Commission into 'poor whiteism', of which he was a leading researcher, Malherbe shocked Afrikaner opinion when he referred to the problem of white poverty as a 'skeleton in our cupboard'. He went on to suggest that whites experienced a psychological 'inferiority complex' as a consequence of an unconscious feeling of vulnerability, which led them to lay the blame for their own social and economic predicament on blacks.[48]

This intervention drew highly favourable comment from some, and was widely seen as a courageous and authoritative intervention into an area of increasing social and economic concern. Dr C. W. Kimmins, president of the Educational Section, resorted to eulogy, while the *Cape Times* called the paper 'brilliant'.[49] Malherbe's mentor and colleague F. W. Clarke (who later became director of the Institute of Education in London) praised his young colleague for displaying: 'Clear-headedness, courage & a genuine patriotism that was ready to face martyrdom .... you honoured the Association by giving it the fearless truth. As a result you have leapt to fame at one bound.' But he went on to warn Malherbe that the 'pinnacle is a dangerous one'.[50]

By genuine patriotism, Clarke meant Malherbe's willingness to speak as a 'good South African' rather than from a sectarian standpoint. Yet his assessment of the attendant dangers proved correct when the Afrikaans press tore into Malherbe for betraying his Afrikaner loyalties. *Ons Vaderland*, mouthpiece of the Transvaal Nationalists, accused Malherbe of 'professorial superficiality' and 'thoughtless stupidity'. The attack was ostensibly focused on Malherbe's methodology and the quality of his research, but the newspaper was clearly stung by the suggestion that whites were somehow psychologically fearful of blacks, and highly sensitive about his calumnious airing in public of a problem which reflected negatively on the standing of the Afrikaner volk.[51] *Die Burger* joined in with an editorial titled 'Pure Nonsense', and *Die Volksblad* proved similarly hostile.[52] Underlying the

[47] S. Swart, 'The Ant of the White Soul: Popular Natural History, the Politics of Afrikaner Identity, and the Entomological Writings of Eugène Marais', in W. Beinart and J. McGregor (eds.), *Social History and African Environments* (Oxford, Ohio, and Cape Town, 2003); see also Leon Rousseau's evocative biography, *The Dark Stream: The Story of Eugène Marais* (Johannesburg, 1982), 70, 404 ff.

[48] *Cape Times*, 24 July 1929. See also E. G. Malherbe, 'Education and the Poor White', *SAJS* 26 (1929).

[49] *Cape Times*, 24 July 1929; 24 July 1929 (editorial); 25 July 1929.

[50] KCM 56973 (47), Fred Clarke to E. G. Malherbe, 24 July 1929.

[51] *Ons Vaderland*, 26 July 1929 (editorial).

[52] *Die Burger*, 26 July 1929 (editorial); *Die Volksblad*, 30 July 1929, (editorial).

negative reaction lay an even deeper fear: that Afrikaner poor whites were becoming biologically and mentally enfeebled as a result of eugenic deterioration.

Greatly concerned that the Carnegie inquiry was becoming politicized, a statement was communicated to the press by its leading investigators, all of whom respected Afrikaner academics, stressing that their primary task was objective 'fact finding' and carefully affirming the commission's political neutrality. Malherbe was forced into a convoluted attempt to defuse the tension by distancing himself from the interpretation that had been placed on his words. This 'clarificatory' explanation was accepted by *Die Burger*. It duly responded with an editorial acknowledging the importance of the commission's work and expressing confidence in its commitment to 'bring the facts to light'.[53] But *Ons Vaderland* remained uncompromisingly hostile. It was especially concerned at the negative impression that would be created by Malherbe's presentation of so-called 'facts' to the British Association as well as to his American backers.[54]

Political controversy thus intruded into the 1929 meeting, though it did not overwhelm the event. In the many reflections and tributes to the success of the 1929 meeting, the British and the South African Associations expressed themselves well pleased with the way in which things had gone. A special meeting to reflect on the proceedings was convened by the Royal Empire Society in December at which the editor of *Nature*, Sir Richard Gregory, reiterated the modern utilitarian role of science in promoting imperial cooperation, economic development and human well-being.[55] In South Africa, a genuine mood of buoyant confidence in the country's achievements was evident. The *Cape Times* captured the spirit when it proclaimed in an editorial titled 'Savants and Servants':

> When last the British Association met in Cape Town its delegates were the guests of the Government of the Cape Colony, then a separate State in the South African group of colonies. But science knew no political boundaries, and in 1905, as in 1929, it held its meetings in the chief centres of what now constitutes the Union, and in doing so it contributed its share to that strong sense of South African unity which was then beginning to take shape and form.[56]

The idea that science 'knows no politics' was, needless to say, a fiction, as Malherbe's intervention into the causes of poor-whiteism and other spats so clearly showed. Racial politics were also conspicuously featured in a number of papers dealing directly or indirectly with eugenics. The interest in physical anthropology and human origins, the heated public debate on the origins of Great Zimbabwe, as well as papers dealing with comparative intelligence quotients and the purpose of native education, all indicated the extent to which questions of race were actively being debated at the 1929 meeting. Moreover, science was frequently invoked as a means of finding an objective 'solution' to the 'native question'.

[53] *Die Burger*, 30 July 1929. [54] *Ons Vaderland*, 7 Aug. 1929 (editorial).

[55] BAAS box 261, 'Science and the Empire', address given by Sir Richard Gregory on 3 Dec. 1929. [56] *Cape Times*, 22 July 1929 (editorial).

As Hofmeyr pointed out, it had an important role in 'determining the lines along which white and coloured races can best live together in harmony and to their common advantage'.[57]

It was not only the message of science but also its personnel that came under scrutiny. *Die Burger*, authoritative voice of Cape Afrikaner nationalism, greeted the 1929 visit of the British Association coolly. It chose to emphasize the European, rather than British, profile of the visitors, and drew attention to the presence of individuals like Professor Freudenberg of the University of Heidelberg. In subdued tones of selfless patriotism, the paper expressed a willingness to learn from the country's overseas guests, observing that the visit was in South Africa's best interests and that it would improve its international status.[58] It was on this basis, too, that the influential organ of the Dutch Reformed Church, *Die Kerkbode*, came to terms with the Association's visit. *Die Kerkbode* was unhappy that there was no explicit theological dimension to the meeting, but took comfort in observing that religion was an intrinsic part of everyday intellectual life and existence.

Strong exception was taken, however, to Hofmeyr's reference to the exciting possibility that Africa might turn out to be 'the scene of Nature's greatest creative effort', namely, the evolution of mankind—because this amounted to an acceptance of evolutionist theory. The unease with which Christian-nationalists viewed Darwinian ideas, especially its atheist connotations, was reflected in cautious discussions in the *Tydskrif vir Wetenskap en Kuns* and in responses to the neo-Calvinist Potchefstroom philosopher, H. G. Stoker, who wrote a book in 1929 which attempted to develop a theist account of natural science.[59] Notwithstanding its qualms about the implications of evolution, *Die Kerkbode* was pleased to welcome the overseas scientists: it warmed to the promise that science would help to advance South Africa's agriculture and industry, and noted with satisfaction Thomas Holland's efforts to speak in Afrikaans.[60]

If Afrikaner opinion-formers were inclined to accept the British Association, they were nonetheless highly sensitive to any advantage that might accrue to Smuts's South African Party, and indeed to anything that could be interpreted as a slight on the Afrikaner volk. Opportunities for political sniping were not wanting. In anticipation of the British Association's arrival, the SAP-supporting *Cape Times* published a cartoon by Wyndam Robinson depicting two of Hertzog's ministers, F. W. Beyers and D. F. Malan, dressed as cave-dwellers. Beyers was shown chewing on a chunk of meat inscribed 'hatred'. The caption read: 'Those members of the

57 Hofmeyr, 'Africa and Science', 16.

58 *Die Burger*, 19 July 1929; 20 July 1929 ('Suid-Afrika en die wetenskap').

59 See e.g. the debate between E. C. N. van Hoepen and D. J. du Plessis on 'Die ewolusieleer in die geologie' (geology and evolution) in *Tydskrif vir Wetenskap en Kuns*, 5 and 6 (1926–7 and 1927–8), and also the contributions of van Hoepen, G. H. T. Malan, and H. G. Stoker on the 'Idee van 'n Teistiese Wetenskap' (the idea of a theistic science) in *Tydskrif*, 8 and 9 (1929–30 and 1930–1); see also A. C. Hoffman, 'Charles Darwin en die onstaan van die idee van ewolusie', *Tydskrif*, 10 (1931–2).

60 *Die Kerkbode*, 31 July 1929 ('Die "British Association"').

British Association especially interested in relics of the Stone Age will not have far to look.'[61]

This provocation did not go unnoticed. *Die Burger* pronounced itself scandalized, in an editorial penned by the nationalist ideologue A. L. Geyer. But when addressing itself to the learned visitors, the paper generously sought to assuage any embarrassment they might suffer; the *Cape Times*, it loftily assured them, was a jingoistic newspaper which could not best claim to represent British values and culture. Nor should the *Cape Times* be seen as the yardstick of South African culture.[62] Geyer's editorial skilfully inverted accusations of race hatred by blaming the English-speaking press in an attempt to drive a wedge between the British Association and the South African Party. This tactic was endorsed in an article by C. J. Langenhoven, the noted Afrikaner poet and critic, who had so percipiently forecast a rise in 'racial' tensions in 1910. He, too, claimed the high ground of cultural and intellectual achievement for the National Party, attacking the view of Smuts as a great world statesman who had suffered from an unappreciative nation of 'backvelders' and blinkered race-haters.[63]

It is worth noting that the hostility of the Afrikaans press was specifically directed. It did not extend to an attack on the British Association itself, or indeed on western scientific traditions. Pride in South African scientific achievement was widely shared and actively courted.[64] The problem had more to do with who stood to gain politically. The Afrikaans press was irked by the ease with which South African Party leaders like Smuts and Hofmeyr were accepted as part of the international community of scholars. Its anger was thus targeted against supporters of the South African Party or untrustworthy Afrikaners like Malherbe. In the space created by a retreating imperialism and an advancing nationalism, there was growing pressure to reposition South Africanism on the Hertzogite end of the spectrum.

SOCIAL SCIENCE AND THE CARNEGIE CORPORATION

From the late 1920s the politics of knowledge entered a new phase. The rise of social science—ranging from statistics and psychometric testing to educational philosophy and sociology—underwent professionalization and sought new institutional moorings. This process can be traced within the expanding university sector, which was somewhat belatedly beginning to build a research capacity. The state also took a growing interest in scientific planning and management. Less visible, but no less important, was the internationalization of research networks,

[61] *Cape Times*, 17 July 1929.

[62] *Die Burger*, 18 July 1929 ('Skandelik', editorial by A. L. Geyer).

[63] C. J. Langenhoven, 'Aan Stille Waters', *Die Burger* (I have been unable to find the date of this article).

[64] In the apartheid era, as the final chapter shows, Afrikaner nationalism reclaimed science and technology and situated itself squarely on the side of 'western' civilization.

conferences, and expertise, as funding institutions, the Carnegie Corporation in particular, looked overseas.[65] Applied knowledge in fields like education, social anthropology, and sociology attracted particular interest. These soon became highly contested areas of debate, and they provided a growing corps of trained intellectuals with the means to discuss matters of national importance—race and ethnicity especially—in a comparative, multinational context.

E. G. Malherbe was a key figure in these developments. In 1921 he had launched his career as a public intellectual by writing an article in the *Cape Times*. Here he pressed for a 'National Bureau of Education' based within the Union government. Echoing the spirit of scientific governmentality, Malherbe argued that the First World War had demonstrated that the supreme duty of a nation was 'the conservation and development of its human resources'. If South Africans aspired to 'national greatness', it was essential to 'take stock' and to focus upon their country's greatest natural resource: its (white) children. In South Africa there were over 100,000 poor whites who constituted a 'menace to the self-preservation and prestige of our white people, living as we do in the midst of a native population which outnumbers us 6 to 1'. In order to diagnose and remedy the causes of this social malady it was therefore necessary to 'get down to the very bed-rock of facts'. Rather than indulge in 'sporadic outbursts of philanthropy', what was required was a 'national project' of research conducted by trained psychologists and sociologists.[66]

Malherbe envisaged his Bureau as cutting across government departments and provincial boundaries. Amongst its many functions, it would undertake systematic and scientific social research in education, act as a clearing-house for new ideas, and collect social and educational statistics for government commissions of inquiry. By acting in cooperation with universities the Bureau could help to frame and standardize group intelligence and educational tests designed for South African conditions. It might also initiate research into the school curriculum to discover which subjects were most liked, how they could be taught most effectively, investigate the effects of language medium on subject choice, and analyse the extent of 'retardation' in South African schools.[67]

Malherbe's article amounted to a manifesto, and in many respects its key objectives were fulfilled. As a well-connected young Afrikaner with strong roots in the western Cape (he was especially proud of his Huguenot ancestry), Malherbe was ideally positioned to champion this project. His mother was linked to the Smuts family and his father had been a Dutch Reformed Church pastor who ministered to Boer troops after the destruction of his church during the South African War. Educated at Stellenbosch, just as Victoria College was gaining university status, Malherbe was part of an aspiring generation of tertiary-educated Afrikaner graduates who were poised to take advantage of career opportunities that were opening

[65] M. Lawn, 'Circulations and Exchanges: Emergence of Scientific Cosmopolitanism in Educational Research', Congress of Historical Sciences, Sydney, 2005.

[66] *Cape Times*, 18 June 1921, 'Educational Research. A Plea for a Bureau', by E. G. Malherbe.

[67] Ibid.

up in politics, industry, and the professions.[68] Instead of studying in England or Scotland, Malherbe enrolled as a doctoral student at Columbia University's Teacher's College, which resulted in his still-unsurpassed study of the history of education in South Africa. Columbia was an important centre for statistically oriented social science, and Malherbe, always an assiduous social networker, came into close contact with several of its leading teachers, including John Dewey, the philosopher of pragmatism, and the psychologist E. L. Thorndike.

The decision to study at an American rather than a British or European university proved a smart choice. Forward-looking, dynamic, and untainted by imperialist associations, America emerged from the First World War as a new international power. Many of the problems addressed by American 'progressivism', the social reform movement that developed in response to the challenges posed by a rapidly industrializing mass society, were recognizable in South Africa. Moreover, the ethnic and racial composition of the southern United States bore a closer resemblance to South Africa's than it did to any other country in the British Commonwealth. In politics and in education the experience of the segregated American South had been eagerly studied from the early twentieth century. South African agricultural experts and environmentalists, like H. S. du Toit, looked to Theodore Roosevelt's conservationist policies and, later, to the experience of the American dustbowl in the 1930s, for solutions to soil denudation and drought.[69] Another practical-minded Afrikaner who spent time working in the United States at this time was the talented Stellenbosch- and German-trained physicist H. J. van der Bijl.

The United States' experience of rapid industrialization and its participation in the First World War stimulated a 'new sense of discontinuity with the past and . . . accelerating historical transformation'. The response of many American social scientists, several of whom were based at Columbia University, was to explore ways in which firm technological control could be established over society.[70] Malherbe was greatly influenced by the apparent certainty offered by empiricist social science and psychology, as well as by the pragmatic and professional approach of American educators and social reformers. He felt sure that outcomes and methods were much more important than theories and dogma. South Africa, like America, was a young, pioneering country, 'bristling with new and unique problems', which could not afford to be 'bound down by traditions'.[71]

[68] E. G. Malherbe, *Never a Dull Moment* (Cape Town, 1981), ch. 3. Malherbe's closest friends at university—his wife Janie teasingly referred to them as the 'Heavenly Quartet'—were J. L. B. Smith, the discoverer of the coelacanth; Frikkie Meyer, later head of ISCOR; and the mining geologist W. Kupferburger. Other friends and associates included H. J. van Eck, T. B. Davie, and Paul Sauer. See also KCM 56973 (11 and 12) for biographical details.

[69] W. Beinart and P. Coates, *Environment and History: The Taming of Nature in the USA and South Africa* (London, 1995), 61–9.

[70] D. Ross, *The Origins of American Social Science* (Cambridge, 1991), pp. xv, 311, 404.

[71] E. G. Malherbe, 'Experimental Research in S. African Education', *New Era*, 8: 3 (1927), in KCM 56973 (144)b.

After his return to South Africa in 1924 Malherbe became a lecturer in the University of Cape Town's education department. Here he was influenced by Fred Clarke, as well as by H. A. Reyburn, professor of logic and psychology and an acknowledged expert in psychometric testing. Both were powerful figures within the university as well as being active in the wider world of education policy. Amongst the students who took courses in education at this time were N. P. van Wyk Louw and I. D. du Plessis, who later became prominent Afrikaner literary figures. Another was P. A. W. Cook, who subsequently worked as Malherbe's deputy in the National Bureau of Educational Research.[72]

Malherbe's big break came in 1927 with the African tour of the Carnegie Corporation of New York (following on the earlier African visit of the Phelps Stokes Fund).[73] The visit was prompted by Andrew Carnegie's establishment in 1912 of a special $10 million fund for exclusive use in the British dominions and colonies.[74] The Corporation gave extensive support to libraries and museums in South Africa, Australia, and New Zealand. It helped to establish national research institutions in the white Commonwealth and inaugurated an extensive programme of US travel and research grants. Carnegie money was vital in the formation of the South African Institute of Race Relations (1929). In sub-Saharan Africa more generally, it assisted in the funding of Lord Hailey's Africa Survey, promoted Jeanes schools, and supported agricultural research in East Africa.[75] Such initiatives were indicative of America's growing interest in parts of the world beyond its backyard.

The Corporation's venture into Africa was led by Frederick Keppel, an urbane patrician who helped to define its enlightened conservative outlook. During his long tenure as head of the Corporation, considerable material and intellectual

[72] Malherbe, *Never a Dull Moment*, 115–16.

[73] Patrick Duncan papers, University of Cape Town, BC294 C1.440, *Carnegie Corporation of New York: Report of the President & the Secretary as to an Educational Program in Africa* (New York, 1927). There is a substantial literature on the Carnegie and other major American philanthropic institutions. See e.g. E. C. Lagemann, *The Politics of Knowledge: The Carnegie Corporation, Philanthropy, and Public Policy* (Middletown, Conn., 1989); W. A. Nielsen, *The Big Foundations* (New York, 1972); R. F. Arnove (ed.), *Philanthropy and Cultural Imperialism: The Foundations at Home and Abroad* (Bloomington, 1982). On the Carnegie Corporation in South Africa see M. Bell, 'American Philanthropy, the Carnegie Corporation and Poverty in South Africa', *JSAS* 26: 3 (2000); C. J. Groenewald, 'The Methodology of Poverty Research in South Africa: The Case of the First Carnegie Investigation 1929–1932', *Social Dynamics*, 13: 2 (1987); Miller, 'Social Science, Philanthropy, and Politics', unpub. Seminar paper, Sussex University (n.d.).

[74] On the history of the Carnegie Corporation's involvement with South Africa see e.g. Malherbe, *Never a Dull Moment*, ch. 11; C. J. Groenewald, 'The Methodology of Poverty Research in South Africa'; Miller, 'Social Science, Philanthropy, and Politics'.

[75] KCM 56983 32b, 519/2, 'Memorandum on the Activities of Carnegie Corporation in the British Colonies and Southern Dominions', by F. P. Keppel, 26 Jan. 1942. This reveals that, between 1925 and 1942, Carnegie spent some $4.8 m in the 'southern hemisphere'. Africa (including South Africa) accounted for nearly $2 m of this, Australia $1 m, and New Zealand $700,000. According to C. Dollard, *Memorandum on Grants to Individuals* (Carnegie Corporation of New York, 1938), 112 South Africans had visited the USA by 1938, of whom 49 were 'educators' and 25 from branches of the government.

resources were made available for what Lagemann usefully characterizes as 'cultural' or 'scientific' philanthropy.[76] The Corporation's espousal of progressivist ideas was marked by a commitment to the preservation of Anglo-Saxon Protestant institutions which, in the case of several of its influential trustees during the 1920s, was inflected by eugenic anxieties about race deterioration.[77] Such concerns were easily translated into an interest in white poverty and social degradation, issues with obvious parallels in America, as were the similarities between racial segregation in South Africa and the American South.

Some critics have traced the rise of the Carnegie, Rockefeller, and Ford Foundations in terms of the rise of American corporate power and cultural imperialism, as well as a means by which northern United States industrialists sought to influence foreign as well as domestic policy.[78] But this view tends to oversimplify the workings of complex organizations through assuming an immediate connection between the sponsorship and control of knowledge on the one hand, and its outcomes on the other. As Morag Bell has pointed out in her study of the Carnegie Corporation's interventions into poverty-alleviation in South Africa, such models of cultural imperialism overlook the sense in which complementarity, reciprocity, and mutual exchange of ideas characterized the relationship between donors and recipients.[79]

The Carnegie Corporation exuded a progressive and strikingly modern developmentalist ethos that reflected the rise of the New World over the Old and presented the United States as an alternative international force to Britain. Unencumbered by accusations of imperialism, the Corporation travelled abroad with light ideological baggage; it was eagerly embraced by ambitious educationists like Malherbe, Charles Templeman Loram, and P. A. W. Cook, who all sought scientific solutions to South Africa's urgent social and political problems.[80]

Keppel took credit for the idea of the Poor White Commission himself, but several thinkers were already working along these lines. One of these was Malherbe, who identified poor-whiteism as the most urgent social issue facing South Africa when he met with Keppel in Cape Town. At the latter's invitation Malherbe proceeded to sketch out the draft of a research project covering the economic, psychological, educational, health, and sociological dimensions of the problem. This, in essence, became the framework of the landmark five-volume Carnegie Commission Report which emerged in 1932 and remains an outstanding example of applied sociology; it gives Malherbe some considerable claim to have been an originator of the inquiry as well as one of its chief researchers.[81]

[76] Nielsen, *The Big Foundations*, 38–9; Lagemann, *Politics of Knowledge*, 7, 95, 100.

[77] Lagemann, *The Politics of Knowledge*, 30, 80–1.

[78] Arnove (ed.), *Philanthropy and Cultural Imperialism*.

[79] Bell, 'American Philanthropy'.

[80] J. D. Shingler, 'Education and the Political Order in South Africa, 1902–1961', Ph.D thesis, Yale University (1973), 184.

[81] Malherbe, *Never a Dull Moment*, 119–20; KCM 56983 32b, 519/2, 'Memorandum on the Activities of Carnegie Corporation in the British Colonies and Southern Dominions', by F. P. Keppel, 26 Jan. 1942, p. 4.

In allocating a sum of $500,000 for a South African research programme extending over five years—of which the Poor White Commission was the centrepiece—Keppel was careful to solicit powerful local support and to avoid accusations of interfering in domestic politics.[82] To assist with its management he created a group of senior trustees, chaired by Smuts's political deputy, Patrick Duncan, a past member of the Milner Kindergarten and a future governor-general of South Africa.[83] Two Afrikaners, Senator F. S. Malan and R. W. Wilcocks, the Stellenbosch academic and senior Carnegie investigator, were given executive responsibility for its administration.

Writing to Duncan in 1928 Keppel invited his 'African colleagues' to exercise the 'widest discretion', not only because of distance and expense of cabling, 'but because this is in accord with our general policy': 'You are on the spot and understand, far better than we would if we were there, what the next step in each enterprise should be. You know, for example, whether any action, which might seem to us perfectly harmless, would appear to the South African public to be unwarranted political activity on our part.'[84] Keppel's decision to include Afrikaner intellectuals and community leaders within the organization and planning of the Commission was carefully considered. He made deliberate use of Afrikaans-speaking graduates of Columbia University Teacher's College (of which he was, coincidentally, an ex-dean), and he was determined to secure the support of the Dutch Reformed Church (DRC), which had already taken a lead by holding conferences on white poverty and organizing poor-relief schemes.[85] The DRC also played a significant role in establishing and legitimizing the Carnegie Commission, several of whose investigators, including the Revd J. R. Albertyn, the Revd T. F. Cronje, and the writer M. E. Rothmann, were drawn from its church and affiliated welfare organizations.[86] Kenyon Butterfield and C. W. Coulter, both specialist American sociologists, were recruited to assist the Commission's researches because of the dearth of sociological expertise in South Africa. They were likewise attuned to the growing influence of young Afrikaner intellectuals in matters of social policy. As Butterfield noted in a letter to Malherbe: 'If you will permit me to say so I think it is from an increasing group of men like yourself, with the Dutch background, but with the broad point of view, who are to save the situation in South Africa. It is fairly clear to me that the Dutch rather than the British will dominate the politics

[82] Duncan Papers, BC294, C1.119, Keppel to Duncan, 16 May 1929.

[83] The other trustees were Carruthers 'Jock' Beattie and C. T. Loram.

[84] Duncan papers, C1.69 Keppel to Duncan, 4 May 1928. See also C1.11, Keppel to Duncan, 28 Oct. 1928; C1.15, Duncan to Keppel, 23 Nov. 1932, and C1.415 Keppel to Duncan, 20 Dec. 1932. A precedent for this form of organization was established in 1928, when a South African committee chaired by C. T. Loram was created to administer the Carnegie Visitors Grants programme.

[85] One of these was Dr M. Boehmke, of Teachers' College, Oudtshoorn. In 'Some Social Implications of the Poor White Problem', *SAJS* 25 (1928), Boehmke proposed the use of community-based social-survey techniques so as to analyse and solve the poor-white problem within a generation. See also Groenewald, 'The Methodology of Poverty Research', 65–6.

[86] Foreword to *The Poor White Problem in South Africa. Report of the Carnegie Commission*, vol. 1 (Stellenbosch, 1932), p. i.

of Africa for an indefinite future, and it is of importance that the leadership shall lie with the progressives.'[87]

As was signalled by Malherbe's controversial 1929 intervention, the issue of white poverty and its growing visibility was a highly sensitive question that crystallized the growing divide between Afrikaner moderates and militants in the 1930s. Afrikaner community leaders were agreed that the solution of the poor-white problem offered an opportunity for collective social 'uplift' and salvation. But the more radical amongst them—most notably the Stellenbosch psychologist and future prime minister H. F. Verwoerd—were inclined to blame the phenomenon of poor-whiteism on the socially exclusionary effects of a rapacious capitalist system (dominated by English and Jewish foreigners) that allowed blacks to 'undercut' the economic status of Afrikaner workers. The issue of poor-whiteism henceforth became a core mobilizing issue for the Afrikaner nationalist movement.

EDUCATIONAL SCIENCE AND SOCIAL RESEARCH

In 1929, as a direct result of his involvement in the Carnegie investigation, Malherbe was appointed to establish a National Bureau of Educational Research within the Department of Education. This agency played a formative role in endowing the state with enhanced social-science research capacity. But the constraints on state expenditure brought about by the world economic depression, coupled with Malherbe's concurrent involvement with the Carnegie investigation, meant that the Bureau was initially little more than a shell organization. The situation altered after 1933, when he successfully persuaded the Carnegie Corporation to grant £12,500 over five years to fund a Council for Educational and Social Research within the education ministry. For its part, the government made available an annual grant of £6,000 towards the running costs and salaries of the National Bureau, which in 1934 was enlarged to include a social-work division. It was renamed the National Bureau of Educational and Social Research.

The reconstituted agency was charged to act as an independent clearing-house for information bearing on all aspects of education and social work, to liaise between relevant departments of state, to maintain links with international bodies, and to conduct research and statistic-gathering activities.[88] In effect, it was the organization that Malherbe had envisaged in his 1921 *Cape Times* article. The Bureau's

[87] KCM 56979(268) 477/9, Kenyon Butterfield to Malherbe, 22 Oct. 1929. See also C. W. Coulter, 'Shifting the Emphasis in Social Work', *Social and Industrial Review*, 5 Jan. 1930, p. 6.

[88] KCM 56973 (130), 'Notes on Activities of National Bureau of Education and Social Research'; Malherbe, *Educational and Social Research*, 34, 49–50. Though formally separate, the relationship between the National Bureau and the South African Council for Educational and Social Research was close. The Bureau acted as the executive office of the Council and also received research grants from the Council. The Council was served by an advisory board, members of which included senior civil servants and provincial educationists, as well as representatives drawn from the South African Institute of Race Relations, teaching associations, universities, and so on.

structure and orientation were also considerably influenced by the American Social Science Research Council (est. 1923) as well as the Carnegie-funded New Zealand and Australian Councils of Educational Research.[89]

The Bureau was not the first grant-giving body in the South African state. In 1918 a Research Grants Board had been established within the Department of Mines and Industries, charged with stimulating applied scientific research and addressing wartime concerns about industrial self-sufficiency.[90] But, unlike this Board, which merely doled out money to support university-based researchers, and was strongly oriented towards the physical and natural sciences, the Bureau had the opportunity to generate social knowledge and the potential to influence government policies. This potential was not fully realized, but it represented an important statement of intent.

Malherbe was the driving force of the Bureau, and he was relatively successful in building up its research establishment between 1934 and 1938.[91] Ever the skilled interdepartmental operator and self-publicist, he used his position as a base from which he could coordinate and influence research activities within a number of state departments, principally those of Census and Statistics, Social Welfare, and Native Affairs. Malherbe sat on a number of interdepartmental commissions and maintained close contact with English and Afrikaans universities, the Inter-University Committee for African Studies, and the Institute of Race Relations.[92]

Malherbe's ability to operate across ethnic and institutional lines is a reminder that, in spite of growing strains, cooperation between anglophone and Afrikaans-speaking organizations and agencies was still possible during this period;[93] it was upon foundations such as these that Malherbe, Hofmeyr, and Smuts sought to entrench the ideology of South Africanism. A group photograph taken of Hofmeyr, Smuts, and Hendrik van der Bijl as the respective chancellors of the Wits, Cape Town, and Pretoria Universities, provides a telling image of this South Africanist concordat.

One of Malherbe's proudest achievements at the Bureau was the 1934 international New Education Fellowship (NEF) Conference, which he convened in association with J. D. Rheinallt Jones of the South African Institute of Race Relations. Like the 1929 BAAS meeting, the Conference took the form of a

[89] Malherbe, *Educational and Social Research*, 37–8, 52. Unlike the Australian and New Zealand Councils, however, the South African Bureau was only partly funded by Carnegie.

[90] The Research Grants Board developed out of the Industries Advisory Board, set up in 1916 and charged with helping to produce war *matériel National Council for Social Research. Annual Report for 1946* (GP-S.13811–1947-8, Pretoria, 1948).

[91] By the late 1930s the Bureau employed about 30 people and consisted of five 'divisions' : educational research and statistics, social research, psychological research, film, and library. See Malherbe, *Educational and Social Research*, 53–5. M. C. Botha, the education secretary, referred to him not only as 'the architect of the Bureau, but in a very real sense . . . the contractor and the clerk of works'. KCM 56973 (130), M. C. Botha to Sec. Public Service Commission, 21 Oct. 1939.

[92] KCM 65980 (242) 481/4, 'Educational and Social Research. Summary'.

[93] C. Kros, 'Economic, Political and Intellectual Origins of Bantu Education, 1926–51', doctoral dissertation, University of the Witwatersrand (1996), ch. 4.

semi-official jamboree. It was presided over by Hertzog, Smuts, and Hofmeyr, who were once more members of the same political party. Over 300 formal addresses were delivered to some 4,000 local and international delegates during the course of a month's proceedings in Cape Town and Johannesburg. Bronislaw Malinowski, John Dewey, Pierre Bovet, Mabel Carney, Fred Clarke, Beatrice Ensor, and William Boyd all featured as part of a glittering array of overseas guest speakers.[94]

The theme of the conference, 'educational adaptations in a changing society', was intended to reflect the progressive Deweyite philosophy of 'education as growth'.[95] But Dewey's potentially radical vision of education as an essential pre-condition for the shaping of a common democratic culture was countered by more prescriptive views of the need for education to serve the interests of society—or sections thereof. The deliberately non-sectarian approach of the New Education Fellowship meant that both these philosophies of education were fully represented.[96] Thus, 'new methods' centring on the needs of the individual child were evident in discussions about the uses of psychoanalysis and art in education, as well as in considerations of personality growth and the negative consequences of coercion in the classroom.

These progressive tendencies were overshadowed by an ethos of socially prescriptive scientism that was manifest in sessions devoted to testing and measurement, vocational guidance, and juvenile delinquency. Concentration on the requirements and maintenance of the prevailing social order emerged clearly in the (separate) section devoted to African education, where anthropological theories of cultural relativism and scientific attempts to measure the intellectual abilities of blacks were debated. Here, indeed, the politics of knowledge became highly instrumental.

The claims of racial science were brought into sharp political focus in debates about the 'educability' of Africans and the nature of 'Bantu mentality'. This discussion centred on research presented by M. L. Fick which purported to demonstrate that blacks were innately less intelligent than whites.[97] Also contentious, as several

[94] E. G. Malherbe, *Educational Adaptations in a Changing Society* (Cape Town, 1937), pp. iii, 540–5; KCM 57012 (16), 505/3/2, 'An International Conference of Education and Social Work in South Africa', July 1934. The local contingent comprised most of the South African educational and social-work establishment—English and Afrikaans, black and white—such as Simon Biesheuvel, Edgar Brookes, J. Chris Coetzee, G. R. Dent, M. L. Fick, A. W. and R. F. Hoernlé, D. D. T. Jabavu, H. P. Junod, I. D. MacCrone, E. B. M. Morake, K. T. Motseti, W. Nicol, H. Verwoerd, G. H. Welsh, W. A. Willemse, and A. B. Xuma.

[95] Malherbe, *Educational Adaptations*, p. v. The NEF was founded in 1915 by Beatrice Ensor. It was 'founded on the belief that education is the key to to-morrow, and that only by organising in a co-operative way all the formative influences which bear on the child through the home, the school, and other social agencies, can we equip the child to solve many of the baffling social and economic problems which to-day confront us and which may be still more complex when our children grow up' (p. xvi).

[96] As Beatrice Ensor, international chair of the NEF, said of its philosophy: 'It was begun and has continued a non-political and non-sectarian movement. It has no formal creed, and its catholicity is such that it embraces the extreme experimentalist and the cautious progressivist.' Ibid. 7

[97] Ibid., ch. 18.

writers have shown, was the notion of fashioning a system of 'social education' specially 'adapted' to African needs and expectations—an idea that was challenged by African teachers, independent mission educationists, and evangelical exponents of the 'social gospel'. Led by Victor Murray, this liberal-oriented Christian network challenged the growing scientific authority of segregationists and self-proclaimed anthropological 'experts' who wished to draw African education within the regulatory domain of the Native Affairs Department.[98]

Considered as public spectacle, the NEF Conference's overriding significance lay in the manner in which South African educational and social debates were projected in a national as well as an international context. The desire of its organizers to express South Africanism in the language of world citizenship and extend its remit beyond the boundaries of Empire/Commonwealth was striking. Remarkably, the NEF Conference proved successful in bringing together theorists and practitioners who were representative of different ideological tendencies. Its timing, after the worst of the economic depression, and at a moment when concerted efforts were being made to show that the ruling United Party was not misnamed, was fortunate. A conference on this scale could not easily have been convened a decade earlier (given the lack of professional expertise and institutional capacity), much less a decade later (when the acrimonious tone of white politics would have made such an eclectic gathering inconceivable).

The question of South Africa's relationship with the outside world was an important point of reference. Amongst the recurring themes of the conference were discussions about internationalism, world cooperation, and the challenges of democracy and industrial modernity. A number of visiting speakers celebrated the appearence in South Africa of a newly emerging white nation, and pointed out affinities between it and their own societies. The problematic relationship between internationalism and nationalism was extensively discussed. Beatrice Ensor, international chair of the New Education Fellowship, spoke enthusiastically about the 'World Commonwealth', while assuring her audience that the Fellowship did not 'favour a form of World Citizenship at the expense of the proper kind of National Citizenship'.[99] Fred Clarke observed that 'Western civilisation no longer stands single, confident and unchallenged'; in his view, the task of safeguarding the 'great Graeco-Roman inheritance' and of refreshing old Anglo-Saxon institutions could best be undertaken by 'new countries' such as the British dominions and, especially, the United States. In his exploration of patriotism, nationalism, and internationalism, Pierre Bovet maintained that a 'double allegiance is possible to one's country and to humanity, which ought not

[98] S. Krige, 'Segregation, Science and Commissions of Enquiry: The Contestation over Native Education Policy in South Africa, 1930–1936', *JSAS* 23: 3 (1997); B. D. Fleisch, 'The Teachers College Club: American Educational Discourse and the Origins of Bantu Education in South Africa, 1914–1951', Ph.D diss., Columbia University (1994), 172 ff.; Shingler, 'Education and the Political Order in South Africa', 128. Shingler's unpublished thesis remains one of the best treatments of education and apartheid.

[99] Malherbe, *Educational Adaptations*, 7.

to be difficult to South Africans'. To support this contention, Bovet took delight in quoting from the Afrikaans poet Louis Leipoldt: 'He is the greatest patriot who loves all mankind.'[100]

Such sentiments would have been welcomed by Smuts and Hofmeyr, who were keen to present South Africa as a dynamic new country with a distinctive role in world affairs. Hofmeyr argued that the country was changing from an economically self-supporting and isolated society with 'a limited vision' to 'a South Africa conscious of its integration with a world family of nations'. As before, he sought to advertise South Africa as a 'microcosm' in which world-scale problems could be studied with unusual clarity. In a global environment of increasing integration and competition, it was necessary to train 'men and women who will at once be quick with the realisation of their South African citizenship, and ready to meet the obligations which rest upon them as citizens of the world'.[101]

In a similar vein, Malherbe placed considerable emphasis on the country's unique status as a focus for worldwide social and scientific enquiry. His introduction to the New Education Fellowship Conference proceedings typified this outgoing approach to South Africa's domestic problems, and shows the confidence and pride which he took in the country's youth and vitality. At the same time it reveals his decidedly narrow conception of race relations:

> South Africa is a new country and, though it simply bristles with problems, the issues (which are really world issues) are thrown into relief more clearly here than in the older countries where they are often obscured by the complexities of tradition and deep underlying prejudices which parade under the name of Western Civilization. As an experimental station and laboratory in racial and cultural relations South Africa is almost unique in the world to-day. A small white population of less than two millions, situated at the tip of a vast continent of 140 millions of black people, is entrusted with the task of bearing the torch of European civilization undimmed. Moreover there is also the clash of cultures between the two white races of English and Dutch descent, each clinging tenaciously to its own cultural distinctiveness.[102]

In the period after the 1934 NEF Conference Malherbe turned his attention to getting research projects within the Bureau under way.[103] Yet, during the period of the Council's and Bureau's greatest activity (1935–8) Malherbe devoted rather more attention to selling the idea of research and to establishing the national and international reputations of these organizations than in monitoring or shaping the quality of their output. Arising out of the Carnegie investigation, resources were targeted towards studies of education in socially deprived white schools. Efforts to standardize individual intelligence tests were given particular emphasis, as were

[100] Ibid. 43, 50, 53. [101] Ibid. 2.

[102] Ibid., p. iv. Cf. KCM 57012 (16), 505/3/2, 'An International Conference of Education and Social Work in South Africa', July 1934, pp. 2–3.

[103] The Bureau also played an important role in organizing the 1946 National Conference on Social Work in Johannesburg. See F. Brummer (ed.), *Report of the National Conference on Social Work* (Pretoria, 1937).

studies into the uses of vocational guidance, juvenile delinquency, the nature of the school curriculum, the educational possibilities of film, the implications of bilingualism, and the effects of nutrition on physical development.[104]

From the mid-1930s Malherbe attached increasing weight to research into 'problems affecting the non-European population'.[105] If, however, he meant that the interests of blacks were being served by such attention, he was being disingenuous: most of the work sponsored by the Council was conditioned by the unstated but automatic assumption that a greater knowledge of problems affecting blacks was needed in order to secure the well-being of whites. In short, it was not so much the problems affecting Africans that needed attention, as Africans themselves who constituted the problem. Unsurprisingly, much of the applied research work commissioned by the Council to seek a solution to such problems—comparative racial intelligence testing in particular—fed directly into segregationist and, later, apartheid discourse.[106]

Two young tyros of apartheid volkekunde anthropology, P. J. Coertze and P. J. Schoeman, were recipients of Council research grants. Students of the Stellenbosch anthropologist-turned-native administrator Werner Eiselen, they were to play significant roles in providing apartheid ethnology with a theoretical underpinning.[107] The unspoken and unquestioned assumption of the Council and Board was that the social and economic interests of whites were of foremost concern. Yet, it would be mistaken to conclude that the Council was possessed of a clear research agenda that consciously strove to entrench racial segregation, or that it entirely shut out contrary views. It is worth noting that a number of grants were made to academics opposed to segregation and doctrinaire Afrikaner nationalism.[108] Inspection of the list of grants made by the Research Grants Board in the period 1919–34 corroborates the view that funds were disbursed on a broadly non-partisan basis. (It is somehow pleasing to note that in 1929 the Board managed to turn down research applications by J. C. Smuts for a study of grasses as well by as H. F. Verwoerd on 'Crime and Punishment in South Africa'.[109])

[104] Malherbe, *Educational and Social Research*, 68–71. The Bureau conducted most of this research but Council grants were also given to individual researchers.

[105] Ibid. 68.

[106] See e.g. M. L. Fick, *The Educability of the South African Native* (Pretoria, 1939); J. A. J. van Rensburg, *The Learning Ability of the South African Native* (Pretoria, 1938). Cf. G. R. Dent, *Zulu Intelligence Test* (Natal, 1938); P. A. W. Cook, *The Native Standard VI Pupil* (Pretoria, 1939). The studies by Dent and Cook were rather more sympathetic to the environmental, cultural, and social factors affecting African education than the racial determinism of Fick and van Rensburg, but Cook's 'culturalist' influence on the 'Bantu' education policy in the 1940s and 1950s was strong. See e.g. M. A. Beale, 'Apartheid and University Education, 1948–1970', Ph.D thesis, University of the Witwatersrand (1998), 191–6, 226–7; Fleisch, 'Social Scientists', 360–1.

[107] R. J. Gordon, 'Serving the Volk with Volkekunde—On the Rise of South African Anthropology', in J. D. Jansen (ed.), *Knowledge and Power in South Africa* (Johannesburg, 1991).

[108] These included H. M. Gluckman, Ellen Hellmann, Simon Biesheuvel, B. W. Vilakazi, R. T. Caluza, T. J. Haarhoff, and Sheila van der Horst. Malherbe, *Educational and Social Research*, 68–71; 'Educational and Social Research', KCM 56980 (242), 481/4.

[109] State Archives, Pretoria, HEN 3313 505/1/13 and MNW 973 917/29, part 2.

Malherbe did not reflect deeply on the substance of the intellectual enquiries that he sponsored. His major concern was to enhance the sociological and educational research capacity of the country, whose deficiencies he ascribed to a lack of resources and to the relative immaturity of the university system. This fluidity, coupled with the non-sectarian manner in which research grants were made, suggests that, notwithstanding political tensions, the politics of knowledge were still relatively open in the 1930s: the Bureau and Council were principally concerned to broaden the base of social research, and national research was conceived as part of the process of constructing nationhood. Malherbe's efforts to expand and coordinate the country's research base—without particular regard to who was conducting this research and why—can perhaps best be explained in terms of his urge to promote the development of South Africa through the ideology of South Africanism. A similar spirit was present in the federal University of South Africa which, during the inter-war years, still managed to 'reconcile a growing South Africanism, a narrower Afrikaner nationalism and the mellow imperialism of those autumn days of empire'.[110]

By 1937 weaknesses in the Bureau were becoming more apparent, notwithstanding—or perhaps as a result of—its rapid growth. The boundaries of Malherbe's research empire were ill-defined, and there are indications that the scope of its activities and ambitions was resented by competing government departments. The position of the old Research Grants Board was also up for review. Since 1920 it had expanded its role by disbursing grants to universities and museums, and from 1928 the Research Grants Board was charged with administering an annual grant of £2,000 provided by the Carnegie Corporation to promote promising South African research work.[111]

Informally constituted and run by an honorary committee of government officials and senior academics, the Research Grants Board had proved reasonably successful in directing funds to projects deemed to be worthwhile, with a minimum of administrative overheads.[112] But the broadening scope of its grant-giving powers now rendered its location in the Department of Mines and Industries increasingly anomalous (a point that the Education Department was particularly insistent upon).[113] In addition, much of the Board's work—as well as the source

[110] M. Boucher, *The University of the Cape of Good Hope and the University of South Africa, 1873–1946* (Pretoria, 1974), 260.

[111] For a survey of the Research Grants Board and a list of the projects funded by it see UG 13-'36, *Report of the Research Grant Board 1918–1935*. Hofmeyr reviewed the reasons for its disbandment in *Cape Times*, 26 July 1938. See also KCM 56980 (168), Sec. Commerce and Industries to J. H. Hofmeyr, 20 Nov. 1936, for detailed proposals to disband the Board.

[112] The Research Board was highly praised in an independent report by H. Miers and S. F. Markham on *Museums and Art Galleries of British Africa* (Edinburgh, 1932), 38. Amongst the projects refused funding in 1929 by the Research Grants Board was research by J. C. Smuts into grasses and by H. F. Verwoerd on 'Crime and Punishment in South Africa'. See State Archives, Pretoria, MNW 973 917/29, part 2.

[113] See e.g. State Archives, Pretoria, UOD 719 E18/12pt.2, Under-Sec. Education to Sec. Education, 25 Feb. 1927.

of its funding—overlapped with Malherbe's Bureau. A further unsettling factor was the Carnegie Corporation's reluctance to continue funding the Council for Educational and Social Research on a permanent basis. In a damning indictment, Keppel wrote in 1942 that 'no outstanding success' had resulted from its heavy investment in the South African government's research programme, though he thought the output to have been of 'creditable quality' and suggested that it had exerted an important indirect influence.[114]

All these factors invited calls for reform and rationalization. Following a lengthy consultative process, the minister of education, J. H. Hofmeyr, finally announced in July 1938 the creation of a newly constituted National Research Council and National Research Board. Chaired by the Wits geologist R. B. Young, it was intended to subsume and expand the activities of the Research Grants Board. Preference was to be given to investigations which 'have a South African bearing' or which could be carried out 'with distinct advantage' in the country.[115] Membership of the new body included representatives from government departments, specialist research institutes, and the universities. The *Cape Times* described the process with a mixture of delicacy and overstatement as a 'peaceful revolution in the control of research work'.[116]

The launch of the new Research Council and Board provided an occasion for taking stock of the country's intellectual development. In public, an upbeat Hofmeyr made much of the 'national prestige' and 'national honour' that flowed from internationally significant scientific discoveries. Hofmeyr envisaged the National Research Council as 'our South African Parliament of Research'. Young spoke proudly of the 'radical change' that had taken place over the past century: 'Our earliest investigators were inquisitive visitors, but the majority today were born, educated and trained in South Africa.' The tributes paid by Hofmeyr and Young to the Carnegie Corporation for its ongoing support of South African research since 1927 further reinforced the trope of increasing national self-reliance. Hofmeyr promised that although the phase of external funding was coming to a natural end, the government intended to increase the overall resources available to the Research Council. An ambitious programme of research initiatives was duly announced.[117]

114 KCM 56983 (32b), 519/2, 'Memorandum on the Activities of Carnegie Corporation in the British Colonies and Southern Dominions', 26 Jan. 1942, p. 4

115 KCM 56980 (171), 481/1, Memo on 'Organisation of Research Work', signed G. W. Eybers, for Secretary for Education, 21 Nov. 1937. The Council was intended to be an advisory and policy committee while the Board acted in an executive role. Young had previously been chair of the Research Grants Board. Although the Carnegie Corporation withdrew from funding government-controlled research, it provided over $50,000 to the new Bernard Price Institute for Geophysical Research at Wits, established in 1938, under the directorship of Basil Schonland.

116 *Cape Times*, 26 July 1938 (editorial).

117 Ibid., 26 July 1938. See also State Archives, Pretoria, UOD 1954 E184/2/2 vol. 1, Minutes of Inaugural Meeting of National Research Council and Board, 25 July 1938, p. 2: 'Mr Hofmeyr had referred to the Carnegie Corporation of New York as the fairy god-mother of the late Research Grant and Board and he [Spencer Lister] thought Mr. Hofmeyr might be referred to as the Prince Charming

THE STATE AND WARTIME RESEARCH

We have seen how, during the 1930s, the South African state (or elements within it) moved to assert centralized control over the domain of knowledge production: this included coordination of existing research work in ministerial departments, funding research work in universities, and taking a more visible role in promoting studies deemed to be of national importance. The Carnegie Corporation's withdrawal of support to the Educational Research Bureau forced the government to prove its self-sufficiency as a national knowledge-producer and manager, and ended its increasingly anomalous reliance on overseas foundation funding. But, despite Hofmeyr's stated intentions, his promises to the National Research Council were not realized.

Expectations that the loss of the Carnegie subvention would be adequately compensated for were soon dashed, as was a proposal, apparently supported by Minister Fagan, to establish a National Research Fund with an initial capital of £100,000. One problem was that many researchers diverted their energies into war work; another was that Hofmeyr's peacetime promise of additional funds did not materialize. University research grants and scholarships, for instance, were pegged back to a total of £1,500 per annum. And there was little to suggest that the monies that were handed out were used effectively. The minutes of the ninth general meeting of the Research Board, held in 1940, for example, record an uninspiring record of individual grants (typically £50–£100), some of which were unused. For example, Professor C. S. Richards's research on 'Company Promotion in South Africa' was discontinued; Dr T. W. B. Osborn repaid his grant of £50 for research into the 'cause and prevention of dental caries' after enrolling for active service; Dr A. A. van Schalkwyk was recorded as owing £15 for an uncompleted project; Mr Language, who had overspent £10, proposed to purchase his bicycle and 'partially worn out' flash-light for £3. 10*s*. 0*d*. himself, on the grounds that he had had no success in securing a buyer. This lame suggestion was accepted.[118]

In 1939 the Research Council initiated a comprehensive survey of ongoing research work in the sciences, social sciences, and humanities, with the aim of prioritizing areas deemed worthy of coordinated research support. Rather than merely responding to requests for grants as before, it now hoped to provide more in the way of strategic direction.[119] Yet there was little to indicate that the Board

who had rescued Cinderella and solved our problems. Mr Hofmeyr had to leave the meeting at this point.' KCM 56980 (244) 'Report of Survey Research Committee', n.d. [1940]. A number of subcommittees were constituted to take stock and recommend research agendas in fields across the sciences and humanities. For replies, see KCM 56980 (194–215).

[118] UOD 1954 E184/2, vol. 3, 'Minutes of 9th general meeting of the National Research Board', 4th Nov. 1940.

[119] E. G. Malherbe papers, KCM 56980 (244), 'Report of Survey Research Committee', n.d. In the field of social research, for instance, it recommended research into South Africa's national income, the 'impact of European civilisation on non-European life in the Union, and vice versa', and

was able to realize this aim, and several of its own members, like the botanist Illtyd Pole-Evans, expressed intense frustration.[120] The Natal university economist H. R. Burrows argued in 1941 that there was no reason for the continued existence of the Research Council if all it did was distribute 'doles' to research workers. Drawing on the discourse of wartime mobilization, he argued that a democracy which remained 'haphazard, unplanned and uncoordinated' could not hope to 'subdue a ruthless totalitarian regime which mobilises all effort, all thought, and all will'. Instead, Burrows outlined a 'somewhat visionary' scheme for a 'really National Research Council' which would help to translate the experience of organizing for war to the solution of longer-range problems in a post-war world with regard to agriculture, urban poverty, and 'inter racial maladjustments'. This would entail not merely responding to requests for financial support but actively encouraging new research, in particular into post-war questions of economic and social change.[121]

Whatever internal weaknesses it suffered from, the Research Board's efforts were stymied by Hofmeyr, who refused an application to allow unused monies from the Council's £4,000 annual research grant-in-aid to be held over to enable it to accumulate resources and fund large research programmes. Hofmeyr insisted that it was unsound, as a matter of principle, 'that in wartime we should be building up what is virtually a hidden reserve', notwithstanding the small amounts of money involved.[122] This petty parsimony, or sanctimony, was at odds with Hofmeyr's own professed enthusiasm for the National Research Council, and also with the tenor of wartime support for government-led planning and social reconstruction. His refusal may have reflected disappointment with the Research Board's mediocre record, but it was fully consonant with his desire to maintain tight control of treasury purse-strings. This is only one instance among many where the government's rhetorical support for ambitious wartime social programmes was tempered by a reluctance, or inability, to back such ideas in practice.[123]

Pressures to expand state support of research work did, however, continue to gain momentum. Participation in the war against fascism raised hopes and expectations of a better future, while the optimistic language of citizenship, welfare, and reconstruction encouraged the growth of a progressive, interventionist state.

'the increasing industrialisation and urbanisation of the Union's population (European and non-European)'. In the field of geophysics, research areas in terrestrial magnetism, atmospheric electricity, wireless and the ionosphere, and the earth's heat were identified. Support for the Bureau of Archaeology, a detailed survey of rock paintings, compilation of a Zulu–English dictionary, and the production of 'electrically recorded gramophone records of Bantu music' were some of the activities deemed worthy of the Council's attention with respect to archaeology and social anthropology.

[120] Brian Austin, *Schonland: Scientist and Soldier* (Bristol and Philadelphia, 2001), 302.

[121] H. R. Burrows, 'Suggested War-Time Functions of National Research Council', 3/2/41, KCM, KCM 56980 (247), 481.

[122] UOD 1954 E184/2/2, vol. 2, Smeath-Thomas to Hofmeyr, 14 Oct. 1943, and reply, 20 Oct. 1943; Smeath-Thomas to Hofmeyr, 2 Nov. 1943.

[123] See e.g. S. Dubow and A. Jeeves (eds.), *South Africa's 1940s: Worlds of Possibilities* (Cape Town, 2005).

Rapid economic expansion, coupled with the urgent need for munitions and military equipment to support the war effort, gave a major boost to manufacturing industry. This, in turn, underlined the need for more intensive research and greater coordination in the fields of pure science and technology.[124] By the end of the war, new state-sponsored scientific and technical organizations, such as the South African Bureau of Standards and the National Institute of Personnel Research, were established in an attempt to improve efficiency and enhance the country's manufacturing capacity.[125]

In 1941 the Industrial and Agricultural Requirements Commission recommended the establishment of a body which, a year later, came to be known as the Social and Economic Planning Council. Chaired by the influential and ubiquitous industrialist H. J. van Eck, this high-level body was given a wide remit, which centred on providing the government with advice on economic and social policy and helping to coordinate and better utilize the country's resources.[126] A year later the South African Association for the Advancement of Science organized a special symposium on 'Science and Post-War Reconstruction'. Hofmeyr presided as chairman, and Smuts sent a message of support in which he made the point that scientific planning would have to play an important part in helping to build a new world on the ruins of the old.[127]

The principal speakers at this symposium included H. J. van der Bijl, director-general of War Supplies since 1939, and H. J. van Eck, director of the newly formed Industrial Development Corporation (of which van der Bijl was chairman). This ambitious and far-seeing pair were the leading figures in South Africa's process of industrial expansion. Through their stewardship of major public-private ventures in electricity and iron and steel production, and much else besides, they exerted a powerful influence on wartime economic development and planning.[128]

As director of War Supplies (in which capacity he coordinated 600 factories and built forty more[129]), and as first chairman of the Industrial Development Corporation (est. 1940), van der Bijl provided critical support to the Smuts government. Production of army boots and blankets, tyres, munitions, armoured cars, precision instruments, and explosives, were all managed by the director of War Supplies and used in operations in North Africa and the Middle East.[130] South Africa's war production was coordinated with that of India, Australia,

[124] H. Mews, 'The Coordination of Scientific Research in Africa South of the Sahara, With Special Reference to Information Services', *South African Libraries*, 21: 2 (1953), 3.

[125] J. Lewis, *Industrialisation and Trade Union Organisation in South Africa, 1924–55* (Cambridge, 1984), 113, 120–1.

[126] Malherbe Papers, KCM 56980 (295), 'Social and Economic Planning Council' [n.d., *c.* 1949]

[127] 'Message from the Prime Minister, the Rt. Hon. J. C. Smuts, F.M.', *SAJS* 39 (1942), 313.

[128] N. L. Clark, *Manufacturing Apartheid: State Corporations in South Africa* (New Haven, 1994).

[129] R. Christie, *Electricity, Industry and Class in South Africa* (London, 1984), 120.

[130] D. G. Kingwill, *The CSIR: The First 40 Years* (Pretoria, 1990), 6–7; A. Jacobs, *South African Heritage: A Biography of H. J. van der Bijl* (Pietermaritzburg, 1948), 159 ff.

and New Zealand. Hofmeyr's shrewd management of the macroeconomy was a vital aspect of South Africa's wartime prosperity. But, to a far greater extent than has been acknowledged by historians, it was van der Bijl's endeavours in the industrial sphere that directly underpinned its contribution to the overall war effort.

Van der Bijl personified the scientism and planning ethos that started to emerge from the First World War and came to fruition in the Second. His career, incidentally, has several parallels with that of E. G. Malherbe. An eighth-generation Afrikaner, whose father was on friendly terms with Smuts, van der Bijl grew up on a farm in the western Cape and graduated in physics from Stellenbosch University. He took a doctorate at Leipzig in 1913, where he came to the attention of the Nobel Prize-winning physicist R. A. Millikan. Through Millikan, van der Bijl was offered a position at the embryonic Bell Telephone Laboratories in New York, where he conducted specialist research in thermionics. He was invited to return to South Africa by Smuts at the end of the First World War, to become the government's first scientific and technical adviser in 1920. Like Malherbe, who returned from Columbia at this time, van der Bijl relished the opportunity to apply modern American industrial and technical methods in South Africa. He wrote a paper in 1919 on 'Scientific Research and Industrial Development', in which he advocated the use of scientific methods to promote secondary industrialization in South Africa. Van der Bijl deployed his considerable managerial skills to coordinate the resources of state and industry, steering clear of party politics wherever possible. As director of War Supplies he declined a salary as well as a cabinet position. His vision of South Africa's long-term national interest was broadly socially inclusive, and shaped by a political pragmatism that projected itself as non-ideological and was shaped by a corporatist view of collective endeavour.[131]

Van der Bijl was an Afrikaner patriot in the Smutsian tradition. But, above all, he was a scientist and technocrat who considered that the expansion of industry was the key to South Africa's future. Self-sufficiency in power, steel, and transport was the essential building-blocks of economic growth. Partnership between the private sector and the state was the preferred industrial model, and he jealously guarded Escom's statutory independence while enforcing its advantages as monopoly. Van der Bijl's optimistic belief in the progressive, socially transforming power of cheap electricity distributed on a wide scale was monumentalized in Escom House, a towering, American-influenced modernist skyscraper in central Johannesburg with a fine art deco interior, whose overall design van der Bijl planned in meticulous detail. The inscription on the central panel of its 'Hall of Achievement' was: 'Dedicated to the ideal of cementing together by common

[131] For details on van der Bijl, see Clark, *Manufacturing Apartheid*; A.Jacobs, *South African Heritage*; 'Hendrik Johannes van der Bijl, 1887–1948', by B. F. J. Schonland, *Obituary Notices of Fellows of the Royal Society*, 7 (1950–1), 27–34.

Fig 9. View of Electricity Supply Commission House, Johannesburg. At the time of its completion in 1937 it stood 21 storeys high and was the tallest building in Africa. (Courtesy of Eskom)

endeavour for achievement, all the peoples of South Africa, regardless of race or creed, into a brotherhood of mutual trust and goodwill for the welfare of our country and the glory of Almighty God.'[132]

On the occasion of the opening of the building in 1937 van der Bijl credited Smuts with the inspirational idea of Escom, whose purpose was nothing less than 'to preserve our civilisation in Southern Africa, to improve the living conditions of our people and to make South Africa a still better country to live in'. In response,

132 When completed in 1937 Escom House was the tallest building in South Africa. Aesthetically pleasing and intended to facilitate 'a maximum of work with a minimum of effort', it was technically sophisticated, air-conditioned, and finished to a high standard. The building was stepped back from the street so that every office received direct sunlight. www.eskom.co.za/heritage/escom house/escom house.htm; Jacobs, *South African Heritage*, 93–5. Cf. Christie, *Electricity, Industry and Class*, who stresses the role of electricity in maintaining South Africa's class and racial hierarchies.

Smuts lauded Escom, referring to it as a form of 'state capitalism'.[133] Van der Bijl was also directly responsible for the construction, in the mid-1940s, of a model industrial city to the south of Johannesburg, which significantly expanded the Witwatersrand's economic base. Vanderbijlpark, as it was called, exemplified key aspects of the collectivist and welfarist ethos that infused wartime state planning and industrial paternalism. Van der Bijl envisaged that this 'ideal city' would be based 'purely upon sociological lines', and that it would 'be guided by legislation primarily intended for the benefit of the masses'. Albeit segregated, the provisions to house black workers on site with their families had the merit of recognizing African urbanization as a fact—at the same time, it ensured their relative isolation from the direct influence of militant Johannesburg-based unions.[134]

In addressing the 'Science and Post-War Reconstruction conference on the industrial aspects of the war', van der Bijl and van Eck together laid stress on the need to manage human and material resources effectively and efficiently. These themes were further reiterated in presentations on agriculture, nutrition, education, housing, manpower utilization, and welfare provision. Productivity was an important theme. But this was balanced by attention to the importance of social inclusivity, most notably in talks delivered by Margaret Ballinger on 'the native and post-war reconstruction' and A. J. Downes on trade-unionism.[135]

Claims for the significance of applied scientific knowledge in welfare and planning were gaining ground in colonial Africa too, as the British government came to realize that the maintenance of imperial ties depended on greater investment and enhanced welfare measures. Smuts's conception of Africa as a 'living laboratory', as outlined in his 1929 Rhodes Memorial Lectures in Oxford, became the 'framing motif', in Helen Tilley's words, of the major research survey of colonial institutions and the 'machinery of knowledge', led by Lord Hailey.[136] Supported by the Colonial Office and sponsored by the Carnegie Corporation, this comprehensive imperial audit culminated in a linked series of volumes, published in 1938.[137] These studies highlighted common problems in colonial administration and contributed, in turn, to a greater readiness to bolster imperial

[133] Cited in Christie, *Electricity, Industry and Class*, 129. Jacobs reports in *South African Heritage*, 221, that van der Bijl was asked by an interviewer in 1938 whether he was 'by any chance a Socialist'. He replied that Escom was 'a form of socialism' but that he did not believe in 'changing our order of things by revolutionary methods. It may generally be admitted that the social structure of the world is far from satisfactory, but it is very closely interlocked with the world's economic machine—a very complicated affair.'

[134] Jacobs, *South African* Heritage, 184–95. On van der Bijl's labour strategies see N. Clark, 'The Limits of Industrialisation Under Apartheid', in P. Bonner *et al.*, *Apartheid's Genesis 1935–1962* (Johannesburg, 1993).

[135] *SAJS* 39 (1942), 313–81.

[136] Helen Tilley, 'Africa as a "Living Laboratory": The African Research Survey and the British Colonial Empire: Consolidating Environmental, Medical, and Anthropological Debates, 1929–40', D.Phil. thesis, Oxford University (2001), 6, 36, 78.

[137] The Survey's official published volumes were: Lord Hailey, *An African Survey* (London, 1938); E. B. Worthington, *Science in Africa* (London, 1938); S. H. Frankel, *Capital Investment in Africa* (London, 1938).

power through the adoption of large-scale development programmes. The 1940 Colonial Welfare and Development Act was one important expression of such intent.

One of the most important outcomes of the African Research Survey was increasing recognition of the need for more research facilities based in Africa itself.[138] Similar requirements were identified in South Africa, but the context here was distinctive. Unlike 'dependent' colonial Africa (British, French, and Portuguese), the Union already possessed a relatively developed scientific and industrial framework. Another difference was that South Africa, though acting in support of the Empire in the war against fascism, was in a position to put its own national interests first; whatever strains of idealism or humanism are detectable in the African Research Survey, the maintenance and consolidation of colonial power through the principle of trusteeship remained its underlying motivation.[139]

From 1943, with the Axis powers in retreat and domestic Afrikaner opposition embroiled in furious internecine conflict, the United Party was in a unique position to capitalize on its more secure political position. It had also just been re-elected to government with a large majority. In social and economic policy several ambitious programmes were mooted, though most came to little.[140] In the field of science, however, reconstruction was carried through with considerable efficacy, borne upon a wave of wartime expectancy, industrial need, and a commitment to societal transformation. The apparent neutrality of technology and applied science meant that ambitious programmes could be put into effect more easily than was the case in the highly contested domain of social policy. It was thus in the context of what Edgerton refers to, with respect to Britain, as a warfare as well as a welfare state, that South African science made its greatest advances.[141]

In 1944 Smuts recruited Basil Schonland, his wartime scientific adviser, to establish a new umbrella scientific and industrial research organization, the Council for Scientific and Industrial Research (CSIR). Schonland was ideally suited to the role of enterprising scientific statesman. After studying under Ernest Rutherford and James Chadwick at the Cavendish Laboratory at Cambridge, Schonland took up a lectureship in the department of physics at the University of Cape Town in 1922, remaining there until 1937 when he became the founding professor of geophysics and director of the newly created Bernard Price Institute at Wits University. A year later he was elected as a Fellow of the Royal Society in recognition of his pioneering work on lightning and atmospheric electricity (the first Wits faculty member to be so honoured). In response to an urgent directive from Smuts in 1939, Schonland secretly used the resources of the Bernard

138 Tilley, 'Africa as a "Living Laboratory" ', 266

139 Anker, *Imperial Ecology*, 218.

140 Dubow and Jeeves (eds.), *South Africa's 1940s*.

141 David Edgerton, 'Science and the Nation: Towards New Histories of Twentieth-Century Britain', inaugural lecture, Imperial College, London, 2002; id. *Welfare State: Britain, 1920–1970* (Cambridge, 2005).

Price Institute to develop from first principles, and at great speed, a system of radar designed to defend the east and southern African coastline from German attack.[142]

In addition to his considerable scientific pedigree—he was by now South Africa's pre-eminent scientist—Schonland enjoyed a distinguished career as a soldier, serving in the Royal Engineers in the First World War (he received an OBE) and as Field Marshal Montgomery's official scientific adviser (with the rank of brigadier) during the allied invasion of France. Schonland's political outlook was liberal and internationalist. He combined a strong sense of South African patriotism with a commitment to the Commonwealth. He moved with ease within the British scientific establishment and maintained close links with peers in Canada, Australia, and New Zealand. These credentials made Schonland the ideal candidate to become Smuts's scientific supremo, but his elevation was also smoothed (as it was in the case of van der Bijl) by personal ties: on the one hand through his father, Selmar, the distinguished Grahamstown botanist who had been a friend of Smuts, and on the other through his wife, Ismay, to the Hofmeyr family.[143]

Armed with Smuts's patronage and support, Schonland worked swiftly to bring the Council for Scientific and Industrial Research into being. The organization's framing document stressed the importance of applied science in all aspects of the country's industrial and social life; it made the case for greater co-ordination of research; and argued that existing state and private support of scientific research was woefully inadequate. Schonland based the CSIR on existing British, Canadian, and especially, Australian models (American institutions were also considered), so as to endow the new organization with autonomy as well as authority. He later claimed that it embodied the best of their virtues and none of their defects.[144]

The cumulative record of the Research Grant Board, the Educational Bureau, and the Research Council and Board over a period of twenty years demonstrated the problems of placing research within the bureaucratic constraints of a government

[142] B. K. Murray, *Wits: The 'Open' Years, 1939–1959* (Johannesburg, 1997), 22–3; G. R. Bozzoli, *A Vice-Chancellor Remembers: The Memoirs of Professor G. R. Bozzoli* (Randburg, 1997).

[143] Austin, *Schonland, passim*, also foreword by Maurice Wilkes, pp. ix–x. Schonland left South Africa in 1954, partly on account of his disaffection with apartheid, and rose to the position of director of atomic energy research at Harwell in 1958, one of the most prestigious scientific positions in Britain. In 1960 he was knighted for services to British science. He maintained connections with South African science and was made Rhodes's first chancellor in 1951. He resigned this position in quiet protest at Rhodes's decision to award an honorary doctorate to the state president, C. R. Swart, who was closely associated with the imposition of apartheid in university education. See Austin, *Schonland*, 405.

[144] CSIR, *Objects and Policy of the CSIR (An Initial Statement)* (Pretoria, 1945), 5–6. This document estimated that the maximum amount voted by parliament for all scientific research in 1944 was £618,000, one-sixth of the 'generally agreed' desirable figure—1% of national income. In Britain a Department of Scientific and Industrial Research was established in 1915. Comparable organizations were established in Australia (1920), Canada (1916–17), and New Zealand (1926). H. J. van der Bijl recommended the creation of a government research institute in 1921, but the proposals lapsed after the election of 1924. See D. G. Kingwill, *The CSIR*, 5–6; also Churchill Archives, Cambridge University, SCHO1/2, Letter from Schonland to Sir Keith Hancock, 2 Jan. 1957.

department. Schonland therefore insisted that the CSIR had to be constituted as a semi-independent 'body corporate', beyond the reach of the civil service and with direct responsibility to the prime minister. He also ensured that power was effectively concentrated: as president of the Council, Schonland was simultaneously its chairman and chief executive officer. By these constitutional means, the CSIR was able to enjoy considerable freedom of movement while benefiting at the same time from direct prime-ministerial access and patronage.[145]

Schonland was also insistent that the CSIR should either absorb or establish effective working relationships with the plethora of research organizations and institutes that already existed within government departments and the universities, ranging from well-established organizations like the Medical Research Institute and Onderstepoort, to disparate laboratories devoted to researching products such as sugar (1925), fuel (1930), and leather (1941). Despite Schonland's preference for avoiding duplication of activities, some of these, like the South African Bureau of Standards and the Metallurgical Research Laboratory, held out as independent entities.[146] The CSIR was formally constituted by parliament in 1945—a decision that commanded a surprising degree of cross-party assent. It commenced existence with four national laboratories or institutes devoted, respectively, to Physical, Chemical, Building, and Personnel research. A fifth, the Telecommunications Research Laboratory, located at Wits University, incorporated the work of Schonland's radar research team.[147]

The interests of the social sciences and humanities were taken care of by a parallel organization, the National Council for Social Research (est. 1946). This body subsumed the functions of the Council for Educational and Social Research as well as the National Research Council and Board. Like its predecessors, the primary function of the National Council for Social Research was to support university research workers, but it also initiated two major 'co-ordinated' research projects. The first of these was a study of social conditions in a sample 'native rural area'. This emerged, under the direction of Lindsay Robb of Rhodes University, as the four-volume multidisciplinary Keiskammahoek Rural Survey (1947–52), whose final report was written by the doyen of post-war liberal anthropology, Monica Wilson. The Social Research Council's companion project was a 'many-sided' study of urban African existence in Atteridgeville, Pretoria. It was initially led by Werner Eiselen, until his elevation to Secretary for Native Affairs. P. J. Coertze subsequently assumed the directorship of the Survey. Coertze was the central figure in the development of volkekunde studies at the University of Pretoria and he exerted an unrivalled and enduring influence on Afrikaner anthropology. The project was eventually taken over by his son, R. D. Coertze. It spawned

145 Austin, *Schonland*, 303–6, 328; *Objects and Policy of the CSIR*, 4–5.

146 Austin, *Schonland*, 307, 326. Kingwill, *The CSIR*, 16–17, 23–4. The SABS lost its independence in 1956 but was re-established as a separate body in 1962.

147 B. J. F. Schonland, 'The Organisation of Scientific Research in the Union of South Africa', paper delivered to the African Regional Scientific Conference, Johannesburg, 1949.

a number of MA theses and resulted in several works on aspects of African family life in Atteridgeville, the most important of which emerged in 1969.[148]

There is curious symmetry in the Social Research Council's inaugural research projects—a large-scale, anthropologically informed, treatment of rural life published in English on the one hand, and an Afrikaner nationalist-inflected series of ethnographic studies of African urban life on the other. The manner in which these projects were initiated, on the eve of apartheid, recalls the studied even-handedness in disbursing funds practised by Malherbe's Bureau in the 1930s; their respective outcomes neatly reflect the post-war divisions that emerged in white South African intellectual and political cultures. Although liberal social anthropology and volkekunde diverged sharply from the 1950s, these traditions shared a considerable amount in terms of institutional, and even intellectual, ancestry—probably more than either would have wished to admit.

Unlike the CSIR, the National Council for Social Research had no statutory basis and was merely an advisory body responsible to the minister of education. It eventually transmuted into the Human Sciences Research Council (1969). The CSIR, by contrast, marked a significant new departure in the scale of government-supported scientific research. It conceived of itself as the topmost point of a tetrahedron or pyramid, from which it presided over a three-cornered base representing university, government, and industrial research respectively.[149] Its primary orientation was applied science, though pure research was not neglected. Basic revenues were provided by parliament (£375,000 in 1945) and these were supplemented by substantial contracts gathered from industry. Considerable sums were involved: whereas the Research Grant Board dispensed a total of £25,000 in the period 1919–36 (approx. £1,300 p.a.), the CSIR funded about £50,000 worth of research grants to universities on an annual basis, while the National Council for Social Research made available a further £20,000 per annum to support the social sciences and humanities.[150]

[148] GP-S. 13811–1947–8, *National Council for Social Research. Annual Report for 1946* (Pretoria, 1948); UG 55–1951, *Report of the National Council for Social Research 1949–50* (Pretoria, 1951). The National Council for Social Research was originally constituted in April 1946 as the South African Council for Educational, Sociological and Humanistic Research, but changed its name a year later. Its first chairman was A. A. Roberts, secretary of the Education Department. See also Government Archives, Pretoria, BAO 7992 838/400, 'The South African Council for Educational, Sociological and Humanistic Research', 25/5/46. For the Keiskammahoek survey, see W. D. Hammond Tooke, *Imperfect Interpreters: South Africa's Anthropologists 1920–1990* (Johannesburg, 1997), 102, and on Coertze see pp. 126 ff. I am grateful to Kees van der Waal and Laureen Rushby for pointing out Coertze's publications on Atteridgeville and to Rob Gordon for further comment about the Atteridgeville output. See esp. R. D. Coertze, *Atteridgeville: 'n Stedelike Bantoewoonbuurt* (Pretoria, 1969), produced under the direction of P. J. Coertze.

[149] CSIR, *Science in South Africa* (Pretoria, 1949), 21.

[150] E. G. Malherbe, 'Our Universities and the Advancement of Science', *SAJS* (Sept. 1951), 58; Schonland, 'The South African Association for the Advancement of Science', *SAJS* 49 (1952), 65; CSIR, *The South African Council for Scientific and Industrial Research* (Pretoria, 1955), 4, 6. These figures, which are not strictly comparable, should be seen as indicative only. According to

CONCLUSION

We have seen in this chapter that the administration and a co-ordination of state-based knowledge was framed by the country's participation in two world wars. As was the case in Britain and other dominions, it was in response to the industrial demands of the First World War, and the vogue for scientific management and expertise encouraged by wartime planning, that the Research Grant Board was first created. The 1920s and 1930s saw a succession of organizations and entities which, with varying degrees of success, attempted to establish a permanent research infrastructure and to promote the country's national profile.

It was only during the Second World War, however, that an enduring and relatively well-endowed research structure was established to cater, separately, for the natural sciences and the humanities. Strong economic growth, wartime mobilization, opportunities for import substitution, and a political environment broadly sympathetic to state intervention and social reform all made this possible. The CSIR was undoubtedly the senior partner in the new arrangement. Modelled on comparable organizations in the British Commonwealth, it was led by an outward-looking, well-connected soldier-scientist who fully comprehended the need to service the demands of growing, large-scale industrial corporations and to cater as well to the fast-expanding university sector.

The CSIR was directly a product of the Smutsian warfare state, and it inaugurated the era of state-sponsored 'big science' in post-war South Africa.[151] Schonland found Smuts's support invaluable and appreciated the prime minister's hands-off approach when it came to running the CSIR. In Schonland's view, Smuts had little genuine understanding about fundamental science or technology. He was not equipped to comprehend 'the technical world of 1946 or of modern physics' and was not at ease when visiting modern laboratories.[152] Yet, Smuts retained a good sense of the larger picture. As a member of the War Cabinet, Smuts had been in on the secret of the atomic bomb as early as 1943, and he responded positively to the British government's request a year later that South African uranium (the presence of which had recently been detected in Witwatersrand gold ore tailings) be available for the Manhattan project (responsible for the construction of the world's first nuclear bomb). Schonland, who

the *Official Yearbook for the Republic of South Africa* (1974), by 1972/3 the CSIR was around ten times the size of the HSRC. It employed around 3,000 people and had a budget of R27 m (60% of which was voted by parliament). By contrast, the HSRC employed around 350 staff and had a budget of R3m.

[151] The term 'big science' was coined by Alvin Weinberg in 1961. It typically refers to government-funded and resource-intensive scientific research in fields like high-energy physics, space research, the human genome project, and so on. The Manhattan Project was the prototype and exemplar of 'big science'.

[152] Churchill Archives, Cambridge University, SCHO1/2, letter from Schonland to Sir Keith Hancock, 2/1/1957.

ensured that research into applied radioactivity and uranium-refining was part of the CSIR's brief, also had contact with scientists associated with the Manhattan project. In 1946 he chaired the first meeting of South Africa's Uranium Research Committee, and he had a private meeting with Lord Portal, Britain's controller of atomic energy, when he represented South Africa at the 1946 Empire Scientific Conference in London.[153]

From the outset the CSIR was committed to national advancement as well as ongoing international exchange. It bore the unmistakable imprint of Smuts's ambitions and marked the highpoint of his wartime South Africanist vision. Schonland considered, with good reason, that the prime minister had given 'a revolutionary new deal to science in South Africa and to South African men of science'.[154] But, as he feared, within a short space of time the CSIR's affiliations and outlook would be tested by the inward-looking and suspicious nationalist forces who took over government in 1948. This is, briefly, the subject of the concluding chapter.

153 A. R. Newby-Fraser, *Chain Reaction: Twenty Years of Nuclear Research and Development in South Africa* (Pretoria, 1979), 20–1; 26–7; Hancock, *Smuts: The Fields of Force*, 433; Austin, *Schonland*, 343. The Atomic Energy Board was created in 1949.

154 Hancock, *Smuts: The Fields of Force*, 506 n.

6

Conclusion: The Renationalization of Knowledge?

THE ECLIPSE OF SMUTSIAN SOUTH AFRICANISM

In October 1949, a year after the election which brought the National Party to power with its slogan of 'apartheid', Prime Minister D. F. Malan was called upon to open the African Regional Scientific Conference in Johannesburg. The event attracted around 100 delegates from Europe, the United States, and colonial Africa, and there were high hopes that it would inaugurate a new era of cooperation in the subcontinent. Basil Schonland was closely involved with the arrangements and, indeed, had been one of its progenitors. At the 1946 British Commonwealth Scientific Conference in London, where he proudly led the South African delegation (and was showered with personal honours), Schonland had called for the establishment of an African Research Committee to look closely at the requirements of the continent.[1] His views were closely in keeping with the pan-African ambitions of Smuts and also with Lord Hailey's vision of continental-wide scientific cooperation as the basis of colonial trusteeship. The 1949 conference was notable as the first generalist scientific conference of this sort, and it could be claimed that this was the most prestigious gathering of scientists in South Africa since the 1929 visit of the British Association. There were other precedents that helped to ensure a sense of unfolding continuity with the African continent: the 1929 pan-African meeting on veterinary and agricultural science held in Pretoria, for instance, and more recently, the first pan-African congress on prehistory, convened in Nairobi in 1947.[2]

Had Smuts still been prime minister he would have revelled in the opportunity to play once more the role of scientific statesman and seer. Many would have regarded this as his entitlement, for the conference could be seen as the culmination of the views he expressed as long ago as his 1929 Oxford Lectures. In all probability Smuts would have reiterated his weary view of white South Africa's unique position and mission as the bearer of civilization in Africa. Malan, if not

[1] B. Austin, *Schonland: Scientist and Soldier* (Bristol and Philadelphia, 2001), 338–40, 344.

[2] 'Speech of the Prime Minister at the Opening of the African Regional Scientific Conference, October 17th 1949', in collection of the Gubbins Library, Witwatersrand University.

actually a reluctant host, was evidently uncomfortable. His opening words—'As head of the Government of the Union of South Africa I have many important and at the same time interesting functions to perform...'—appears to suggest that he needed to remind himself of his prime-ministerial duties.[3] The message that followed, albeit dutiful and polite, was bland to the point of perfunctoriness.

Malan's discomfort is understandable. He was head of a new, exclusively Afrikaner government, that was determinedly insular and wholly focused on the survival of white South Africa. Since 1939 Afrikaner nationalists had been agreed on little more than their hostility to fighting on the side of the Empire and their desire to create a republic outside of it. Science and internationalism were associated with the despised figure of Smuts, the 'handyman of empire', who had overlooked the needs of his own people in his self-aggrandizing role as international statesman. Scientific research had long been dominated by an anglophone elite who maintained strong imperial connections. Moreover, the universalist and secularist traditions of international science posed a threat to the Christian-national theologians—Malan was himself a *dominee* (minister)—whose influence on the new government was powerful and growing.

Yet it was not only domestic political change in South Africa that affected the 1949 scientific conference. The context of international opinion was changing too. Three years earlier, when the conference was mooted in London, South Africa's external reputation was at an all-time high. Smuts was revered in Britain, and much of white South Africa was happy to bask in the glow of military victory and warmth of economic prosperity.[4] Late in 1946 the government's complacency was disturbed when India denounced South Africa's unjust race policies in the General Assembly of the United Nations, the preamble to whose Charter Smuts had himself helped to write. This was an important signal of changing times. Although the international 'retreat' from race was anything but smooth, white supremacy was coming under increasingly critical scrutiny, prompted in particular by the realization of the role played by racial science in Nazi Germany. The nature of empire was also being actively reassessed in the light of a bipolar, nuclear world, as well as the ambitions expressed by the newly independent Asian subcontinent.

[3] 'Speech of the Prime Minister at the Opening of the African Regional Scientific Conference, October 17th 1949', Austin, *Schonland*, 386–7, comments on Malan and Schonland's joint entry to the Scientific Conference thus: 'It was a short but symbolic journey that they took together up those steps beneath the colonnades of the Great Hall. Two men of Cape stock: one, from its western reaches, a former *predikant* turned politician of marked anti-British sentiment; the other, a 53-year-old Anglophile born in the very heart of British South Africa in the eastern Cape, with liberal views and a personal philosophy of scientific rationalism. Their closeness as they walked together that day was a massive contradiction. Schonland, the visionary who saw his country as the pilot of Africa, able to offer it the means to a better future through scientific cooperation; and Malan, a bigot whose view was of a South Africa turning in on itself...'

[4] The period 1936–51 was an era of unprecedented economic growth, especially in the manufacturing sector. See J. Nattrass, *The South African Economy: Its Growth and Change,* 2nd edn. (Cape Town, 1988), 164–5.

The tensions between the pre- and post-war worlds could readily be discerned at the 1949 African Regional Conference. Delegates had been chosen by the colonial powers (Britain, France, Belgium, and Portugal—German representatives were excluded) as well as the white settler states of Southern Rhodesia and South Africa. Observers were also present from affiliates to the United Nations (Unesco, the World Health Organization, and the Food and Agricultural Organization); their presence suggested that internationalism was taking on new forms.[5] In 1945 the landmark sixth Pan-African Congress in Manchester had been attended by many of the continent's future African leaders. To anyone aware of this, or indeed the history of the fifty-year-old pan-African movement, Malan's casual use of the term 'pan-African' to describe a conference of white scientists in apartheid South Africa must have struck a discordant note. Coverage by the local press trumpeted white supremacy in complacent headlines and editorials, with strap-lines like 'Science plans to dispel Africa's darkness' and 'The Backward Continent'.[6] Such attitudes evidently worried the visiting Kenyan anthropologist Louis Leakey, who warned his colleagues at the 1949 scientific meeting that 'anthropologists trained in South Africa were unlikely to be acceptable to Africans in other parts . . .'.[7]

In retrospect, Leakey may be seen as a harbinger of South Africa's growing isolation from international contacts, but his was not yet a dominant voice and the country was by no means the pariah state that it would become a decade or more later. In South Africa the late 1940s were an interregnum. Liberal post-war optimism was rapidly dissipating, but the intentions and strength of the new apartheid government still remained uncertain. Historians are generally agreed that apartheid was more like a slogan than a detailed programme in the years immediately after 1948.[8] Surprised, like most observers, by its narrow electoral victory, the National Party was unsure of its political base and could not take its longevity in government for granted. Administrative inexperience was compounded by the fact that the country's industrial base had undergone enormous expansion during the war, while the apparatus of government planning and management had become commensurately more complex. By the end of the war substantial elements of the economy were controlled by state-run corporations.[9] It would take a while before the apartheid administration became sufficiently confident to seize the enhanced levers of bureaucratic and industrial power it

[5] *African Regional Scientific Conference. Detailed Programme of Meetings* (Johannesburg, 1949), Gubbins Library, Wits University, pam G. 1718.

[6] Austin, *Schonland*, 387–8.

[7] Leakey added somewhat ambivalently, 'Scientists knew, however, that such [South African] anthropologists were satisfactory'. 'Minutes of African Regional Scientific Conference. Section E&D. Ethnography & Psychology', 21st October 1949, Gubbins Library, Witwatersrand University.

[8] D. Posel, *The Making of Apartheid, 1948–1961* (Oxford, 1991); J. Lazar, 'Conformity and Conflict: Afrikaner Nationalist Politics, 1948–1961', D.Phil. thesis, Oxford University (1987).

[9] Christie, *Electricity, Industry and Class*, 142. These included iron and steel, electricity, harbour boards, the railways, agricultural control boards, and the Industrial Development Corporation.

inherited; but as soon as the government realized the power these mechanisms could unlock, it moved swiftly to exploit their potential.

The technocratic dimension of the Smuts government's legacy has tended to be overlooked by historians, who have underestimated, or merely taken for granted, the role played by scientifically trained industrialists and managers in South Africa's burgeoning war economy. Although the country's rapid economic growth is widely noted, most account has been taken of the social and political impact this had on a massively expanded African population in the cities. Recently, attention has begun to focus on the incipiently welfarist nature of the wartime administration, and also on how Smuts's conservatism acted to constrain those liberal and progressive voices who were actively pressing for the provision of universal health schemes, social security, and improved housing and services for all South Africa's citizens.[10] Smuts's failure to carry through promises of social reconstruction, and his habit of promoting mediocre and superannuated politicians to positions of power and influence—the patriarchal 'oubaas' (old master) habitually gave preferment to those whom he knew through family ties—created torpor in his government and was a source of frustration for promising young liberals and reformists.[11]

This record needs to be qualified in respect of Smuts's greater success in selecting scientists and technocrats. Malherbe was one of these, though he was rather less effective than either Hendrik van der Bijl, Basil Schonland, or H. J. van Eck, all of whom occupied key scientific and industrial positions.[12] These men gave outstanding service to the wartime administration and formed part of an unofficial scientific cabinet. Van der Bijl's death in December 1948 (the day before Jan Hofmeyr's) marked the end of an era and proved a severe personal loss to Smuts, who described van der Bijl 'as our greatest industrialist'.[13] Basil Schonland wrote an appreciative obituary for the Royal Society in London that neatly summarized his scientific achievements as well as his contribution to the country's industrial development. E. G. Malherbe mentioned Schonland to illustrate the point that Smuts was 'not always a poor picker of men'.[14]

The transition to a new relationship between science and state in the post-1948 period is evident in the career of van der Bijl's protégé H. J. van Eck, who graduated in physics and chemistry at Stellenbosch University and went on to

[10] S. Dubow and A. Jeeves (eds.), *South Africa's 1940s: Worlds of Possibilities* (Cape Town, 2005).

[11] See B. Friedman, *Smuts: A Reappraisal* (London, 1975), 159–60.

[12] One reason why Malherbe was less effective was that he was much more obviously 'political' in his opposition to Afrikaner nationalism, in particular because of his role in monitoring the Afrikaner Broederbond and the public position he took in favour of bilingualism. Nor did he command the big budgets or enjoy the influence that state planners and scientists possessed.

[13] J. van der Poel (ed.), *Selections from the Smuts Papers, Vol. VII* (Cambridge, 1973), doc. 829 'Speech (1948)' and doc. 830, letter to M. C. Gillett, 6 Dec. 1948; A. Paton, *Hofmeyr* (London, 1964), 528.

[14] E. G. Malherbe, *Never a Dull Moment* (Cape Town, 1981), 278; Hendrik Johannes van der Bijl, 1887–1948', by B. F. J. Schonland, *Obituary Notices of the Royal Society*, 7 (1950–1), 27–34. Van der Bijl was elected a Fellow of the Royal Society in 1944. His nominees were Smuts and Schonland.

study in Leipzig and Berlin, where he gained a chemical engineering doctorate in 1927. Like van der Bijl, van Eck was an influential figure in the wartime scientific establishment. He was a member of the National Provisions Council (chaired by Smuts), and had an important role in ensuring the production of arms and ammunition. Yet van Eck's career extended well into the apartheid era and was coloured by association with it.[15]

Van Eck's experience at Escom, Iscor, and his chairmanship of the Social and Economic Planning Council (1942–52) provided the grounding for major state-sponsored scientific and industrial ventures in the post-war period. It also gave him the opportunity to put into practice the principles of integrated economic planning that he had articulated in a major government report of 1940–1.[16] Over the course of a thirty-year career as director or chair of the powerful Industrial Development Corporation, van Eck helped to create a series of new parastatals and factories devoted to the production of oil-from-coal (utilizing the Fischer-Tropsch process),[17] phosphate and cellulose, textiles and aluminium, military hardware, hydroelectricity, and nuclear energy. Van Eck's close links to government and his championship of industrial decentralization and export-oriented policies inevitably drew him close to the heart of the apartheid regime (he was said to have had a direct telephone line from his office at the IDC to Verwoerd),[18] and he became a symbol of Afrikaner technical and industrial prowess for a new generation of scientists.[19]

This shift in allegiance is remarkable for someone whose route to advancement came through Smuts's wartime administration. In the mid-1940s van Eck could still be mentioned by Leo Marquard as one of the South African industrialists potentially sympathetic to the formation of a new 'liberal-left' political party capable of keeping the Nationalists at bay.[20] Marquard's hopes may have been

[15] On van Eck, see entry in *DSAB*. Also Clark, *Manufacturing Apartheid: State Corporations in South Africa* (New Haven, 1994), and K. Middlemas, *Cabora Bassa: Engineering and Politics in Southern Africa* (London, 1975), 22–3, 29–33.

[16] *Third Interim Report of the Industrial and Agricultural Requirements Commission* (UG40-'41), known as the Van Eck Commission.

[17] The South African Coal, Oil and Gas Corporation (SASOL) has been described as 'an exercise in courage' (see E. A. Bunt, 'Some Highlights of Engineering Research in South Africa', in A. C. Brown (ed.), *A History of Scientific Endeavour in South Africa* (Cape Town, 1977). There is an interesting comparison to be made here with the synthetic fuels programme in Spain, where Franco pursued a very expensive state-led policy of energy self-sufficiency, based on Nazi oil-from-coal technology. See E. San Román and C. Sudrià, 'Synthetic Fuels in Spain, 1942–66: The Failure of Franco's Autarkic Dream', *Business History*, 4 (2003).

[18] Middlemas, *Cabora Bassa*, 36.

[19] Clark, *Manufacturing Apartheid*, ch. 6, provides a good discussion of state corporations in the apartheid era and points out how van Eck's post-war goal of creating consumer-oriented manufacturing industry that could provide skills and employment came to be realized in the form of large-scale industrial development based on cheap labour. Van Eck's important role in the implementation of industrial decentralization policies, utilizing the resources of the Industrial Development Corporation, was a key element of Verwoerdian 'separate development'.

[20] Leo Marquard Papers, University of Cape Town, C71.2, Marquard to J. H. Hofmeyr 25/9/46 (personal and confidential). Van Eck and van der Bijl both gave lectures to the Army Education Service, established by Marquard and Malherbe during the war to support troop morale and democratic ideals.

misplaced, but it was reasonable at the time to assume the anti-Nationalist political sympathies of industrialists and planners like van der Bijl and van Eck whose politics were resoundingly centrist. During the first years of Nationalist rule the state corporations created by van der Bijl and van Eck were treated with suspicion and hostility on account of their historic links to the mining industry and 'foreign' capital. Such suspicion was compounded by deep-seated resentment that state corporations had not done enough to protect the interests and job security of semi-skilled white workers.[21] Yet, after a brief hiatus, the increasingly confident apartheid government came to appreciate how state corporations could be used to cement its power, pursue its overall political objectives, and guarantee national self-sufficiency and security.

APARTHEID SCIENCE

The warfare state endowed by the wartime government of Smuts provided a technocratic template for social engineering which the apartheid government realized in earnest during the 1960s. In the process the ethos of scientific internationalism and the spirit of genuine enquiry which Smuts did so much to cultivate yielded to a much more insular technicism. The underlying institutional continuities represented by parastatal corporations and related scientific institutions are not as well understood as developments in the public realm of legislation and ideology.[22] But their importance in understanding the transition to apartheid should not be underestimated.

Although widely noted in anecdotal evidence, memoirs, and biographies, there has been no full study of the manner in which successive Afrikaner administrations moved to replace Smuts appointees and anglophone civil servants by Nationalist loyalists.[23] Deliberate placement of political loyalists in key positions was perfected as a strategy by the clandestine and increasingly influential Afrikaner Broederbond. With the burgeoning of the apartheid state bureaucracy the scope for preferment expanded hugely. Within the civil service it seems likely that transition was effected more quickly in the upper reaches of political departments like justice, education, and 'native affairs' than in their technical, diplomatic, or scientific equivalents. In semi-autonomous or independent organizations like universities, the emphasis was rather on building up parallel Afrikaner institutions.

[21] Clark, *Manufacturing Apartheid*, 71–2, 109, 148–9.

[22] An exception is N. Clark, 'The Limits of Industrialisation Under Apartheid', in P. Bonner *et al.*, *Apartheid's Genesis 1935–1962* (Johannesburg, 1993).

[23] But see D. Posel, 'Whiteness and Power in the South African Civil Service: Paradoxes of the Apartheid State', *JSAS* 25: 1 (1999). Posel records (p. 105) that 'By 1959, "Afrikanerization" had advanced to such an extent that of more than 40 government departments and sub-departments, only six had English-speaking heads'. A comparison with the transition to a post-apartheid state is obviously due.

The nationalist makeover is clearly discernible in the newly created National Council for Social Research, whose annual reports after 1948 make increasing use of Afrikaans as the medium of expression and reveal a rapid turnover in membership. English-speakers representing the leading universities were not excluded immediately, though their influence was soon curtailed. In 1949 the Council's chairman and secretary for education, A. A. Roberts, was replaced by H. S. van der Walt and rewarded by the high commissionership of Canada.[24] A year later the Council's expert committee structure was significantly overhauled so as to consolidate Afrikaner nationalist representation. E. G. Malherbe was one of those notables who lost his position on the Council's governing committee; Werner Eiselen, anthropologist, leading expert on 'native education', and new secretary for native affairs, was equally notable as an entrant.[25]

By virtue of its constitution and scientific standing the CSIR was more independent than the Social Research Council, yet it was by no means immune to pressure. Basil Schonland was acutely aware of this and, having anticipated the possibility of a nationalist political victory well before the event, he began to plan accordingly. In 1945 he had met the nationalist trio of D. F. Malan, A. J. Stals, and Eric Louw to ask whether, in the event of a change of government, a new administration would support the CSIR as a 'national institution'. The discussion was conducted in both official languages. Malan replied in Afrikaans that the CSIR would have to adopt ''n Suid-Afrikaanse kleur' (South African colouring), and that it should demonstrate its dedication to South Africa rather than any other country. So long as this was assured the CSIR would have his support and its staff would enjoy security of tenure similar to a university faculty. As Schonland's biographer, Brian Austin, puts it: 'From both the tenor of their remarks and the vehemence of their delivery it was clear that Schonland had touched a fundamental tenet of Nationalist philosophy: not only South Africa first, but South Africa alone, if needs be.'[26]

In the appointment of senior CSIR staff, Schonland placed a premium on choosing outstanding scientific talent, yet he was shrewd enough to maintain political balance wherever possible. On the one hand he selected Simon Biesheuvel, the liberal-inclined psychologist and critic of race bias in intelligence testing, to head up the CSIR's Institute for Personnel Research. On the other, he approached Stefan Meiring Naudé, the highly regarded German- and American-trained Stellenbosch professor of physics, to form the Physical Laboratory.[27] A committed Afrikaner nationalist, though not a parochial one—and a scientist first

[24] Deon Geldenhuys, *The Diplomacy of Isolation: South African Foreign Policy Making* (Johannesburg, 1984), 16, points out that unwanted English-speaking civil servants were often moved to overseas diplomatic missions.

[25] *Reports of the National Council for Social Research for 1946* and for *1949–1950* (GP-S. 13811–1947–8) and UG '55–1951.

[26] Austin, *Schonland*, 311.

[27] Ibid. 314–5. Naudé turned town a Rhodes scholarship to Oxford and chose instead to study in Germany, where he gained his doctorate, *cum laude*, in 1928. He also worked in Chicago with the Nobel laureate R. A. Millikan.

and foremost—Naudé rose to become head of the CSIR in 1952. He also served as president of the (South African) Royal Society (1960–1), chairman of the Akademie vir Wetenskap en Kuns (1967–9), and in retirement worked as the prime minister's scientific adviser (1971–6).

Schonland meanwhile prepared an exit strategy for himself. Fearing unwelcome government intrusion into scientific research, and into the CSIR in particular (the appointment of the fascist-sympathizer Eric Louw as minister of Economic Affairs was a particular worry), Schonland returned to the Bernard Price Institute at Wits and resumed his research activity there until leaving in 1954 for Harwell in England. Schonland's departure from the CSIR in 1950 was of his own volition, and he was replaced as its head by P. J. du Toit, the long-term director of Onderstepoort. Malan made clear that he would like to keep Schonland in post, at least in the short term. The government, which had recently come under fire for replacing the general manager of the South African Railways, W. Marshall Clark, was evidently keen to avert further controversy and to maintain continuity and public confidence in the prized CSIR.[28]

The growing influence of Afrikaners in science was reflected in the history of the Akademie vir Wetenskap en Kuns (Academy for Science and Arts). Originally founded in 1909 as the Academy of Language Literature and Art, it was intended to further the cause of Dutch and Afrikaans culture. The august Akademie participated fully, if discreetly, in the struggle for language equality, but for some years it was consumed at least as much by internal battles over the relative virtues of Afrikaans and Dutch as it was in countering the supremacy of English. In 1921 the Akademie was incorporated by an act of parliament which gave it quasi-official status.[29] Until the middle of that decade there was little input from scientists, though its associated journal, the *Tydskrif vir Wetenskap en Kuns*, edited by D. F. Malherbe, included science in its title and carried occasional discussions on scientific topics. These were mostly devoted to broad philosophical or applied matters.

[28] Austin, *Schonland*, chs. 16 and 18. On his reappointment as president of the CSIR for one more year in 1948, Austin comments: 'The new Prime Minister [Malan] saw no sense nor political capital in losing the services of the man who had done so much for South African science, while Schonland himself, though more and more attracted by the scientific challenges at the BPI, was prepared to see how the new government's policies, both towards science and in a broader sense, affected him personally before opting to return to research or, indeed, to explore other pastures' (p. 362). D. C. Kingwill, *The CSIR: The First 40 Years* (Pretoria, 1990) 18, records that after Schonland's departure responsibility for the CSIR was indeed transferred to the Ministry of Economic Affairs, headed by Louw, a move 'which inevitably diminished the status of the CSIR as a co-ordinating and advisory body to the Government on all matters affecting scientific and industrial research'. The extent to which the CSIR was subject to direct government interference is unclear. A full history of this organization is badly needed.

[29] P. J. Nienaber, *Op Brandwag vir ons Taal. Die Geskiedenis en Werksaamhede van die Afrikaansse Akademie vir Wetenskap en Kuns* (Pretoria, 1950). Nienaber (p. 10) points out that in the early years the Akademie was caught up not only with the struggle for language recognition, but also in internal arguments as to whether Dutch or Afrikaans should be given official status. The Akademie's name was initially rendered in Dutch.

From the late-1920s efforts were made to attract Afrikaner scientists into the Akademie so as to prevent their being 'lost' to 'English' organizations, but the organization's leaders proved reluctant or slow to formalize their inclusion. The Bloemfontein National Museum curator E. C. N. van Hoepen was deputed in 1938 to investigate the creation of a scientific section. However, institutional conservatism blocked any rapid progress. Significantly, the impetus to establish a fully constituted body came from without, prompted by the formation of the Vereniging vir Wetenskap and Wysbegeerte (Society for Science and Philosophy) by a group of Wits students in 1937, along with a journal, *Die Denker* (The Thinker). The chairman of the Vereniging and editor of *Die Denker* was A. J. A. Roux, later the country's most prominent nuclear physicist. This organization survived only until 1939, but a year later several of those involved in its creation formed a successor body, the Afrikaanse Akademie vir Wetenskap en Tegniek (Afrikaans Academy for Science and Technology).[30]

The Akademie vir Wetenskap en Kuns, which had hitherto been slow to take up the cause of science, now responded to these moves by reviving van Hoepen's initiative. In 1942 it carried through a major restructuring process. This resulted in the creation of a fully fledged 'Faculty of Science and Technology', whose agenda included promoting the professional interests of Afrikaner scientists, honouring their achievements, and developing the technical and scientific vocabulary of the Afrikaans language. By the end of the decade specialist subsections (engineering, medicine, chemistry, agriculture, mathematics, agriculture, and economics) had been constituted.[31] A rising generation of Afrikaner scientists was now positively encouraged by an organization that had hitherto focused almost entirely on literary and cultural affairs. Whether it was desirable or, indeed, possible to stimulate a distinctively Afrikaans scientific culture is not discussed in official histories;[32] nevertheless, it is clear that science was rapidly being absorbed into the broader nationalist struggle.

The 1948 election gave new momentum to champions of Afrikaner science, which rapidly established a foothold in state-funded institutions. A small but symbolic victory for the Akademie was its success in persuading Malan to rename the Department of Education as the Department of Education, Arts, and Science (1949).[33] The expanded Department chose to highlight the role of science at the 1952 Van Riebeeck Festival Fair, a major government-sponsored event which used

[30] A. J. A. Roux, 'Die Konsolidering van die Afrikaner in die Natuurwetenskappe', in *Feesalbum 1909–1959*, SAAWK (Pretoria, 1959), 51–3; C. A. Engelbrecht, 'Natuurwetenskap, Tegniek en die Akademie', in *Akademie 75* (Pretoria, 1984), 130–3. Those responsible for creating the 1940 Akademie included A. J. A. Roux (secretary), R. L. Straszacker, C. V. van Abo, M. C. van Schoor, and P. J. Meyer.

[31] Engelbrecht, 'Natuurwetenskap', 134–5; Roux, 'Die Konsolidering', 52–3.

[32] According to Beale, it was Afrikaanse Studentebond 'standing policy' that science ought to be practised 'in the light of God's word', a doctrine that was reiterated in 1963 when the student organization decided they were in favour of the principle of Christian science in all branches of the sciences'. M. A. Beale, 'Apartheid and University Education, 1948–1970', Ph.D thesis, Wits University (1998), 385 and n.

[33] F. C. Fensham, 'Die Pad van die Akademie (1909–1984)', in *Akademie 75*, 13.

the tercentenary of white settlement in South Africa to celebrate the triumph of Afrikaner nationalism. In the 'Hall of Science' visitors were encouraged to reflect on the beneficial effects on national development of technological and industrial progress. The CSIR figured prominently in the display, and considerable pride was taken in its activities. According to Leslie Witz, mention of the theory of evolution was barred from the scientific exhibit—apparently with the consent of the South African Archaeological Society. This concession to religious and creationist sensibilities did not, however, prevent the depiction of human origins by means of prominent showcases which 'constructed an imagined racial evolutionary scale within the "human race" '.[34]

Political pressure was increasingly experienced (or anticipated) by the anglophone South African Association for the Advancement of Science, which gave its journal a shared Afrikaans title in 1943 and embraced bilingualism from the end of that decade.[35] Having previously ignored the Akademie, the S2A3 referred to an official meeting with the Akademie's leaders in 1950 and conveyed that organization's 'greetings' to its membership. The existence of 'goodwill' between the two organizations was duly noted and reference was made to a proposal that scientific organizations in the country be federated.[36]

Basil Schonland was acutely concerned about creeping nationalism. In a speech delivered to the Pretoria Women's Club in 1947 he insisted that scientists should be left alone to do their research and, in a playful image freighted with meaning, he characterized scientists as 'the new pioneers of South Africa, travelling with an electron-microscope instead of an ox-wagon'.[37] After 1948 Schonland was forced onto the defensive. In his 1952 presidential address to the S2A3 he observed that South Africa was a bilingual country and that, although not an internationally recognized scientific language, there should be 'no hindrance' to publication in Afrikaans. Reiterating the S2A3's long-held desire to be a broadly representative and generalist rather than specialist scientific organization, he deftly deflected the attentions or overtures of any particularist constituency: '. . . our Association has no right and no historical claim to LEAD science in this country. Such a claim could only be advanced by the Royal Society of South Africa and the Akademie, who will, I feel sure, one day amalgamate to form a single body'.[38]

Although hardline nationalists were active in the Akademie, as an elite and genteel organization representing the great and the good it chose to eschew crude populist politics.[39] Instead, the domain of 'practical' cultural mobilization was

[34] L. Witz, *Apartheid's Festival: Contesting South Africa's National Pasts* (Bloomington, Ind., 2003), 201 and 191–3. See also *Science Serves South Africa: An Introduction to the Hall of Science Exhibition* (Cape Town, 1952); *Exhibit of the CSIR: Van Riebeeck Festival* (Pretoria, 1952).

[35] C. A. Englebrecht, 'Natuurwetenskap, Tegniek en die Akademie', in *Akademie 75*, 131.

[36] *SAJS* 47 (Sept. 1950), 27. See also 46 (Nov. 1949), 105.

[37] Cited in Austin, *Schonland*, 353.

[38] B. J. F. Schonland, 'The South African Association for the Advancement of Science, its Past and Future', *SAJS* 49 (Oct./Nov. 1952), 64–5.

[39] This may be one reason why the Akademie has been virtually ignored in the critical literature on Afrikaner nationalism.

franchised out to the Broederbond-directed FAK (Federation of Afrikaans Cultural Organizations), established in 1929.[40] Science proved rather more resistant to being captured on an ethnic basis, though latent nationalist pressures were building. This much was clear in an article written in 1950 by S. M. Meiring Naudé, who reflected on a half-century of Afrikaner involvement in the physical and technical sciences. Naudé's survey played to ethnic pride by summarizing the achievements of Afrikaner scientists and traced their growing representation in the academic institutions. Yet, as he must have been aware, many of the individuals he mentioned would have eschewed exclusivist definitions of Afrikanerdom, and some were only Afrikaners in name. Significantly, Naudé made much of the fact that leading Afrikaner scientific researchers (himself included) had pursued their studies overseas, often in Germany, after initial training in South Africa. His underlying motivation was thus to encourage young Afrikaners to become scientists in the new South Africa while ensuring that their intellectual horizons remained open.[41]

In the first decade of apartheid, pragmatism, combined with Malan's often cautious approach to institutional change, characterized the National Party's approach to scientific matters. Historians of foreign affairs have remarked on the premium that was placed on maintaining a presence in scientific and technical organizations such as the Scientific Council for Africa South of the Sahara, whose formation, in 1950, was a direct outcome of the 1949 African Regional Scientific Conference.[42] At a time of growing isolation from international associations, the apartheid state valued its participation in organizations dedicated to technical cooperation, not least because they offered quasi-diplomatic channels of communication.

Membership of inter-African technical bodies was also commensurate with Malan's 1952 'African Charter', in which the prime minister portrayed South Africa as an independent European state in Africa, dedicated, like the colonial powers, to resisting decolonization and countering communist influence. The Strijdom and Verwoerd administrations continued to mouth platitudes about South Africa's role as an 'African power' and reiterated its commitment to technical and scientific cooperation, but by 1965 diplomatic isolation ensured South Africa's exclusion from virtually all inter-African organizations.[43]

[40] Fensham, 'Die Pad van die Akademie', 6, 11, 13. Many Broederbonders were active in the Akademie, though they do not appear to have controlled it, and links with the FAK were close. I. M. Lombard, for instance, who served terms as secretary of the Broederbond and the FAK, was part-time secretary of the Akademie in the years 1934–48.

[41] S. Meiring Naudé, 'Die Afrikaner en die Ontwikkeling op Fisies-Wetenskaplike en Tegniese Gebied Gedurende die Afgelope 50 Jaar', *Tydskrif vir Wetenskap en Kuns*, 10 (1950).

[42] Its first chair was P. J. du Toit, the ex-director of Onderstepoort, who succeeded Schonland as president of the CSIR (1950–2) and had played an important role in convening the 1949 Regional Scientific Conference.

[43] Geldenhuys, *The Diplomacy of Isolation*, 12–14; J. Spence, 'South Africa in the Modern World', in M. Wilson and L. Thompson (eds.), *The Oxford History of South Africa, Vol. 2* (Oxford, 1971), 494. The Scientific Council for the Africa South of the Sahara was tied to the Commission for Technical Cooperation in Africa South of the Sahara (CCTA) and the Inter-African Bureau of Soil Conservation and Land Utilisation. Regarded as vestiges of colonialism, by the mid-1960s they were dismantled or reconfigured within the Organisation of African Unity.

It was the Sharpeville massacre in 1960, followed by the country's withdrawal under pressure from the Commonwealth, that radically altered the apartheid state's view of its place in the world. Under severe internal and external pressure, the government moved decisively to crush all political opposition. Verwoerd's mood of truculent defiance, coupled with his intellectual self-confidence and conspicuous lack of religiosity, persuaded him to seek rational scientific answers to the predicament of white supremacy. The growing authority of the Afrikaner scientific establishment helped to make this possible. By the end of the 1970s science and technology were firmly invoked in support of an ideology of modernizing techno-nationalism which ensured that white—and particularly Afrikaner—intellectual prowess was celebrated as a key weapon in the fight against communism and African nationalism. As H. L. de Waal proclaimed, in an Akademie-sponsored discussion on future scientific research in the universities which took place shortly after Sharpeville: '*The power of a nation (volk) has always depended on the level of technological progress in its military, agricultural, and industrial spheres.*'[44]

Massive engineering projects, such as the construction of dams, may serve to illustrate this point. In the mid-nineteenth century farmers in the dry Cape midlands applied resources and ingenuity in order to construct small earth-and-stone dams on their farms. Hailed as the epitome of agricultural progress and improvement, such productive endeavour was often associated with English settler initiative.[45] Half-a-century on, as we have already seen, Percy FitzPatrick demonstrated the exemplary qualities of British South Africanism through his large personal investment in irrigation along the Sundays River. Seen in this light, the use made by Afrikaner nationalists of large dam construction as a trope of national self-sufficiency and modernity in the latter part of the twentieth century is of particular interest, though of course it was not exceptional. Similar ambitious schemes were paralleled elsewhere in Africa (the Aswan High Dam in upper Egypt for example, which became so closely identified with Nasserite ideology), and also in developing countries in Asia. As Sunil Khilnani has shown, India 'fell in love with the idea of concrete' in the 1950s and constructed massive dams to demonstrate its commitment to a developmental-based vision of modernity.[46]

The prospect of diverting some of the Orange River through a series of tunnels in order to enhance the flow of water through the Great Fish River valley in the eastern Cape had been envisaged as early as 1928 by a government hydrologist, A. D. Lewis. This idea was repeatedly pressed by eastern Cape farming representatives, but made no impact on government. As late as 1959 it was rejected as far too expensive by the Nationalists. However, as H. J. Simons remarked in

[44] H. L. de Waal, 'Natuurwetenskaplike navorsing aan ons Universiteite—Waarheen?', paper delivered to 'Die Suid-Afrikaanse Akademie vir Wetenskap en Kuns. Simposium gehou in die Universiteit van Pretoria, 20 Oktober 1960, Gubbins collection, Wits University, pam. 4950. The Afrikaans original reads: 'Die krag van 'n volk was altyd afhanklik van die niveau van tegnologiese vooruitgang in sy militêre, landboukundige en industriële sfere.' (Emphasis in original).

[45] W. Beinart, *The Rise of Conservation in South Africa: Settlers, Livestock, and the Environment 1770–1950* (Oxford, 2003), 91–4.

[46] S. Khilnani, *The Idea of India* (London, 1977), 61–2.

Fig 10. View of Hendrik Verwoerd (now Gariep) Dam, which at the time of its completion in 1972 was the largest construction in South Africa. (Courtesy of Eskom)

a perceptive study of the embryonic Orange River Project, government policy 'changed dramatically after the Sharpeville massacre of 1960 and during the state of national emergency that followed'.[47]

Military and strategic, rather than economic, calculations proved decisive, though the booming economy of the 1960s provided the means to achieve grandiose ambitions. In May 1960 it was announced in parliament that the Orange River Project would go ahead, and that it would help to bolster white settlement in the north-western as well as the eastern Cape. In 1966, at the newly created settlement of Oranjekrag (Orange force), Prime Minister B. J. Vorster triggered explosives to inaugurate work on the Hendrik Verwoerd Dam, the huge gravity-arch construction that underpinned the entire scheme. It was completed in 1972.[48] A concrete watercourse (the 'longest continuous water tunnel in the world') designed to convey water for over 80 kilometres to the Great Fish and Sundays rivers was put into commission in 1975.[49] In Christie's words, the Orange River Project was 'designed to "inject morale" into the ruling groups after the state

[47] H. J. Simons, 'Harnessing the Orange River', in N. Rubin and W. M. Warren (eds.), *Dams in Africa: An Interdisciplinary Study of Man-made Lakes in Africa* (London, 1968), 132–3.

[48] *Cape Times*, 19 Nov. 1966. It has now been renamed the Gariep Dam.

[49] H. Olivier, *Great Dams in Southern Africa* (Cape Town, [1976]), 154–68. The P. K. Le Roux hydroelectric dam (named after a minister of water affairs) is the third component of the Orange River scheme. Its function is to control water supply to the lower reaches of the Orange.

of emergency; to "say it with concrete"; to emphasise the rulers' determination to fight'.[50]

Saying it with concrete, more specifically reinforced concrete, is more apt than the familiar but misleading 'granite' metaphor so often used to characterize the apartheid state's supposed rock-like intransigence.[51] The massive grain silos peppering the countryside, the national roads and militarily important motorways that linked regions of the country, the new urban public architecture of banks, corporations, universities, and public memorials, all featured the use of concrete, often in the neo-brutalist idiom of 1960s modernism.[52] David Goldblatt's brilliant photographs of Protestant church architecture have drawn attention to the paradox that the ultra-conservative religious message of the Dutch Reformed churches was preached in new brick and concrete buildings whose 'Vertical, bold, often powerfully triangulated' lines aggressively redefined the landscape in order to proclaim a 'modern message with a radical core: apartheid'.[53] A similar point is made by David Bunn, who draws a neat distinction between the use of dressed stone in Herbert Baker's organic imperial architecture, and the prominence of reinforced concrete in modern Afrikaner nationalist memorials and monuments to symbolize vigorous 'forward movement and protective enclosure'.[54]

No structure built for the use of humans can compare in scale and massified use of concrete with dams. And modern South African dams say a great deal besides. Henry Olivier, one of the leading consultant engineers involved in the Orange River Project, was able to draw directly on his extensive international experience of large-scale dam-building in post-colonial Africa, Iran, and Pakistan as well as his extensive contacts with British and European contractors. The account which Olivier provides in his semi-autobiographical book, suitably entitled *Damit*, is richly illustrative of the huge technical and financial demands involved in the construction of such prestige national projects. Numbers and statistics—including the precise volume of concrete poured in each case—are displayed in clear tabular form, reminding us of the vogue for quantification shared by colonial and post-colonial modernizing states alike.[55]

The political and ideological dimensions of these dams, though unstated, are plain enough. Olivier's lavishly illustrated *Great Dams in Southern Africa* is

[50] Christie, *Electricity, Industry and Class*, 167–8 (citing several sources).

[51] See e.g. J. Botha, *Verwoerd is Dead* (Cape Town, 1967), ch. 2, 'Hendrik Verwoerd, Man of Granite'; A.Hepple *Verwoerd* (Harmondsworth, 1967), 157.

[52] There are intriguing parallels with the Nazi case where, notwithstanding violently anti-modernist propaganda, modernism in architecture and design often received state support. See P. Betts, *The Authority of Everyday Objects: A Cultural History of West German Industrial Design* (Berkeley and Los Angeles, 2004).

[53] D. Goldblatt, 'Introduction', *South Africa: The Structure of Things Then* (Cape Town, 1998), 17, 18.

[54] D. Bunn, 'Whited Sepulchres: On the Reluctance of Monuments', in H. Judin and I. Vladislavic (eds.), *Blank Architecture, Apartheid and After* (Rotterdam, 1998), 96, 106.

[55] See e.g. A. Appandurai, 'Number in the Colonial Imagination', in C. A. Breckenridge and P. van der Veer (eds.), *Orientalism and the Postcolonial Predicament* (Philadelphia, 1993).

dedicated to 'All the Countries of Africa'. It bears a foreword by the prime minister, B. J. Vorster, with an accompanying epigraph from Francis Bacon ('The people's welfare is the supreme law'). Both books appeared on the eve of the Soweto uprising—a moment when Vorster's pragmatic embrace of 'détente' with neighbouring African states, coupled with independence for South Africa's Bantustans, was favoured by apologists or fellow-travellers of the apartheid government. Here we have an unmistakable reminder of South Africa's technical might, albeit softened by its willingness to assist pliant African states with economic development in the spirit of 'partnership'.

Olivier's detailed account of the Orange River Project is yet more revealing of the state of domestic politics. His description of Verwoerd's keen involvement in the scheme, of H. J. van Eck's critical participation through the Industrial Development Corporation, as well as the experience of close cooperation with the Anglo American Corporation and British and French international consortia, are together indicative of the merging of powerful international political and economic interests.[56] Closely linked was the contemporaneous Cabora Bassa scheme on the Zambezi River, a massive hydroelectric project which was explicitly directed to the bolstering of faltering Portuguese rule in Mozambique and the defence of settler power in the subcontinent.[57] Scientific and technical cooperation, nowhere shown to better effect than in grandiose water and hydroelectric schemes, was perfectly suited to the growing influence of *verligte* (enlightened) factions within the cabinet and the apartheid government's shift to an 'outward policy' from the late-1960s. Its restatement of Smutsian expansionism was imbued by Vorster with a sense of menace: 'We have a sense of mission in respect of Africa. In addition, providence has been very good to us in Africa and we want to return to Africa something of this. This is the spirit that inspires us—and this is the spirit that will conquer Africa.'[58]

The success of the Orange River scheme in economic terms was in doubt from the outset, which is why its viability depended upon combining hydroelectric generation with irrigation.[59] In the case of Cabora Bassa, connection to a 'pan-African water and power grid' represented an attempt to secure South Africa as the region's hegemon.[60] Within the country, a pumped storage hydroelectric scheme in the western Cape was associated with the building, from 1976, of the continent's first (and only) nuclear power station at Koeberg to the north of Cape Town.[61]

[56] H. Olivier, *Damit: Great Dams in Southern Africa* (Johannesburg, 1975).

[57] K. Middlemas, *Cabora Bassa.*

[58] B. J. Vorster, *The Star*, 4 Nov. 1968, cited in M. P. Mitchell, 'The Politics of Water', B.Phil. diss., University of York (1976), 25. See also B. J. Vorster, 'South Africa's Outward Policy', in *South Africa in the World: Lectures Delivered at Annual Meeting of Suid-Afrikaanse Akademie vir Wetenskap en Kuns* (Cape Town, [1969]).

[59] Olivier, *Damit*, 114; Christie, *Electricity, Industry and Class*, 167–8.

[60] Christie, *Electricity Industry and Class*, 166. By the time the delayed Cabora Bassa power plant entered commission, the Frelimo nationalist movement had taken power from the Portuguese.

[61] Koeberg is a pressurized water reactor. It was 'synchronized' to the electricity grid in 1984.

This facility brought to fruition the use of nuclear power for civilian purposes, a process which began with the initiatives taken by Smuts and Schonland immediately after the Second World War. It was given fresh impetus and new direction from the early-1960s under the leadership of A. J. A. (Ampie) Roux, a pioneer of the Afrikaner presence in science, who left his senior position at the CSIR in order to head up research at the increasingly powerful Atomic Energy Board.[62] Electricity generation was, however, only one of the Atomic Board's interests. With barely concealed glee, Vorster announced to the world in 1970 that the country was about to embark on a programme of uranium enrichment using a unique process at a pilot research station—the so-called 'Y Plant' at Valindaba, Pretoria: 'South African scientists have thus again added to the prestige of their country in no uncertain terms. In the past they have made lasting contributions to science, but perhaps the achievement that I am announcing to-day is unequalled in the history of our country.'[63]

The military implications of this development, albeit clothed in the deepest secrecy and deliberately confused by misinformation, censorship, and denial, was self-evident: it is broadly accepted that South Africa embarked upon a nuclear-weapons programme from the early 1970s, and that by end of the decade it posed a credible nuclear threat. When Vorster warned the world in 1977, 'So far and no further; do your damnedest if you wish', it was clear what sort of retaliation he had in mind.[64]

This was the ultimate weapon of last resort and it added immeasurably to South Africa's formidable capacity to produce conventional military hardware and ammunition—ranging from jet fighters and armoured cars to self-propelled cannon, anti-aircraft missiles, and multiple rocket launchers. The effort to circumvent international armament sanctions and to achieve a degree of military self-sufficiency was led by Armscor, a generously funded parastatal which, by the 1980s, had become a major international arms exporter. Armscor's wares were advertised to developing countries with the macho boast: 'Born in necessity, tested under fire.'[65] For a government that was deliberately cultivating a siege mentality in response to growing domestic opposition and international pressure,

[62] A. R. Newby-Fraser, *Chain Reaction*, ch. 4. Roux left the CSIR in 1959 for the Atomic Energy Board, and persuaded the government of the need for a separate national nuclear research programme. This controversial decision ended an important dimension of the CSIR's activities. See Kingwill, *The CSIR*, 23. The decision to pursue nuclear power was taken when R. L. Straszacker, an old friend and colleague of Roux's, was chairman of Escom. Needless to say, both were members of the Afrikaner Broederbond.

[63] Vorster, cited in Newby-Fraser, *Chain Reaction*, 92. In 1969 the Uranium Enrichment Corporation was established under the leadership of W. L. Grant and A. J. A. Roux.

[64] J. W. de Villiers, R. Jardine, and M. Reiss, 'Why South Africa Gave Up the Bomb', *Foreign Affairs*, 72: 5 (1993); J. Barber and J. Barratt, *South Africa's Foreign Policy* (Cambridge, 1990), 241; http://cns.miis.edu/research/safrica/. South Africa remains the only country to have voluntarily relinquished its nuclear weapons capacity.

[65] P. L. Moorcraft, *African Nemesis: War and Revolution in Southern Africa 1945–2010* (London, 1994), 51; Barber and Barratt, *South Africa's Foreign Policy*, 101–3.

and which was increasingly dominated by a tight-knit network of 'securocrats', the availability of high technology to ensure strategic power and national security had incalculable ideological as well as military importance.

Vorster's suggestion that uranium enrichment marked the apogee of South African scientific achievement says much about the value placed by Afrikaner nationalist leaders on state-controlled prestige projects and trophy technology. A little over two years before, in December 1967, Vorster telegrammed Dr Christiaan Barnard to congratulate the surgeon who had just performed the first ever heart transplant on a 'world achievement'. Shortly afterwards the prime minister hosted a dinner at his official residence to honour the team of doctors and University of Cape Town dignitaries.[66] The unique triumph of the heart transplant provided South Africa with international acclaim and attention. It also transformed the photogenic Chris Barnard into a medical superstar. We cannot know how Vorster rated nuclear capability against medical science, but it would not be surprising if enriched uranium was closer to his own heart.

Analysis of the series of transplants performed by Barnard in 1967–8 still awaits detailed analysis by historians. Yet even a cursory reading of the papers of the period suggests that the Afrikaner nationalist establishment may not have been as fulsome in its enthusiasm as might be expected. Peter Hawthorne, the freelance British journalist whose instant book, *The Transplanted Heart*, was serialized in the *Cape Argus* within a month of the operation, did more than anyone else to shape the way in which the news was produced and received. His description of the surgeon's 'almost impoverished' life in a 'sheep-farming district in the arid heart of South Africa' was especially resonant. Barnard's father, readers are reminded, served as a missionary to the coloured people of Beaufort West. As Christiaan recalled: 'We were poor. We never had shoes. Even in winter we walked barefoot'.[67]

Encapsulated in this account we have the romantic story of Afrikaner survival, piety, and triumph against the odds. Yet, by the 1970s, this narrative may have had more sentimental appeal to an international audience than it had to sleek leaders of the modern *volk*, for whom reminders of poor whiteism were becoming embarrassing. The fact that the team of University of Cape Town doctors included English-speakers and Jews, and that the first two heart recipients, Louis Washkansky and Philip Blaaiberg, were Jewish as well, may also have been the cause of some ambivalence in official quarters. Far more controversially, a subsequent transplant performed by Barnard was clumsily revealed to have involved

[66] *Cape Argus*, 4 Dec. 1967; 18 Jan. 1968. The dinner invitation was extended to members of the Groote Schuur medical team as well as the principal of the University of Cape Town, Sir Richard Luyt. Intriguingly, a separate dinner was scheduled at a later date for Prof. H. B. Thom of the University of Stellenbosch and its medical faculty.

[67] Peter Hawthorne, *The Transplanted Heart* (Johannesburg, 1968), 70–1, 72. Cf. *Cape Argus*, 12 Jan. 1968. Albert Grundlingh points out that Barnard was given assistance by the Helpmekaar movement to study medicine—which makes him a perfect, if wayward, example of Afrikaner upliftment.

the use of an organ belonging to an unnamed black woman—a fact that proved shocking on both sides of the racial divide.[68] That South Africa's leading role in heart transplantation was a source of considerable national pride for all whites is not in doubt. But the complexities of race and ethnicity meant that the warm glow of satisfaction was most naturally experienced by white liberals (especially those associated with the University of Cape Town) who regarded themselves as being above sectional politics.[69]

Sectionalism in the world of tertiary education was by now, however, deeply entrenched and growing. We have seen that the ethos of inclusive South Africanism proved resilient in the face of nationalist pressures through the inter-war years. By the end of the 1940s it was exhausted. The pretensions of the University of Cape Town and Witwatersrand to represent the interests of all white South Africans were routinely dismissed by Afrikaner cultural nationalists, who had long regarded broad South Africanism and claims to universalism as a mere cloak for liberal anglocentrism.[70] Afrikaner resistance to English or imperial domination in the university sector had been signalled by the formation of the Christian-national ANSB (Afrikaner National Student Union) in 1933. The Akademie vir Wetenskap en Kuns offered a separate sphere for Afrikaner intellectual activity, and was growing in confidence. Yet it would be mistaken to overrate the power or influence of such organizations. Until the capture of state power in 1948, Afrikaner nationalist cultural politics was largely contained within its natural constituencies and was often more marginal than either its champions or opponents chose to assume.

In the apartheid years huge resources were deployed to establish parity, if not outright predominance, in the tertiary educational sector. However, in the first decade after 1948 institutional change proved slow (as it always is in universities). Vociferous opposition to the landmark university apartheid legislation of 1959 was directed from liberal or English-speaking quarters to the exclusion of black students from 'open' institutions; significantly, there was also considerable opposition from within the Afrikaner nationalist movement to the imposition of greater state control over universities and loss of academic and religious autonomy

[68] *Cape Times*, 14 Sept. 1968. Other ethical questions, such as determination of the moment of death and the right to donor privacy, were also vigorously debated. In 1987 Chris Barnard revealed that the first operation might have taken place two weeks earlier had a decision not been made to exclude black donors or recipients for reasons of political sensitivity. See *Die Burger*, 17 Dec. 1987.

[69] See e.g. *The Argus*, 9 Dec. 1967, 'Cause for Pride' (editorial): 'Let us say frankly that it was time for some brilliant achievement to bring into the clear light of day the real character and meaning of the University of Cape Town, so often unjustly belittled of late years because of the political activities of a handful of students. It is not a political institution but a seat of learning which devotes itself to the pursuit of knowledge and the discovery of truth, together with the training of students to become members of the learned professions, of which medicine is one.' Cf. the opinion piece by the 'coloured' intellectual Adam Small, who argued that 'truly scientifically-people' were irked by the exaggerated attention given to the operation and expressed unease at the emphasis placed on 'South Africa first!', *Cape Times*, 5 Feb. 1968.

[70] M. A. Beale, 'Apartheid and University Education', 5, 63, 92–3.

Fig 11. View of Rand Afrikaans University (now University of Johannesburg), The inward-facing semicircular campus buildings, suggestive of a concrete 'laager', were designed by Jan van Wyk and opened officially in 1975 (Courtesy Neil McCarthy)

(the 'conscience clause' was a particular sticking point).[71] By the 1960s the power of Christian-nationalist advocacy reached the point where the very notion of universities as institutions serving universal ends could no longer be tolerated. The debate now shifted from a general desire to Afrikanerize existing universities to the more ambitious idea of creating entirely new institutions to serve the particularist needs of the *volk*.

The practical response to this shift in emphasis was the creation of two new universities: the University of Port Elizabeth (1964) and the Rand Afrikaans University (1966). Their avowed purpose was to counter the alleged liberal influence of, respectively, Rhodes in the eastern Cape and Wits in Johannesburg. With strong support from the Broederbond, the Rand Afrikaans University became one of the centrepieces of modernizing Afrikanerdom at its most bombastic.[72] Sculpted in soaring concrete, and arranged in semicircular form, the RAU campus resembled an urban laager in the midst of Johannesburg. It may be seen as the

[71] Ibid. 196 ff.

[72] Ibid., ch. 6. Cf the University of South Africa campus in Pretoria, which resembles a vast battleship on the rolling hills of Pretoria.

monumental equivalent, in the cultural and intellectual sphere, of the great dams under construction at the same time. Either way the new campus was a clear statement that Afrikanerdom was securing its position in the country's economic centre. Knowledge was being politicized, institutionalized, and nationalized as never before.

The consolidation of Afrikaner power and the rapid growth of Afrikaner universities encouraged the development of separate or parallel intellectual fields. By the 1970s the social sciences were thoroughly bifurcated. Rival professional organizations were formed on English- and Afrikaans-speaking campuses, corresponding closely to broader political and ethnic divisions within white society. In fields ranging from criminology, anthropology, and sociology, to psychology and education (where 'fundamental pedagogics' was in vogue), specially tailored teaching syllabi were in place catering to the distinctive needs of Christian-national ideology. The historical profession sought to maintain a semblance of unity by reviving the bilingual South African Historical Society in the mid-1960s in order to head off the Christian-nationalist oriented SA Historiese Genootskap (est. 1956). The founding of the SA Historical Society represented a carefully choreographed attempt to enable Afrikaans- and English-speaking academics 'to find each other'.[73] But Afrikaner nationalist and English liberal historiography effectively inhabited separate intellectual universes, as represented by two defining collective statements of the time: C.F.J. Muller's *Five Hundred Years: A History of South Africa* (1969), and the two-volume *Oxford History of South Africa* (1969–71), edited by Monica Wilson and Leonard Thompson.[74]

Perhaps the most remarkable of the home-grown Afrikaner nationalist intellectual traditions was *volkekunde*. This version of cultural anthropology, which developed from the 1930s at Stellenbosch and then spread to Pretoria and Bloemfontein, drew eclectically from German as well as British and American anthropological teachings in order to produce a theoretical framework that conformed to the specific needs of Christian-national theory. Cultural relativism and cultural boundedness were hereby invested with the force of the categorical imperative. The notion that each *volk*, tribe, or racial group possessed a distinct ethnos that was worthy of cultivation and preservation was of central importance, as was the imperative of *volkdiens* or service to the Afrikaner people. In the period of high apartheid the devotees of *volkekunde* provided important legitimation for the idea of separate development, as well as practical guidance for the implementation of the Bantustan policy of tribally based social engineering.

[73] See C. Saunders and B. Le Cordeur, 'The South African Historical Society and its Antecedents', *SAHJ* 18 (1986), 23.

[74] Muller's *History of South Africa* was first published in Afrikaans in 1968. The powerful challenge to the liberal school by Marxist historians from the early 1970s further complicated—and enriched—South African historiography.

Its effects were also to be seen in the elaboration of 'Bantu' education and 'Bantu' law.[75]

One of the reasons for the disproportionate influence of *volkekunde* was the close relationship between Afrikaans universities and the state bureaucracy. The connections were especially close in the field of 'native' or 'Bantu' administration. This department of state experienced enormous expansion and considerable professionalization under the leadership of Verwoerd and Eiselen, both of whom began their careers as senior academics at Stellenbosch University (in applied psychology and anthropology, respectively). As minister of Native Affairs, and later as prime minister, Verwoerd took the lead in conceiving of, and then applying, apartheid as a rigorous and logical system of ordering society. Verwoerd's belief in apartheid was guided less by the religious sentiments of his Christian-nationalist peers than by his training in the American social sciences, which, when combined with his unreflective intellectual arrogance, persuaded him (like many other leaders of developing countries) that there were no limits to the scope of state-centred social engineering. This helps to explain why Verwoerd was at least as much a pragmatist as he was an idealist or 'visionary'.[76] It also draws attention to the importance of administrative method in the madness that was apartheid.

It was indeed in the messy business of putting apartheid into practice that the government was most easily able to flex its modernizing muscles. Ivan Evans has drawn attention to the manner in which the Department of Native Affairs, reinvigorated and strengthened under Verwoerd, combined modern administrative and scientific methods in order to bring apartheid fully to fruition. In its efforts to solve the African housing crisis and to design 'the "properly planned location" ', considerable resources were expended from the early 1950s. An important contribution to the building of mass housing was made by the National Building and Research Institute (est.1946), one of the CSIR's original set of specialist research units. Its architects and engineers, some of whom had been socialist radicals in the 1930s, drew freely on the ideas developed by European modernists like Le Corbusier as well as Ebenezer Howard, the British pioneer of the English 'garden city'. Such expertise was easily adapted to the government's coercive intentions. As Evans mordantly puts it, planted boundary strips in London became machine-gun belts in South Africa.[77]

In a series of illuminating studies on the modernizing apartheid bureaucracy and its 'mania for measurement', Deborah Posel has demonstrated how data was obsessively collated in order to control all aspects of African urbanization and to

[75] J. Sharp, 'The Roots and Development of "Volkekunde" in South Africa', *JSAS* 8: 1 (1981); R. Gordon, 'Serving the Volk with Volkekunde: On the Rise of South African Anthropology', in J. D. Jansen, *Knowledge and Power in South Africa* (Johannesburg, 1991). See also W. D. Hammond-Tooke, *Imperfect Interpreters: South Africa's Anthropologists 1920–1990* (Johannesburg, 1997).

[76] On this point see esp. J. Lazar, 'Verwoerd Versus the Visionaries'; R. B. Miller, 'Science and Society in the Early Career of H. F. Verwoerd', *JSAS* 19: 4 (1993).

[77] I. Evans, *Bureaucracy and Race: Native Administration in South Africa* (Berkeley, 1997), 130.

service the labyrinthine system of labour bureaux that sought to manage the flow of African labour to white employers.[78] Sophisticated techniques of enumeration and biometric taxonomy were adopted to put into effect the landmark 1950 Population Registration Act, which assigned a pre-set racial category to every individual in the country. In Posel's view this project was 'a high-modernist fantasy, a hankering for totalizing order, positioning the eye of the state at the pinnacle of an orderly bureaucracy, with a panoramic view of the racial landscape and everything constructed upon it'.[79]

For a variety of reasons—including Christian-nationalist ideology's preference for dealing in the currency of cultural rather than biological essentialism—racial classification did not ultimately depend on scientific-based typologies: social convention and the authority of lived experience were also key arbiters in defining racial belonging.[80] The implications of universalism (in this case evolutionary thought) sat uneasily with the particularism of apartheid in the social as well as the natural sciences. But this did not in any way constrain the state from adopting the technology and apparatus of modernity to proclaim and secure its political omnipotence.

SCIENCE IN THE 'NEW' SOUTH AFRICA

Apartheid reached its hubristic high-point in the 1960s and early 1970s. After the Soweto uprising of 1976–7 its authority eroded palpably, and the urgency of reform came to be acknowledged as an imperative. An infusion of exciting new intellectual ideas from the beginning of the 1970s helped to open up space for political contestation, while also offering opponents of the regime a deeper understanding of apartheid's underlying structures of exploitation and power. The Black Consciousness movement in particular played a critical role in challenging white hegemony, by identifying the psychological and cultural internalization of inferiority and helplessness that contributed to black subordination. Concurrently, the rise of Marxist theory exposed apartheid as a system of class domination that could only be overthrown through a radical reordering of society. Adherents of Black Consciousness and Marxism were deeply divided on the key question of the salience of 'race' and class, but there was substantial common ground in the view that liberal universalism—or bourgeois ideology—was deeply

[78] D. Posel, 'A Mania for Measurement: Statistics and Statecraft in Apartheid South Africa', in S. Dubow (ed.), *Science and Society in Southern Africa* (Manchester, 2000).

[79] D. Posel, 'Race as Common Sense: Racial Classification in Twentieth Century South Africa', *African Studies Review*, 44: 2 (2001), 100. K. Breckenridge, in 'The Biometric State: The Promise and Peril of Digital Government in the New South Africa', *JSAS* 31: 2 (2005), 280, argues that contemporary South Africa may become the world's 'first genuinely biometric society'.

[80] Posel, 'Race as Common Sense'. For an analysis of Christian-nationalism's ambivalence towards biological determinism and eugenics see also my *Scientific Racism in Modern South Africa* (Cambridge, 1995).

complicit in sustaining racial capitalism as a system, even if it purported to oppose government policy. There was also consensus in the two radical positions around the idea that imperialism, intellectual as well as material, was a malign influence.

The influence of radical thought on anti-apartheid activists and intellectuals, black and white alike, was especially prominent on English-speaking university campuses in the 1980s. Here there were growing calls to 'Africanize' institutions and syllabi, confront 'Eurocentric' assumptions, and develop new modes of thinking to support 'progressive' political movements. Critiques of liberalism abounded, led in particular by historians and sociologists. From law and literature to psychology and education, the ideological structures of racial and capitalist domination were debated intensively. An informative distillation of such thinking appeared in a volume entitled *Knowledge and Power in South Africa: Critical Perspectives Across the Disciplines* (1991), edited by Jonathan D. Jansen. Contributors to this broad-ranging collection sought to expose the role of the social sciences in perpetuating capitalist-based forms of racial domination. The opening sentence of Jansen's own chapter established the tenor of the book's approach, in which Black Consciousness and Marxist thinking were freely combined:

> The decade of the 1980s will be remembered in South African social science as a period of profound epistemological turmoil. Within the small black intellectual community, as well as the dominant white academic sphere, there has emerged a strong discontent with prevailing modes of scientific discourse, its limits and politics.[81]

The impact of this volume was perhaps not as powerful as it might have been. Conceived in the mid-1980s, at a time when mass resistance to apartheid reached a peak, it was published only after the unbanning of the liberation movements—by which time debates about transformation to a post-apartheid society had begun to eclipse critiques of the system itself. Yet, in other respects the book pointed to the future. The first of its stated objectives was to 'put forward black representations of knowledge/power in South Africa'. While taking account of the contribution of Marxist ideas to progressive thought, historical materialists were chided for drowning out emerging radical theories of race: 'We stand at a point in South Africa's intellectual history in which we need to creatively reclaim the radical discourse on race.'[82]

Many white left-wing intellectuals took time to appreciate that the official demise of apartheid, far from solving the problem of race, instead opened up new avenues for its expression. Bishop Tutu's 'rainbow nation', which was so fulsomely endorsed by Nelson Mandela's politics of forgiveness and reconciliation, proved to be only a transient moment in the new South Africa. More overtly nationalist and Africanist sentiments were soon to be heard. In 1996, on the eve of parliament's

[81] J. D. Jansen, 'Knowledge and Power in the World System: The South African Case', in id. (ed.), *Knowledge and Power in South Africa: Critical Perspectives Across the Disciplines* (Johannesburg, 1991), 17.

[82] Ibid. 5, 7.

adoption of South Africa's new constitution, Deputy-President Thabo Mbeki delivered a major address which has come to be known as his 'I am an African' speech. Using rich, poetic language, Mbeki proclaimed that he owed his very being 'to the Khoi and the San whose desolate souls haunt the great expanses of the beautiful Cape'. He decried racism and the denial of human dignity by referring obliquely to the science of race in apartheid South Africa: 'I know what it signifies when race and colour are used to determine who is human and who subhuman.' The speech ended on a determinedly positive note as Mbeki proclaimed his faith in Africa's future, 'however improbable it may sound to the sceptics'.[83]

In subsequent addresses Mbeki elaborated on his theory of the 'African Renaissance', a theme that has since become the leitmotif of his presidency and a key point of reference for much public discussion in South Africa. This idea embraces the view that the African continent is poised to make remarkable progress in political, economic, and spiritual values; it suggests that the process of rebirth and regeneration will deliver the continent from despair and condescension; and it presumes that South Africa has a leading role to play as a catalyst in the dawn of an African century.[84]

Science and technology feature strongly in the concept of the African Renaissance, which is ambitious and forward-looking while invoking the romance of the past. As Vale and Maseko put it, the Renaissance draws at once on a modernist tradition linking South Africa to the African continent through a 'globalist' logic, and also an 'Africanist' ethos which emerges from 'a series of complex constructions around African identity'.[85] Both elements are present in the National Research Foundation's newly designated 'focus area' on Indigenous Knowledge Systems. This proposes, *inter alia*, to 'Decipher and unlock the inherent', 'Shed light on nation-building', and 'Develop research capacity in the field of IK [indigenous knowledge] in South Africa'. Dr Mogege Mosimege, an ethno-mathematician, is currently director of the Indigenous Knowledge Systems Unit within the National Department of Science and Technology. He is an active participant in the wider scientific conference circuit.[86] A new journal, *Indilinga*, is devoted to interdisciplinary research into indigenous African knowledge systems. It has attracted submissions by several well-known scholars, as well as by President Mbeki.

Indigenous knowledge seeks to affirm core African values in the face of globalization. Its proponents are not necessarily opposed to modernization or

[83] Thabo Mbeki, 'I Am an African', in *Africa. The Time has Come: Selected Speeches* (Cape Town, 1998).

[84] See e.g. 'Africa's Time Has Come', 19–22 Apr. 1997, 'The African Renaissance, South Africa and the World', 9 Apr. 1998, in ibid.

[85] P. Vale and S. Maseko, 'South Africa and the African Renaissance', in *South Africa and Africa: Reflections on the African Renaissance*, Foundation for Global Dialogue Occasional Paper no. 17 (Johannesburg, 1998), 8.

[86] www.nrf.ac.za./focusareas/iks/. See also its journal, the *People's Science Voice Newsletter*, in which projects germane to 'IKS' are sketched out.

'western' scientific knowledge. The modernizing dimension of the African Renaissance is indeed frequently expressed through endorsement of the market potential of new technology such as fibre-optic cables, computers, and the world-wide web. South Africa's major businesses and utilities (transport, energy, telephony, entertainment), whose operations are rapidly spreading throughout Africa, have not been slow to take up the challenge. It is possible to detect, in their confident rhetoric, corporate echoes of Rhodes's Cape-to-Cairo vision and Smutsian expansionism.[87]

As well as a commitment to intellectual and spiritual renewal, opportunism is also a feature of the African Renaissance. This was much in evidence at a major conference on the topic of the African Renaissance, convened in 1999, which drew substantial corporate sponsorship from parastatals which were inextricably connected with the old apartheid regime. These included the defence contractor Denel and the electricity company Eskom, for whom the 'African renaissance is not a dream nor a theoretical discourse but an everyday reality'.[88] The conference, organized by the leading medical scientist and university vice-chancellor Malegapuru Makgoba, included a section on science and technology, with contributions that spanned utilitarian as well as intellectual and ideological concerns. The potential of technology in wealth-creation was thereby interwoven with evocations of indigenous African knowledge systems. This theme is most fully developed in a contribution devoted to 'The lost sciences of Africa', which touches on the continent's accomplishments in fields ranging from metallurgy and astronomy to mathematics, engineering, medicine, and writing systems.[89]

A more sustained treatment of these issues, with a particular emphasis on environmental and biological themes, was offered at a conference on 'An African Renais-Science' held in Durban in 2002. Its opening address was delivered by the world-renowned anatomist Phillip Tobias, who spoke on 'Africa: The Cradle of Humanity'. In this brief survey of hominid development and human evolution, Tobias lent his personal authority to what is rapidly becoming part of a key set of tropes in the discourse of the African Renaissance: that *Australopithecus* is the progenitor of all living humans; that 'Africa gave the world its first culture'; and that hominids in southern Africa were 'not a southern African aberration, but a pan-African revelation'.[90]

It is ironical that one of the foundation myths of the new African Renaissance amounts to a neat inversion of what was once a core component of the country's

[87] Tom Lodge, *Politics in South Africa: From Mandela to Mbeki* (Cape Town, 2002), 228. There are compelling parallels between Mbeki and Smuts, not least in the sense that both have taken themselves seriously as philosopher-statesmen and both have taken personal command of a peace-making foreign policy devoted to making the world a better place.

[88] 'Acknowledgements', in Malegapuru William Makgoba (ed.), *African Renaissance: The New Struggle* (Sandton and Cape Town, 1999).

[89] I. van Sertima, 'The Lost Sciences of Africa: An Overview', in ibid.

[90] Phillip V. Tobias, 'Africa: The Cradle of Humanity', in Himansu Bajinath and Yashica Singh (eds.), *Rebirth of Science in Africa: A Shared Vision for Life and Environmental Sciences* (Pretoria, 2002), 6, 10.

tradition of scientific racism.[91] In segregationist South Africa, as we have seen, the discovery of *Australophithecus* and other fossil remains was commonly held up as proof of underlying typological racial difference in modern human beings: it emphasized evolutionary diversity over unity, and it proclaimed scientists like Raymond Dart and Robert Broom as exemplars of colonial South African scientific achievement.

Tobias, who studied under Dart and duly inherited his mentor's mantle, took some time to reconcile fully his uncompromising liberal political views with the racialized methodology of comparative anatomy within which he was trained. While Tobias has forthrightly disowned aspects of this tradition, in particular the association of brain size with intelligence, he has been selective in his representation of the achievements of physical anthropology in South Africa.[92] Dart's revelation that the African continent harboured the origins of humankind has therefore been encouraged to segue into a romantic vision of indigenous initiative and achievement. Paradoxically, the 'back to Africa' model, which portrays the African continent as the original cradle of civilization, now serves, in the words of Dialo Diop, as a key prop of 'the cultural foundation of the African renaissance'. Unesco's proclamation of the fossil-bearing sites at Sterkfontein as a World Heritage Site in 1999 has given an international seal of approval to this narrative, which is being widely marketed as the basis of 'Afrikatourism'.[93]

Appeals to indigenous scientific knowledge systems are another essential aspect of the African Renaissance. In fields ranging from ecological management to sustainable development and medicine, traditional understandings of the environment and of humanity's place within it are avidly discussed. The discovery and exploitation of *Hoodia gordonii*, a bitter cactus plant used by generations of San (Bushmen) people to stave off hunger and thirst, is an excellent example. In 1996 an appetite-suppressant molecule in *Hoodia*, known as P57, was patented by the CSIR, which has conducted research into indigenous plants since the 1960s. The pharmaceutical company Phytopharm, in collaboration with Pfizer, gained the licence to produce a 'natural' anti-obesity drug. But the San were overlooked as financial beneficiaries, because Pfizer said they were 'extinct'.

An energetic campaign ensued, by and on behalf of various San communities who were aggrieved to learn of their non-existence. It was guided by Roger Chennells, a human-rights lawyer and lands-claims specialist, who has pointed to the irony whereby an anti-obesity drug has been produced for westerners utilizing

[91] Dubow, *Scientific Racism in Modern South Africa*.

[92] P. V. Tobias, 'Brain-size, Grey Matter and Race—Fact or Fiction', *American Journal of Physical Anthropology*, 32: 1 (1970). In Tobias's otherwise illuminating 'History of physical anthropology in southern Africa', *Yearbook of Physical Anthropology*, 28 (1985), 18–19, he notes that the 'story of the search for "pure strains", "physical types," and "racial groups" among the numerous Khoisan, Negro, Caucasoid, and mixed populations of Southern Africa is a tortuous one'. But he proceeds to skip over this, 'because so many of the earlier notions have now been laid to rest . . .'.

[93] Dialo Diop, 'Africa: Mankind's Past and Future', 3–4; 8; also Nthobi Moaholi, 'Afrikatourism', in Makgoba (ed.), *African Renaissance*.

the indigenous knowledge of some of the world's hungriest and most deprived people. An apology from the CSIR's bio-prospecting programme for ignoring the San's interests was made at the 1992 World Summit on Sustainable Development in Johannesburg. Tough negotiations followed. The next year, at a ceremony in the Kalahari Desert, San elders and representatives of the CSIR signed an agreement whereby substantial royalties from the drug would be paid into a Hoodia Benefit Trust in recognition of the San's intellectual property rights.[94] This outcome, whereby the interests of an indigenous people have been recognized in a deal brokered between a government agency and a leading international corporation, is a signal example of how the commercialization of indigenous or traditional knowledge by scientific companies may work to mutual advantage in the 'new' South Africa.

In stark contrast is the story of Virodene P058, a highly contentious, locally produced anti-AIDS drug developed by a private company linked to Pretoria University. Official responses to the HIV/AIDS pandemic illustrate broader confusion and ambivalence towards western science, as well as inappropriate reliance on local solutions and remedies. Claims that Virodene could halt or reverse the progress of HIV were presented to the South African cabinet in 1997, along with a request for funding. In the scandal that followed, the health minister, Dr Nkosazana Zuma, was accused of undue interference and of breaching basic medical protocols. It was revealed, *inter alia*, that the drug contained a toxic industrial solvent, that claims for its efficacy as an HIV inhibitor in pre-clinical trials were unwarranted, and that 'sloppy' experiments had been conducted without the permission of the Medicines Control Council, which considered the trials to have been illegal. President Mbeki was himself forced to issue a statement in March 1998 denying unethical conduct, and refuting allegations that the ANC might have had a financial stake in the product.[95]

This scandal is but one episode in the broader failure of the South African government to tackle effectively the rampant problem of HIV/AIDS. Mbeki's eccentric views on the transmission of the disease have greatly exacerbated the crisis, both because of his reluctance to sanction standard drug treatments, and also on account of the support he has lent to the views of an increasingly discredited group of so-called 'dissidents' who refuse to accept a direct link between HIV and AIDS. Mbeki has variously lent support to the view that conditions of poverty rather than the HIV virus are the cause of immune-system collapse,

[94] See e.g. Antony Barnet, 'In Africa the Hoodia Cactus Keeps Men Alive', *Observer*, 17 June 2001; 'South Africa—Marginalised San Win Royalties from Diet Drug *hoodia gordonii*', www.hoodiadietpills.org; 'Native Healers want Drug Company Cooperation', by D. Craft, www.unl.edu/rhames/courses/current/hoodia.htm; minutes of 'Workshop on Benefit Sharing with Developing Countries. The Case of Human Genetic Resources', Cape Town, May 2004, http://www.wits.ac.za/bioethics/AIDS-ELSI/CTBeShaMinutes.pdf. Hoodia pills are now widely advertised on the internet.

[95] Pat Sidley, 'South African Research into AIDS "Cure" Severely Criticised', *British Medical Journal*, 17 Mar. 1997; Thabo Mbeki, 'ANC has no financial stake in Virodene', www.anc.org.za/ancdocs/history/mbeki1998/virodene.html.

and that anti-retroviral drugs are dangerously toxic. His tacit defence of the Virodene researchers presents a marked contrast to his hostile attitude towards recognized internationally produced drug treatments, a position that has plausibly been explained in terms of his desire to show that Africa is itself capable of finding a solution to its own AIDS crisis.[96]

Political, economic, and psychological reasons have all been advanced to explain the president's intransigence, none of them alone fully satisfactory. Whatever his own motivations, it is clear that Mbeki's attitudes are symptomatic of a much wider culture of denial which turns on a complex of historically conditioned attitudes to sex and to gender, as well as the malign interactions of race, science, and imperialism. Steven Robins has attributed AIDS 'denialism' to several different, though related, factors. These include resistance to racialized assumptions about rampant African sexuality, to the presumed authority of mainstream 'western' science, and to a suspicion that international drug companies are seeking to exploit Africa's health crisis for their own commercial advantage.[97]

Questions of race and cultural identity figure prominently in public discourses on AIDS, and the debate has been exhaustively played out in competing understandings of science, citizenship, and popular democracy. It was on the issue of the aetiology of AIDS and the right of scientific professionals to pronounce on these issues that Malegapuru Makgoba, one of the original proponents of the African Renaissance, broke privately as well as publicly with Mbeki. As president of the Medical Research Council, Makgoba recognized that AIDS had become the single biggest cause of death in the country. He was insistent that politicians had no right to interfere with scientific research, and he cited Stalin's championship of Lysenko's biological research by way of warning.[98]

The texture of official resistance to dominant explanations of HIV/AIDS is well illustrated by a peculiarly titled and confusedly argued document, posted on the ANC's website in 2002, which is widely believed to have emanated from the president's office.[99] Styled as a 'dissertation' or 'monograph', its numerous 'chapters' are compiled from an assemblage of quotes drawn from medical and non-medical authorities in order to support Mbeki's (defensible) view that 'The world's biggest killer and the greatest cause of ill health and suffering across the globe, including South Africa, is extreme poverty.' Quotations from non-medical sources—which include references to Shelley and Yeats, Fanon, Marcuse, Le Carré, H. L. Gates,

[96] My thanks to James Myburgh for providing me with his Oxford term paper in which he analyses the development of Mbeki's views on AIDS in some detail.

[97] S. Robins, '"Long Live Zackie, Long Live": AIDS Activism, Science and Citizenship After Apartheid', *JSAS* 30: 3 (2004); see also id., 'AIDS, Science and Citizenship after Apartheid', in M. Leach, *et al.* (eds.), *Science and Citizens: Globalization and the Challenge of Engagement* (London and New York, 2005).

[98] Robins, '"Long Live Zackie, Long Live"', 652–3 and *passim*.

[99] 'Castro Hlongwane, Caravans, Cats, Geese, Foot & Mouth and Statistics. HIV/AIDS and the Struggle for the Humanisation of the African', Mar. 2002. The title derives from the account of how Castro Hlongwane, a 17-year-old schoolboy, was prevented from sharing a picnic with white friends in a caravan park in 2001 because the owners alleged he had AIDS and was a potential rapist.

and Du Bois, among others—abound. Denunciations of racism and imperialism predominate, and phrases like the 'omnipotent apparatus' occur over and again. Whoever wrote or sanctioned the text is not declared, but it reads as a disjointed and derivative cocktail of underdevelopment literature, radical epistemology, and post-colonial theory.

It would be easy to dismiss this document and other similar pronouncements as the misguided and tragic conjoining of pseudo-science and pseudo-intellectualism. Many professionals and activists have taken this view, angered and frustrated at the way in which the government's entry into the AIDS 'debate' has justified inaction and thus caused the loss of an incalculable number of lives. Yet, from the longer historical perspective, the terms in which this debate are carried out bear further scrutiny. In an incisive discussion of the AIDS debate, the rhetorician Yehoshua Gitay has called our attention to the *manner* in which Mbeki and his close supporters have pursued their arguments. At issue is the question as to whether science should be understood as precise and absolute truth, or else as a form of contestable public knowledge upon which politicians and statesmen have the right—and even the duty—to declare. That Mbeki appeals to both views at different times only adds to the state of confusion.[100]

Beneath the president's distant and sometimes irascible front lies an intellectual who purports to take ideas seriously. He sees it as his prerogative to communicate directly with his people, and regards it as his role to make key executive decisions. Mbeki's deep mistrust of the received wisdom of 'western science', and of the assumed authority of professional organizations, is given some credence by current views of scientific fallibility and of knowledge as a form of social power. In South Africa this epistemological relativism acquires a particular twist in the government's rejection of the superimposition of western solutions to what it prefers to see as a distinctively African situation requiring an appropriate African response.[101] It is a matter of savage irony that blacks are the primary victims of a set of policies driven by a passionately ideological technocrat who presumes to speak in their name. But it is perfectly comprehensible in terms of the president's understanding of the pernicious purposes to which scientific knowledge has historically been put to use.

At least some of the resistance to dealing with the problem of AIDS has to be explained by reference to deep sensitivities relating to the politics of the body. In a country where racial hierarchies have been synonymous with physical appearance, where state-backed contraceptive campaigns have been linked to

[100] Yehoshua Gitay, 'The Word, Science, Medicine and Politics', paper presented to Centre for African Studies, University of Cape Town, 2002. As Gitay puts it: 'In short, Mbeki claims that actually there is at the end of the day a scientific truth, but still this truth has not been found. However, he, the Statesman, is able to determine that the so-called "generally accepted" scientific assessment of the African HIV/AIDS situation, is based on Western research, which is false' (p. 16).

[101] In an open letter to world leaders, dated 3 April 2000, President Mbeki drew attention to differences in the spread of the disease in Africa and the west and concluded: 'A simple superimposition of Western experience on African Reality would be absurd and illogical.' Cited in ibid.

demographic fears and the desire to control population growth, and where the symbolic power of blood—'pure', 'mixed', and 'tainted'—runs so deep, it could hardly be otherwise.[102]

Despite its dedication to non-racialism, the field of medicine and health-care, broadly conceived, provides many examples where colour has supervened. It was undeniably a factor in the heart-transplant programme of the late 1960s. The scandal of the 'Biko doctors', and their failure to diagnose or treat fatal brain injuries inflicted on the Black Consciousness leader by security policemen in 1977, raised awkward questions about the professed political neutrality of the medical profession. A recommendation by the South African Medical and Dental Council that no action be taken against the doctors provoked uproar, and was one of the factors leading to the formation of an explicitly anti-apartheid medical organization in 1982.[103]

More recently, there have been shocking revelations of the alleged activities of Dr Wouter Basson, dubbed 'Dr Death' by the media. This cardiologist, who was at one time a personal physician to President P. W. Botha, headed a secret military chemical- and biological-warfare programme (Project Coast) during the 1980s that enjoyed high-level political support and an estimated $10m annual budget. It was gradually wound down only after 1990. Claims surfaced in the media about research into clandestine methods of contraception designed to retard the birth-rate of blacks and, even more bizarrely, the production of a racially specific bacterial 'bomb' to effect this result. Sixty-seven charges were originally levelled at Basson as part of a detailed indictment. As well as multiple counts of fraud, these centred on Basson's role in administering lethal toxins, devising methods of drugging and murdering opponents of the apartheid state, and in conspiracies to murder ANC political leaders such as Pallo Jordan, Ronnie Kasrils, Frank Chikane, and Dullah Omar.[104]

The failure to convict Basson was widely interpreted as a signal of the persistent influence of the old regime in the post-apartheid judiciary. There are many other parts of the state in which old attitudes and practices persist. In 2004 it was revealed that racial profiling was still in operation in the South African blood-transfusion service, apparently because of the blood bank's desire to minimize the risk of HIV transmission. The disclosure of the practice was doubly shocking

[102] Shula Marks makes the additional point that a long legacy of missionary education and reticence on the part of older Africans to deal openly with matters of sex adds to this reluctance to confront the AIDS problem directly.

[103] See e.g. D. Silove, 'Doctors and the State: Lessons from the Biko Case', *Social Science and Medicine*, 30: 4 (1990), 417–29. The anti-apartheid National Medical and Dental Association, formed in 1982, was formed in opposition to the Medical Association of South Africa.

[104] After a massive ten-year investigation and a hugely expensive 30-month trial, the unrepentant Dr Basson was finally cleared of all 67 charges against him in 2002. Judge Hartzenberg was accused by the senior prosecutor of bias in favour of the defendant. For details on Project Coast see S. Burgess and H. Purkitt, *The Rollback of South Africa's Chemical and Biological Warfare Programme* (Air University, Maxwell AirForce Base, Ala., 2001); www.au.af.mil/au/awc/awcgate/cpc-pubs/southafrica.pdf; also hearings of the Truth and Reconciliation Commission.

because of the manner in which it came to light: the unceremonious disposal of a pint of blood publicly donated by President Thabo Mbeki during the country's Christmas holiday season. Mbeki tactfully chose to play down this incident, and the controversy gradually abated amidst allegations of institutional racism and explanations that the policy was soundly based on unavoidable statistical method. One commentator, defending the need for racial profiling, argued that the 'so-called blood transfusion scandal illustrates precisely why we need technocrats and why it can be dangerous to air technocratic arguments in public . . .'.[105]

We have seen in this book how scientific and intellectual culture has been used to support many different political positions: knowledge has always been bound up in views of national identity and belonging. In the nineteenth century scientific and intellectual debate was employed to carve out a domain for the exercise of civil rights and to promote a form of civic nationalism consonant with a liberal Cape colonial identity. The formation and defence of scientific institutions and societies was a key component of this process. During the first half of the twentieth century scientific culture moved beyond its close association with the Cape to become one of the central supports of the ideology of broad South Africanism. While disdaining ethnic nationalism of the Afrikaner variety and proclaiming the virtues of internationalism, South Africanism took for granted the superior attributes of western civilization, rationalism, and progress. Implicitly or explicitly, intellectual achievements were publicly celebrated in order to construct and defend an ethnically inclusive but racially exclusive white nation-state. It is hardly surprising that ideologists in the new South Africa are now reacting against this historical legacy.

Ever since the emancipationist movement of the early nineteenth century, there have been powerful universalist strands in South African intellectual discourse which have sought to include all people, regardless of race, in a common future. The governing African National Congress, with its roots firmly in the inclusivist and progressivist traditions of late Victorian Christianity and liberalism, continues to reflect important aspects of this modernizing, humanitarian impulse. Yet the ANC was also shaped from its outset by more exclusivist Africanist and pan-Africanist strains of thought and sentiment, whose nationalist tone and intensity have varied over time. As the twentieth century unfolded, many ANC intellectuals rejected liberalism in favour of socialism or, alternatively, pan-Africanist and, in turn, Black Consciousness modes of thinking. In the case of Marxism, the supposedly scientific principles of historical materialism encouraged internationalism, non-racialism, and progressive rationalism. By contrast, the subjectivity inherent in Africanist philosophy coexisted uneasily with the putative race-blindness so central to liberal and Marxist ideology.

The terms in which the debate about AIDS has been played out mirror these competing traditions, as does the African Renaissance more generally. In many

[105] Sean Muller, 'Comment', *Weekly Mail and Guardian*, 10–16 Dec. 2004.

ways it is symptomatic of a new society that is still struggling to be born and whose institutions and founding principles remain fluid. With this goes a confidence and dynamism that gives reason for optimism in the country's future. One positive indication is that government resistance to the weight of scientific evidence on the causation of AIDS, and the most efficacious manner of its treatment, has been less obdurate of late. In explaining recent shifts and modifications in official views, the persistence of expert criticism and rise of vigorous citizen activism, as well as independent thinking in the trade-union movement, should not be underestimated. It should be noted that the ANC's embrace of global technology has been driven by concerns for the welfare of the populace as a whole. There is no trace of the rampant militarism that characterized the late-apartheid state.

While it is too early to speculate on the long-term consequences of its transformation of intellectual and scientific institutions, so as better to reflect the complexion of the country, there are few indications that processes of political transformation are motivated by narrowly introspective sectionalism. Intolerance of criticism in public life is, however, a concern, as is a reluctance to appreciate the role of independent institutions in creating and sustaining a vibrant civic consciousness. It is as yet too early to assess the impact of major changes in the funding and organization of scientific bodies in particular, as a result of the creation of the National Research Foundation (1999) and the Academy of Science of South Africa (2002). This book has focused on the 'South Africanization' of knowledge over a considerable span of time. Glimpses have also been offered of its 'Afrikanerization' and, most recently, of its 'Africanization', both of which mark attempts to renationalize knowledge for particular political purposes that closely reflect the politics of identity.

The directions this latest phase will take are as yet unclear. As this book goes to press, there is positive news of South Africa's international bid to host a vastly expensive international astronomical project known as the Square Kilometre Array. This would capitalize on the recently completed Southern African Large Telescope in the Northern Cape, which lays claim to being the largest single optical telescope in the southern hemisphere. The international competition for the prize of hosting the Square Kilometre Array has been likened to the scientific equivalent of gaining the Olympic Games. Significantly, the country's new push for astronomical leadership has benefited from enthusiastic government support, and from President Mbeki himself, who has reportedly instructed all his cabinet ministers to visit the Large Telescope near the remote town of Sutherland. An American astronomer based in Cape Town explains: 'Prestige is . . . part of it. They want a very visible symbol of national pride to show that this is a serious country.'[106] At least for the moment, the dominion of knowledge remains relatively open, globally oriented, and free for contestation.

[106] *Guardian*, 13 Feb. 2005; www.ska.ac.za/index.html; www.engineeringnews.co.2a/?show=80679

Select Bibliography

Alberti, S. J. M. M., 'Placing Nature: Natural History Collections and their Owners in Nineteenth-Century Provincial England', *British Journal for the History of Science*, 35: 3 (2002).

Altick, R. D., *The English Common Reader* (Chicago, 1957).

Anker, P., *Imperial Ecology: Environmental Order in the British Empire, 1895–1945* (Cambridge, Mass., 2001).

Anon., *Life at the Cape a Hundred Years Ago: By a Lady* (Cape Town, 1963).

Arnove, R. F. (ed.), *Philanthropy and Cultural Imperialism: The Foundations at Home and Abroad* (Bloomington, 1982).

Austin, Brian, *Schonland: Scientist and Soldier* (Bristol and Philadelphia, 2001).

Bajinath, Himansu and Singh, Yashica, *Rebirth of Science in Africa: A Shared Vision for Life and Environmental Sciences* (Pretoria, 2002).

Baker, H., 'The Architectural Needs of South Africa', *The State* (May, 1909).

Bank, A., 'Liberals and their Enemies: Racial Ideology at the Cape of Good Hope, 1820–50', Ph.D thesis, Cambridge University (1995).

—— 'Losing Faith in the Civilizing Mission: The Premature Decline of Humanitarian Liberalism at the Cape, 1840–60', in M. Daunton and R. Halpern (eds.), *Empire and Others: British Encounters with Indigenous Peoples, 1600–1850* (London, 1999).

—— 'Evolution and Racial Theory: The Hidden Side of Wilhelm Bleek', *SAHJ* 43 (2000).

Barber, J. and Barratt, J., *South Africa's Foreign Policy* (Cambridge, 1990).

Barnard, W. S., 'Darwin at the Cape', *South African Journal of Science*, 100 (2004).

Basalla, G., 'The Spread of Western Science', *Science*, 5: 156 (1967).

Bayly, C. A., *Imperial Meridian: The British Empire and the World 1780–1830* (Harlow, 1989).

—— *Empire and Information: Intelligence Gathering and Social Communication in India, 1780–1870* (Cambridge, 1996).

Beale, M. A., 'Apartheid and University Education, 1948–1970', Ph.D thesis, University of the Witwatersrand (1998).

Beer, G., *Darwin's Plots*, 2nd edn., (Cambridge, 2000).

Beinart, W., *The Rise of Conservation in South Africa: Settlers, Livestock, and the Environment 1770–1950* (Oxford, 2003).

—— and McGregor, J. (eds.), *Social History and African Environments* (Oxford and Ohio, 2003).

Belich, James, *Making Peoples: A History of New Zealanders* (Auckland, 1996).

Bell, M., 'American Philanthropy, the Carnegie Corporation and Poverty in South Africa', *Journal of Southern African Studies*, 26: 3 (2000).

Benyon, J., *Proconsul and Paramountcy in South Africa* (Pietermaritzburg, 1980).

Berger, C., *Science, God, and Nature in Victorian Canada* (Toronto, 1983).

Betts, P., *The Authority of Everyday Objects: A Cultural History of West German Industrial Design* (Berkeley and Los Angeles, 2004).

Bickford-Smith, V., *Ethnic Pride and Racial Prejudice in Victorian Cape Town* (Cambridge, 1995).
—— 'Leisure and Social Identity in Cape Town, British Cape Colony, 1838–1910', *Kronos*, 25 (1998/9).
Bolus, H., 'The Native Flora of South Africa and its Preservation', *The State*, 2: 7 (1909).
Bonyhady, T., *The Colonial Earth* (Melbourne, 2000).
Botha, H. C., *John Fairbairn in South Africa* (Cape Town, 1984).
Boucher, M., *Spes in Arduis: A History of the University of South Africa* (Pretoria, 1973).
—— *The University of the Cape of Good Hope and the University of South Africa 1873–1946* (Pretoria, 1974).
Bozzoli, B., *The Political Nature of a Ruling Class: Capital and Ideology in South Africa 1890–1933* (London, 1981).
Bozzoli, G. R., *A Vice-Chancellor Remembers: The Memoirs of Professor G. R. Bozzoli* (Randburg, 1997).
Bradlow, E. and F., *Thomas Bowler of the Cape of Good Hope* (Cape Town, 1955).
Bradlow, F., *Thomas Bowler: His Life and Work* (Cape Town, 1967).
Brain, C. K. and Erasmus, M. C., *The Making of the Museum Professions in Southern Africa* (Pretoria, 1986).
Breckenridge, K., 'The Biometric State: The Promise and Peril of Digital Government in the New South Africa', *Journal of Southern African Studies*, 31: 2 (2005).
Bregman, L. D., ' "Snug Little Coteries": A History of Scientific Societies in Early Nineteenth-Century Cape Town, 1824–1835', Ph.D thesis, University of London (2003).
Brookes, E. H., *et al.*, *Coming of Age: Studies in South African Citizenship and Politics* (Cape Town, 1930).
Brown, A. C., (ed.), *A History of Scientific Endeavour in South Africa* (Cape Town, 1977).
Brown, K., 'Progressivism, Agriculture and Conservation in the Cape Colony *circa* 1902–1908', D.Phil. thesis, Oxford University (2002).
—— 'Tropical Medicine and Animal Diseases: Onderstepoort and the Development of Veterinary Science in South Africa 1908–1950', *Journal of Southern African Studies*, 31: 3 (2005).
Buchan, J., *The African Colony: Studies in the Reconstruction* (Edinburgh and London, 1903).
—— *Memory Hold-the-Door* (London, 1940).
Buffetaut, E., *A Short History of Vertebrate Palaeontology* (Beckenham, 1987).
Bull, M., *Abraham de Smidt 1829–1908: Artist and Surveyor-General of the Cape Colony* (Cape Town, 1981).
Bunn, D., 'Whited Sepulchres: On the Reluctance of Monuments', in H. Judin and I. Vladislavic (eds.), *Blank Architecture, Apartheid and After* (Rotterdam, 1998).
Burkitt, M. C., *South Africa's Past in Stone and Paint* (Cambridge, 1928).
Burrow, J. W., *A Liberal Descent: Victorian Historians and the English Past* (Cambridge, 1981).
—— 'Images of Time: From Carlylean Vulcanism to Sedimentary Gradualism', in S. Collini *et al.* (eds.), *History, Religion, and Culture: British Intellectual History 1750–1950* (Cambridge 2000).
Burrows, E. H., *A History of Medicine in South Africa* (Cape Town, 1958).
Buttman, G., *The Shadow of the Telescope: A Biography of John Herschel* (London, 1974).

Cadbury, D., *The Dinosaur Hunters* (London, 2000).
Canny, N. and Pagden, A. (eds.), *Colonial Identity in the Atlantic World, 1500–1800* (Princeton, 1987).
Carman, J., ' "Modern Art for South Africa": The Founding of the Johannesburg Art Gallery', Ph.D thesis, University of the Witwatersrand (2002).
Carruthers, J., *Wildlife and Warfare: The Life of James Stevenson-Hamilton* (Pietermaritzburg, 2001).
—— ' "Our Beautiful and Useful Allies": Aspects of Ornithology in Twentieth Century South Africa', *Historia*, 49: 1 (2004).
Cetina, K. K., *Epistemic Cultures: How the Sciences Make Knowledge* (Boston, 1999).
Chanock, M., *The Making of South African Legal Culture 1902–1936: Fear, Favour and Prejudice* (Cambridge, 2001).
Chant, C. and Fauvel, J. (eds.), *Darwin to Einstein: Historical Studies on Science and Belief* (Harlow, 1980).
Chapman, M., *South African Literatures* (London, 1996).
Christie, R., *Electricity, Industry and Class in South Africa* (London, 1984).
Clark, N. L., *Manufacturing Apartheid: State Corporations in South Africa* (New Haven, 1994).
Clarke, P., *British Clubs and Societies 1580–1800* (Oxford, 2000).
Coetzee, J. M., *White Writing: On the Culture of Letters in South Africa* (New Haven, 1988).
Cohen, A., 'Mary Elizabeth Barber, the Bowkers, and South African Prehistory', *South African Archaeological Bulletin*, 54: 170 (1999).
——'Mary Elizabeth Barber: South Africa's First Lady Natural Historian', *Archives of Natural History*, 27: 2 (2000).
Cole, A. W., *Reminiscences of My Life and of the Cape Bench and Bar* (Cape Town, 1896).
Colley, L., *Britons: Forging the Nation 1707–1837* (New Haven, 1992).
Compton, R. H., *Kirstenbosch, Garden for a Nation* (Cape Town, 1965).
Cook, P. A.W., *The Native Standard VI Pupil* (Pretoria, 1939).
Cope, R., *Ploughshare of War: The Origins of the Anglo-Zulu War of 1879* (Pietermaritzburg, 1999).
Crais, C., *The Making of the Colonial Order: White Supremacy and Black Resistance in the Eastern Cape, 1770–1865* (Johannesburg, 1992).
Davenport, T. R. H., *The Afrikaner Bond* (Cape Town, 1966).
Davidoff, L. and Hall, C., *Family Fortunes: Men and Women of the English Middle Class, 1780–1850* (London, 1987).
Davison, P., 'Material Culture, Context and Meaning: A Critical Investigation of Museum Practice, with Particular Reference to the South African Museum', Ph.D thesis, University of Cape Town (1991).
De Kiewiet, C. W., *A History of South Africa: Social and Economic* (London, 1941).
De Villiers, J. W., Jardine, R., and Reiss, M., 'Why South Africa gave up the Bomb', *Foreign Affairs*, 72: 5 (1993).
Denoon, D., *A Grand Illusion: The Failure of Imperial Policy in the Transvaal Colony During the Period of Reconstruction 1900–1905* (London, 1973).
Dippenaar, N. J., *Staatsmuseum 100* (Pretoria, 1992).
Drayton, R., *Nature's Government: Science, Imperial Britain, and the 'Improvement' of the World* (New Haven, 2000).

Du Plessis, Jean, 'Colonial Progress and Countryside Conservatism: An Essay on the Legacy of van der Lingen of Paarl, 1831–1875', MA thesis, University of Stellenbosch (1988).

Du Toit, A., 'The Cape Afrikaner's Failed Liberal Moment 1850–1870', in J. Butler *et al.*, *Democratic Liberalism in South Africa* (Cape Town and Middletown, 1987).

Du Toit, S. J., *Die Geskiedenis van ons Land in die Taal van ons Volk* (Cape Town, 1877).

Dubow, Jessica, 'Colonial Space, Colonial Identity: Perception and the South African Landscape', Ph.D thesis, University of London (2000).

—— 'Rites of Passage: Travel and the Materiality of Vision at the Cape of Good Hope', in Barbara Bender and Margot Winer (eds.), *Contested Landscapes: Movement, Exile and Place* (Oxford and New York, 2001).

Dubow, S., *Scientific Racism in Modern South Africa* (Cambridge, 1995).

—— 'Human Origins, Race Typology and the Other Raymond Dart', *African Studies*, 55: 1 (1996).

—— 'Colonial Nationalism, the Milner Kindergarten and the Rise of "South Africanism", 1902–10', *History Workshop Journal*, 43 (1997).

—— 'A Commonwealth of Science: The British Association in South Africa, 1905 and 1929', in S. Dubow (ed.), *Science and Society in Southern Africa* (Manchester, 2000).

—— 'Scientism, Social Research and the Limits of "South Africanism": The Case of Ernst Gideon Malherbe', *South African Historical Journal*, 44 (2001).

—— 'Imagining the New South Africa in the Era of Reconstruction', in D. Omissi, and A. S. Thompson (eds.), *The Impact of the South African War* (Houndmills, 2002).

—— 'Earth History, Natural History, and Prehistory at the Cape, 1860–1875', *Comparative Studies in Society and History*, 46 (2004).

—— and Marks, S., 'Patriotism of Place and Race: Keith Hancock on South Africa', in D. A. Low (ed.), *Keith Hancock: The Legacies of a Historian* (Melbourne, 2001).

—— and Jeeves, A. (eds.), *South Africa's 1940s: Worlds of Possibilities* (Cape Town, 2005).

Dunn, E. J., 'On the Stone Implements of South Africa', *Transactions of the South African Philosophical Society*, 2 (1879–81).

Dunn, W. H., *James Anthony Froude: A Biography 1857–1894* (Oxford, 1963).

Eddy, J. and Schreuder, D., *The Rise of Colonial Nationalism* (Sydney, 1988).

Elphick, R. and Giliomee, H. (eds.), *The Shaping of South African Society, 1652–1840*, 2nd edn. (Cape Town and London, 1989).

Erlank, N., 'Writing Women In(to) Early Nineteenth Century Cape Town', *Kronos*, 23 (1996).

Evans, D. S., *et al.*, *Herschel at the Cape: Diaries and Correspondence of Sir John Herschel, 1834–1838* (Austin, Tex., 1969).

Evans, I., *Bureaucracy and Race: Native Administration in South Africa* (Berkeley, 1997).

Fairbridge, D., *Along Cape Roads* (Cape Town, 1928).

Ferguson, W. T. and Immelman, R. F. M., *Sir John Herschel and Education at the Cape 1834–1840* (Cape Town, 1961).

Fick, M. L., *The Educability of the South African Native* (Pretoria, 1939).

Fleisch, B. D., 'Social Scientists as Policy Makers: E. G. Malherbe and the National Bureau for Educational and Social Research, 1929–1943', *Journal of Southern African Studies*, 21: 3 (1995).

—— 'The Teachers College Club: American Educational Discourse and the Origins of Bantu Education in South Africa, 1914–1951', Ph.D dissertation, Columbia University (1994).

Fontana, B., *Rethinking the Politics of Commercial Society* (Cambridge, 1985).

Foster, J., 'The Poetics of Liminal Places: Landscape and the Construction of White Identity in early Twentieth Century South Africa', D.Phil. thesis, University of London (1998).

—— ' "Land of Contrasts" or "Home we have always known"?: The SAR&H and the Imaginary Geography of White South African Nationhood, 1909–1930', *Journal of Southern African Studies*, 29: 3 (2003).

Frankel, S. H., *Capital Investment in Africa* (London, 1938).

Fraser, M. and Jeeves, A., *All that Glittered: Selected Correspondence of Lionel Phillips 1890–1924* (Cape Town, 1977).

Frere, B., 'The Native Races of South Africa', *Transactions of the South African Philosophical Society*, 1 (1877–8).

Friedman, B., *Smuts: A Reappraisal* (London, 1975).

Froude, J. A., *Oceana, or England and her Colonies* (London, 1886).

Gale, A., 'Science at the Margins: The British Association and the Foundations of Canadian Anthropology, 1884–1910', Ph.D dissertation, University of Pennsylvania (1986).

Gibbens, M., 'Two Decades in the Life of a City: Grahamstown 1862–1882', MA thesis, Rhodes University (1982).

Giliomee, H., *The Afrikaners: Biography of a People* (London, 2003).

Gitay, Yehoshua, 'The Word, Science, Medicine and Politics', paper presented to the Centre for African Studies, University of Cape Town (2002).

Godby, M., 'The Colonial Experience of Space: Thomas Bowler's Representation of the Figure in the Landscape in Lithographs and Other Views of the Cape', UCT Centre for African Studies seminar paper (1992).

Godby, M.,'The Art of Charles Bell: An Appraisal', in P. Brooke Simons, *The Life and Work of Charles Bell* (Cape Town, 1998).

Goldblatt, D., *South Africa: The Structure of Things Then* (Cape Town, 1998).

Goodfellow, C. G., *Great Britain and South African Confederation 1870–1881* (Cape Town, 1966).

Goodwin, A. J. H., 'Sir Langham Dale's Collection of Stone Implements', *South African Journal of Science*, 25 (1928).

—— 'A Commentary Upon South African Prehistory up to the Introduction of the New Terminology', *Bantu Studies*, 9: 4 (1935).

—— and Van Riet Lowe, C., 'The Stone Age Cultures of South Africa', *Annals of the South African Museum*, 27 (1929).

Gordon, R. J., *The Bushman Myth* (Boulder, Colo., 1992).

Green, E. H. H., *The Crisis of Conservatism: The Politics, Economics and Ideology of the British Conservative Party, 1880–1914* (London, 1995).

Greswell, W., *Our South African Empire*, vol. 1 (London, 1885).

Griffiths, T., *Hunters and Collectors* (Cambridge, 1996).

—— and Robin, L. (eds.), *Ecology and Empire: Environmental History of Settler Societies* (Edinburgh, 1997).

Groenewald, C. J., 'The Methodology of Poverty Research in South Africa: The Case of the First Carnegie Investigation 1929–1932', *Social Dynamics*, 13: 2 (1987).

GROVE, R., *Green Imperialism: Colonial Expansion, Tropical Island Edens and the Origins of Environmentalism, 1600–1860* (Cambridge, 1995).

—— 'Scotland in South Africa: John Croumbie Brown and the Roots of Settler Environmentalism', in T. Griffiths and L. Robin (eds.), *Ecology and Empire: Environmental History of Settler Societies* (Edinburgh, 1997).

GRUNDLINGH, A., *Fighting Their Own War: South African Blacks and the First World War* (Johannesburg, 1987).

HAAGNER, A. and IVY, R. H., *Sketches of South African Bird-Life* (London, 1908).

HAHLO, H. R. and KAHN, E., *The South African Legal System and its Background* (Cape Town, 1968).

HAILEY, M. (Lord), *An African Survey* (London, 1938).

HALL, N. J., *Trollope: A Biography* (Oxford, 1991).

HAMILTON, C., *Terrific Majesty: The Powers of Shaka Zulu and the Limits of Historical Invention* (Cambridge, Mass., 1998).

HAMMOND TOOKE, W. D., *Imperfect Interpreters: South Africa's Anthropologists 1920–1990* (Johannesburg, 1997).

HANCOCK, W. K., *Smuts: The Sanguine Years 1870–1919* (Cambridge, 1962).

—— *Smuts: The Fields of Force 1919–1950* (Cambridge, 1968).

HARRIES, P., 'Field Sciences in Scientific Fields: Entomology, Botany and the Early Ethnographic Monograph in the Work of H.-A. Junod', in S. Dubow (ed.), *Science and Society in Southern Africa* (Manchester, 2000).

HATTERSLEY, A. F., The *Convict Crisis and the Growth of Unity: Resistance to Transportation in South Africa and Australia 1848–1853* (Pietermaritzburg, 1965).

—— *An Illustrated Social History of South Africa* (Cape Town, 1969).

HAWTHORNE, P., *The Transplanted Heart* (Johannesburg, 1968).

HEADLAM, C. (ed.), *The Milner Papers: South Africa 1899–1905, vol.* II (London, 1933).

HENSHAW, P., 'John Buchan from the "Borders" to the "Berg": Nature, Empire and White South African Identity, 1901–1910', *African Studies*, 62: 1 (2003).

HILTON, B., *The Age of Atonement: The Influence of Evangelicalism on Social and Economic Thought, 1795–1865* (Oxford, 1988).

HINCHLIFFE, P., *The Anglican Church in South Africa* (London, 1963).

HOFMEYR, J. H., 'Africa and Science', *South African Journal of Science*, 26 (1929).

HOWARTH, O. J. R., *The British Association for the Advancement of Science: A Retrospect 1831–1931* (London, 1931).

HYSLOP, J., 'The Imperial Working Class Makes Itself "White": White Labourism in Britain, Australia and South Africa Before the First World War', *Journal of Historical Sociology*, 12: 4 (1999)

—— 'An Anglo-South African Intellectual, the Second World War, and the Coming of Apartheid: Guy Butler in the 1940s', in S. Dubow and A. Jeeves (eds.), *South Africa's 1940s: Worlds of Possibilities* (Cape Town, 2005).

INKSTER, I. and MORRELL, J., *Metropolis and Province: Science in British Culture, 1780–1850* (London, 1983).

INNES, J. ROSE, *Autobiography*, ed. B. A. Tindall (London and Cape Town, 1949).

JACOBS, A., *South African Heritage: A Biography of H. J. van der Bijl* (Pietermaritzburg, 1948).

JACOBS, N. J., *Environment, Power and Injustice* (Cambridge, 2003).

JANSEN, J. D. (ed.), *Knowledge and Power in South Africa* (Johannesburg, 1991).

JEBB, R., *Studies in Colonial Nationalism* (London, 1905).
KEEGAN, T., *Colonial South Africa and the Origins of the Racial Order* (Cape Town, 1996).
KENDALL, F. K., *A Short History of the South African Fine Arts Association* (Cape Town, 1941).
KHILNANI, S., *The Idea of India* (London, 1977).
KILPIN, R., *The Romance of the Cape Parliament* (London, 1930).
KINGWILL, D. G., *The CSIR: The First 40 Years* (Pretoria, 1990).
KIRBY, P. R., *Sir Andrew Smith—His Life, Letters and Works* (Cape Town and Amsterdam, 1965).
KRIGE, S., 'Segregation, Science and Commissions of Enquiry: The Contestation over Native Education Policy in South Africa, 1930–1936', *Journal of Southern African Studies*, 23: 3 (1997).
KRIKLER, J., *Revolution from Above, Rebellion from Below: The Agrarian Transvaal at the Turn of the Century* (Oxford, 1993).
KROS, C., 'Economic, Political and Intellectual Origins of Bantu Education, 1926–51', doctoral dissertation, University of the Witwatersrand (1996).
LAGEMANN, E. C., *The Politics of Knowledge: The Carnegie Corporation, Philanthropy, and Public Policy* (Middletown, Conn., 1989).
LAVIN, D. (ed.), *FitzPatrick's South African Memories: Scraps of History* (Johannesburg, 1979).
LEACH, M., *et al.* (eds.), *Science and Citizens: Globalization and the Challenge of Engagement* (London and New York, 2005).
LEHFELDT, R. A., *The National Resources of South Africa* (Johannesburg, 1922).
LESTER, A., *Imperial Networks: Creating Identities in Nineteenth-Century South Africa and Britain* (London, 2001).
LEVITZ, C., 'The Politics, Ideology and Social Practices of South African Natural Science and Technology Museums', MA dissertation, Wits University (1995).
LEWCOCK, R., *Early Nineteenth Century Architecture in South Africa* (Cape Town, 1963).
LEWIN ROBINSON, A. M., 'The English Periodical Literature of the Cape Colony from its Beginnings (1824) to 1835', Ph.D thesis, University of Cape Town (1961).
—— *Selected Articles from the Cape Monthly Magazine (New Series 1870–76)*, (Cape Town, 1978).
LEWSEN, P., *Selections from the Correspondence of J. X. Merriman 1870–1890* (Cape Town, 1960).
—— *John X. Merriman: Paradoxical South African Statesman* (Cape Town, 1982).
LISTER, A. H., *Journals of Andrew Geddes Bain* (Cape Town, 1949).
LLOYD, A. C., 'Some Secrets of South African Literature', *South African Bookman*, 4 (1911).
LUBBOCK, J., *Prehistoric Races as Illustrated by the Ancient Remains and the Manners and Customs of Modern Savages*, 3rd edn. (London 1872).
McCRACKEN, D. P. and E. M., *The Way to Kirstenbosch* (Cape Town, 1988).
McCRACKEN, J. L., *The Cape Parliament 1854–1910* (Oxford, 1967).
—— *New Light at the Cape of Good Hope: William Porter, the Father of Cape Liberalism* (Belfast, 1993).
MACKENZIE, J. M. (ed.), *Imperialism and the Natural World* (Manchester, 1990).
McKENZIE, K., 'The *South African Commercial Advertiser* and the Making of Middle Class Identity in Early-Nineteenth Century Cape Town', MA thesis, University of Cape Town (1993).

McKenzie, K., 'Gender and Honour in Middle-Class Cape Town: The Making of Colonial Identities 1828–52', D.Phil. thesis, Oxford University (1997).
—— *Scandal in the Colonies* (Melbourne, 2004).
MacLeod, R., 'On Visiting the "Moving Metropolis": Reflections on the Architecture of Imperial Science', in N. Reingold and M. Rothenberg (eds.), *Scientific Colonialism: A Cross-Structural Comparison* (Washington, DC, 1987).
—— and Collins, P. (eds.), *The Parliament of Science: The British Association for the Advancement of Science 1831–1981* (Northwood, 1981).
Makgoba, M. W. (ed.), *African Renaissance: The New Struggle* (Sandton and Cape Town, 1999).
Malherbe, E. G., *Education in South Africa Vol.I:1652–1922* (Cape Town, 1925).
—— 'Education and the Poor White', *South African Journal of Science*, 26 (1929).
—— *Education in South Africa Vol.II: 1923–75* (Cape Town, 1977).
—— *Never a Dull Moment* (Cape Town, 1981).
—— *et al.* (eds.), *Educational Adaptations in a Changing Society* (Cape Town, 1937).
Martin, S. J. R., 'British Images of the Zulu, c.1820–1879', D.Phil. thesis, University of Cambridge (1982).
Mbeki, Thabo, *The Time Has Come: Selected Speeches* (Cape Town, 1998).
Meiring, J., *Thomas Pringle: His Life and Times* (Cape Town, 1968).
Meltzer, J. L., 'The Growth of Cape Town Commerce and the Role of John Fairbairn's *Advertiser* (1835–1859)', MA thesis, University of Cape Town (1989).
Mendelsohn, R., *Sammy Marks: 'The Uncrowned King Of The Transvaal'* (Cape Town, 1991).
Merrington, P., 'Pageantry and Primitivism: Dorothea Fairbridge and the "Aesthetics of Union" ', *Journal of Southern African Studies*, 21: 4 (1995).
Metcalf, T. R., *An Imperial Vision: Indian Architecture and Britain's Raj* (London and Boston, 1989).
Middlemas, K., *Cabora Bassa: Engineering and Politics in Southern Africa* (London, 1975).
Millar, A. K., *Plantagenet in South Africa: Lord Charles Somerset* (London, 1965).
Miller, J. D. B., *Richard Jebb and the Problem of Empire* (London, 1956).
Miller, R. B., 'Science and Society in the Early Career of H. F. Verwoerd', *Journal of Southern African Studies*, 19: 4 (1993).
Moll, J. C. *et al.* (eds.), *Akademie 75* ([Pretoria], 1984).
Molteno, J. T., *Further South African Recollections* (London, 1926).
Molteno, P. A., *A Federal South Africa* (London, 1896).
Morris, R. J., 'Clubs, Societies and Associations', in F. M. L. Thompson (ed.), *Cambridge Social History of Britain 1750–1950*, (Cambridge, 1990).
Murray, B. K., *Wits: The 'Open' Years, 1939–1959* (Johannesburg, 1997).
Murray, Joyce (ed.), *In Mid-Victorian Cape Town: Letters from Miss Rutherfoord* (Cape Town, 1953).
Musselman, Elizabeth Green, 'Swords into Ploughshares: John Herschel's Progressive View of Astronomical and Imperial Governance', *British Journal for the History of Science*, 31 (1998).
—— 'Worlds Displaced: Projecting the Celestial Environment from the Cape Colony', *Kronos*, 29 (2003).
Nasson, Bill, 'Delville Wood and South African Great War Commemoration', *English Historical Review, 119:* 480 (2004).

NAUDÉ, S. MEIRING, 'Die Afrikaner en die Ontwikkeling op Fisies-Wetenskaplike en Tegniese Gebied Gedurende die Afgelope 50 Jaar', *Tydskrif vir Wetenskap en Kuns*, 10 (1950).

NEWBY-FRASER, A. R., *Chain Reaction: Twenty Years of Nuclear Research and Development in South Africa* (Pretoria, 1979).

NEWTON, A. P., *Select Documents Relating to the Unification of South Africa*, (London, 1924).

NIELSEN, W. A., *The Big Foundations* (New York, 1972).

NIENABER, P. J., *Op Brandwag vir ons Taal. Die Geskiedenis en Werksaamhede van die Afrikaansse Akademie vir Wetenskap en Kuns* (Pretoria, 1950).

NIMOCKS, W., *Milner's Young Men: The 'Kindergarten' in Edwardian Imperial Affairs* (Durham, NC, 1968).

NOBLE, JOHN, *South Africa, Past and Present: A Short History of the European Settlements at the Cape* (Cape Town and London, 1877).

NOBLE, RODERICK (ed.), *The Cape and its People and Other Essays: By South African Writers* (Cape Town, 1869).

NUMBERS, R. and STENHOUSE, J. (eds.), *Disseminating Darwin: The Role of Place, Race, Religion, and Gender* (Cambridge 1999).

OLIVIER, H., *Damit: Great Dams in Southern Africa* (Johannesburg, 1975).

—— *Great Dams in Southern Africa* (Cape Town, [1976]).

PATON, A., *Hofmeyr* (London, 1964).

PENN, N., 'Mapping the Cape: John Barrow and the First British Occupation of the Colony, 1795–1803', *Pretexts*, 4: 2 (1993).

PHILLIPS, E. PERCY, 'A Brief Historical Sketch of the Development of Botanical Science in South Africa and the Contribution of South Africa to Botany', *South African Journal of Science*, 27 (1930).

PHILLIPS, L., *Some Reminiscences*, ed. M. Fraser (Johannesburg, 1986).

The Poor White Problem in South Africa. Report of the Carnegie Commission, 5 vols. (Stellenbosch, 1932).

PORTER, R., *The Making of Geology: Earth Science in Britain, 1660–1815* (Cambridge, 1977).

POSEL, D., *The Making of Apartheid, 1948–1961* (Oxford, 1991).

—— 'A Mania for Measurement: Statistics and Statecraft in Apartheid South Africa', in S. Dubow (ed.), *Science and Society in Southern Africa* (Manchester, 2000).

—— 'Race as Common Sense: Racial Classification in Twentieth Century South Africa', *African Studies Review*, 44: 2 (2001).

POTTER, S. J., *News and the British World: The Emergence of an Imperial Press System* (Oxford, 2003).

RAFFLES, H., 'Intimate Knowledge', *International Social Science Journal*, 173 (2002).

Report of the South African Association for the Advancement of Science. First Meeting Cape Town 1903 (Cape Town, 1903).

Report of the SAAAS. 3rd and 4th Meeting 1905–6 (Cape Town, 1906).

RICH, P. B., 'Milnerism and a Ripping Yarn: Transvaal Land Settlement and John Buchan's Novel *Prester John*, 1901–1910', in B. Bozzoli (ed.), *Town and County in the Transvaal* (Johannesburg, 1983).

RITCHIE, W., *The History of the South African College 1829–1918*, vol. 1 (Cape Town, 1918).

ROBINS, S., '"Long live Zackie, long live": AIDS Activism, Science and Citizenship After Apartheid', *Journal of Southern African Studies*, 30: 3 (2004).

Robinson, R. and Gallagher, J., *Africa and the Victorians* (New York and London, 1961).
Ross, D., *The Origins of American Social Science* (Cambridge, 1991).
Ross, R., *Beyond the Pale: Essays on the History of Colonial South Africa* (Johannesburg, 1993).
—— *Status and Respectability in the Cape Colony, 1750–1870* (Cambridge, 1999).
Rousseau, L., *The Dark Stream: The Story of Eugène Marais* (Johannesburg, 1982).
Rubin, N. and Warren, W. M. (eds.), *Dams in Africa: An Interdisciplinary Study of Man-Made Lakes in Africa* (London, 1968).
Rupke, N. A., *The Great Chain of History: William Buckland and the English School of Geology (1814–1849)* (Oxford, 1983).
Ruskin, S., *John Herschel's Cape Voyage: Private Science, Public Imagination and the Ambitions of Empire* (Aldershot and Burlington, 2004).
Rutherford, J., *Sir George Grey* (London, 1961).
Sachs, A., *Justice in South Africa* (London, 1973).
Sank, Yvonne Pauline, 'The Origin and Development of the Cape Progressive Party (1884–1898)', MA thesis, University of Cape Town (1955).
Saunders, C. C., *The Making of the South African Past* (Cape Town, 1988).
—— and Le Cordeur, B., 'The South African Historical Society and its Antecedents', *South African Historical Journal* 18 (1986).
Schlanger, N., 'The Burkitt Affair Revisited: Colonial Implications and Identity Politics in Early South African Prehistoric Research', *Archaeological Dialogues*, 10: 1 (2003).
Schoeman, K. (ed.), *A Debt of Gratitude: Lucy Lloyd and the 'Bushman Work' of G. W. Stow* (Cape Town, 1997).
Schonland, B. J. F., 'The South African Association for the Advancement of Science, its Past and Future', *South African Journal of Science*, 49: 3–4 (1952).
Schreiner, Olive, *Closer Union* (London, 1909).
—— *Thoughts on South Africa* (London,1923).
Schreuder, D. M., 'The Imperial Historian as "Colonial Nationalist": George McCall Theal and the Making of South African History,' in G. Martel (ed.), *Studies in British Imperial History* (London, 1986).
Sharp, J., 'The Roots and Development of "Volkekunde" in South Africa', *Journal of Southern African Studies*, 8: 1 (1981).
Shaw, D. J., 'The Writings of Thomas Pringle', Ph.D thesis, Cambridge University (1996).
Shingler, J. D., 'Education and the Political Order in South Africa, 1902–1961', Ph.D dissertation, Yale University (1973).
Silove, D., 'Doctors and the State: Lessons from the Biko Case', *Social Science and Medicine*, 30: 4 (1990).
Skotnes, P. (ed.), *Miscast: Negotiating the Presence of the Bushmen* (Cape Town, 1996).
Smuts, J. C., 'South Africa in Science', *South African Journal of Science*, 22 (1925).
Solomon, W. E. G., *Saul Solomon: 'The Member for Cape Town'* (Cape Town, 1948).
Stafford, R. A., 'Annexing the Landscapes of the Past: British Imperial Geology in the Nineteenth Century' in J.M. McKenzie (ed.), *Imperialism and the Natural World* (Manchester, 1990).
—— 'Scientific Exploration and Empire', in A. Porter (ed.), *The Oxford History of the British Empire* (Oxford 1999).
Statham, F. R., *Blacks, Boers and British: A Three-Cornered Problem* (London, 1881).
—— *South Africa As It Is* (London, 1897).
Stevenson, M., 'Old Masters and Aspirations: The Randlords, Art and South Africa', Ph.D thesis, University of Cape Town (1997).

Stokes, E., *The English Utilitarians and India* (Oxford, 1959).
Sturgis, J., 'Anglicisation at the Cape of Good Hope in the Early Nineteenth Century', *Journal of Imperial and Commonwealth History*, 11: 1 (1982).
Summers, R. F. H., *A History of the South African Museum* (Cape Town, 1975).
Swart, S., 'The Ant of the White Soul: Popular Natural History, the Politics of Afrikaner Identity, and the Entomological Writings of Eugène Marais', in W. Beinart and J. McGregor (eds.), *Social History and African Environments* (Oxford and Ohio, 2003).
—— ' "An Irritating Pebble in Kruger's Shoe"—Eugène Marais and *Land en Volk* in the ZAR, 1891–1896', *Historia*, 48: 2 (2003).
Tamarkin, M., *Cecil Rhodes and the Cape Afrikaners* (London, 1996).
—— 'The Cape Afrikaners and the British Empire from the Jameson Raid to the South African War', in D. Lowry (ed.), *The South African War Reappraised* (Manchester, 2000).
Taylor, M., 'The 1848 Revolutions and the British Empire', *Past and Present*, 166 (2000).
Thakray, Arnold, 'Natural Knowledge in Cultural Context: The Manchester Model', *American Historical Review*, 79: 3 (1974).
Theal, G. M., *Compendium of South African History and Geography* (Lovedale, 1873).
Thompson, A., 'The Language of Imperialism and the Meanings of Empire: Imperial Discourse in British Politics, 1895–1914', *Journal of British Studies*, 36 (1997).
—— 'The Languages of Loyalism in Southern Africa *c.* 1870–1939', *English Historical Review*, 118: 477 (2003).
—— *The Empire Strikes Back? The Impact of Imperialism on Britain from the Mid-Nineteenth Century* (Harlow, 2005).
Thompson, L., *The Unification of South Africa* (Oxford, 1960).
Thompson, T. W., *James Anthony Froude on Nation and Empire: A Study in Victorian Racialism* (New York and London, 1987).
Tilley, Helen, 'Africa as a "Living Laboratory": The African Research Survey and the British Colonial Empire: Consolidating Environmental, Medical, and Anthropological Debates, 1929–40', D.Phil. thesis, Oxford University (2001).
Tobias, P. V., 'Africa. The Cradle of Humanity', in Himansu Bajinath and Yashica Singh (eds)., *Rebirth of Science in Africa: A Shared Vision for Life and Environmental Sciences* (Pretoria, 2002).
Trollope, A., *South Africa. A Reprint of the 1878 Edition, with an Introduction and Notes by J. H. Davidson* (Cape Town, 1973).
Turrell, R. V., 'A Cape Periodical: The Cape Monthly Magazine 1870–1875', BA (Hons.) thesis, University of Cape Town (1974).
Tyrrell-Glynn, W. H. P. A., 'The History of the South African Public Library, Cape Town, 1830–1961', Ph.D thesis, Stellenbosch University (1983).
Van Jaarsveld, F. A., *The Afrikaner's Interpretation of South African History* (Cape Town, 1964).
Van Onselen, C., 'The Modernization of the Zuid Afrikaansche Republiek: F. E. T. Krause, J. C. Smuts, and the Struggle for the Johannesburg Public Prosecutor's Office, 1898–1899', *Law and History Review*, 21: 3 (2003).
Van Rensburg, J. A. J., *The Learning Ability of the South African Native* (Pretoria, 1938).
Van Sittert, L., 'From "Mere Weeds" and "Bosjes" to a Cape Floral Kingdom: The Reimagining of Indigenous Flora at the Cape, c. 1890–1939', *Kronos*, 28 (2002).
—— 'The Bourgeois Eye Aloft: Table Mountain in the Anglo Urban Middle Class Imagination, c. 1891–1950, *Kronos*, 29 (2003).

VARLEY, D. H., 'The Nineteenth Century and the "Common Reader"', *South African Libraries*, 19: 4 (1952).

VERGUNST, N., *Hoerikwaggo. Images of Table Mountain* (SA National Gallery, 2001).

VERSCHOYLE, F. (Vindex), *Cecil Rhodes. His Political Life and Speeches 1881–1900* (London, 1900).

VOSS, A. E., 'Thomas Pringle and the Image of the "Bushmen"', *English in Africa*, 9: 1(1982).

WAHRMAN, DROR, *Imagining the Middle Class. The Political Representation of Class in Britain, c.1780–1840* (Cambridge, 1995).

WALKER, E. A., *Lord De Villiers and his Times. South Africa 1842–1914* (London, 1925).

—— *A History of South Africa* (London, 1928).

—— *The South African College and the University of Cape Town 1829–1929* (Cape Town, 1929).

—— *W. P. Schreiner. A South African* (London, 1937).

WALLIS, J. P. R., *Fitz. The Story of Sir Percy FitzPatrick* (London, 1955).

WARNER, B., *Astronomers at the Royal Observatory Cape of Good Hope* (Cape Town, 1979).

—— *Royal Observatory, Cape of Good Hope, 1820–1831. The Founding of a Colonial Observatory* (Dordrecht, 1995).

WHITELOCK, D. A., *The Great Tradition: A History of Adult Education in Australia* (Queensland University Press, 1974).

WILMOT, A., *History of the Zulu War* (London, 1880).

—— *The History of South Africa* (London, 1901).

WILSON, M. and THOMPSON, L. (eds.), *The Oxford History of South Africa Vol.II* (Oxford, 1971).

WITZ, L., *Apartheid's Festival: Contesting South Africa's National Pasts* (Indiana, 2003).

WORDEN, N. *et al.*, *Cape Town. The Making of a City* (Cape Town, 1998).

—— and CRAIS, C., *Breaking the Chains: Slavery and its Legacy in the Nineteenth-Century Cape Colony* (Johannesburg, 1994).

WORTHINGTON, E. B., *Science in Africa* (London, 1938).

WRIGHT, H. (ed.), *Sir James Rose Innes Selected Correspondence (1884–1902)* (Cape Town, 1972).

ZIMMERMAN, R. and VISSER, D. (eds.), *Southern Cross: Civil Law and Common Law in South Africa* (Cape Town, 1996).

Index